S0-BEB-752

TANKER SPILLS
PREVENTION BY DESIGN

Committee on Tank Vessel Design

Marine Board

Commission on Engineering and Technical Systems

National Research Council

NATIONAL ACADEMY PRESS
Washington, D.C. 1991

National Academy Press • 2101 Constitution Avenue, N.W. • Washington, D.C. 20418

NOTICE: The project that is the subject of this report was approved by the Governing Board of the National Research Council, whose members are drawn from the councils of the National Academy of Sciences, the National Academy of Engineering, and the Institute of Medicine. The members of the panel responsible for the report were chosen for their special competencies and with regard for appropriate balance.

This report has been reviewed by a group other than the authors according to procedures approved by a Report Review Committee consisting of members of the National Academy of Sciences, the National Academy of Engineering, and the Institute of Medicine.

The program described in this report is supported by Cooperative Agreement No. 14-35-0001-30475 between the Minerals Management Service of the U.S. Department of the Interior and the National Academy of Sciences.

Library of Congress Cataloging-in-Publication Data

National Research Council (U.S.). Committee on Tank Vessel Design.
Tanker spills : prevention by design / Committee on Tank Vessel Design, Marine Board, Commission on Engineering and Technical Systems, National Research Council.
p. cm.
Includes bibliographical references and index.
ISBN 0-309-04377-8
1. Tankers—Accidents. 2. Oil spills. I. Title.
VM455.N35 1991
623.8'245—dc20 91-12489
CIP

Printed in the United States of America

Cover: The Arco Marine, 122,000 deadweight ton, single hulled, tanker *Arco Juneau* serves the U.S. trade, including Alaska and West Coast ports. The *Arco Juneau* is a U.S.-flag ship, classified by the American Bureau of Shipping. With 17 years of service (as of 1991), it is about average in age for the U.S.-flag tanker fleet. Photo copyright © Vince Streano, Streano-Havens, Anacortes, Washington.

COMMITTEE ON TANK VESSEL DESIGN

HENRY S. MARCUS, *Chairman*, Massachusetts Institute of Technology, Cambridge
WILLIAM O. GRAY, *Vice Chairman*, Skaarup Oil Corporation, Greenwich, Connecticut
DAVID M. BOVET, Temple, Barker & Sloane, Inc., Lexington, Massachusetts
J. HUNTLY BOYD, JR., Booz-Allen & Hamilton, Inc., Arlington, Virginia
JOHN W. BOYLSTON, Argent Marine Operations, Inc., Solomons, Maryland
JOHN W. BURKE, Mobil Shipping & Transportation Company, Fairfax, Virginia
THOMAS D. HOPKINS, Rochester Institute of Technology, Rochester, New York
JAMES HORNSBY, Transport Canada (retired), Ottawa, Ontario, Canada
VOJIN JOKSIMOVICH, Accident Prevention Group, San Diego, California
SALLY ANN LENTZ, The Oceanic Society and Friends of the Earth, Washington, D.C.
ROBERT G. LOEWY, NAE, Rensselaer Polytechnic Institute, Troy, New York
TOMASZ WIERZBICKI, Massachusetts Institute of Technology, Cambridge

Liaisons

DONALD LIU, American Bureau of Shipping, Parmamus, New Jersey
JAMES M. MACDONALD, United States Coast Guard, Washington, D.C.
BERT MARSH, United States Navy, Washington, D.C.
FREDERICK SEIBOLD, Maritime Administration, Washington, D.C.
DANIEL F. SHEEHAN, United States Coast Guard, Washington, D.C.

Staff

DONALD W. PERKINS, Associate Director
LAURA OST, Editor
GLORIA GREEN, Project Assistant

The National Academy of Sciences is a private, nonprofit, self-perpetuating society of distinguished scholars engaged in scientific and engineering research, dedicated to the furtherance of science and technology and to their use for the general welfare. Upon the authority of the charter granted to it by the Congress in 1863, the Academy has a mandate that requires it to advise the federal government on scientific and technical matters. Dr. Frank Press is president of the National Academy of Sciences.

The National Academy of Engineering was established in 1964, under the charter of the National Academy of Sciences, as a parallel organization of outstanding engineers. It is autonomous in its administration and in the selection of its members, sharing with the National Academy of Sciences the responsibility for advising the federal government. The National Academy of Engineering also sponsors engineering programs aimed at meeting national needs, encourages education and research, and recognizes the superior achievements of engineers. Dr. Robert M. White is president of the National Academy of Engineering.

The Institute of Medicine was established in 1970 by the National Academy of Sciences to secure the services of eminent members of appropriate professions in the examination of policy matters pertaining to the health of the public. The Institute acts under the responsibility given to the National Academy of Sciences by its congressional charter to be an adviser to the federal government and, upon its own initiative, to identify issues of medical care, research, and education. Dr. Samuel O. Thier is president of the Institute of Medicine.

The National Research Council was organized by the National Academy of Sciences in 1916 to associate the broad community of science and technology with the Academy's purposes of furthering knowledge and advising the federal government. Functioning in accordance with general policies determined by the Academy, the Council has become the principal operating agency of both the National Academy of Sciences and the National Academy of Engineering in providing services to the government, the public, and the scientific and engineering communities. The Council is administered jointly by both Academies and the Institute of Medicine. Dr. Frank Press and Dr. Robert M. White are chairman and vice-chairman, respectively, of the National Research Council.

Preface

Each year thousands of tons of crude oil and petroleum products are spilled in U.S. waters as a result of many minor, and a few major, tank vessel collisions, groundings, and other accidents. Although the average annual spillage represents 1/500th of 1 percent of the total amount of oil moved through U.S. waters, the effects can be costly, as well as devastating to the environment. Reducing the risk of oil spills is critical to continued public acceptance of tank vessel operations, particularly in light of the nation's growing consumption of oil and projections for increased seaborne imports.

As part of the overall effort to promote maritime safety and environmental protection, the U.S. Coast Guard regulates some aspects of U.S.-flag tankship design, crew qualifications and training, and vessel traffic and routing. The Coast Guard also enforces national and international shipping regulations in U.S. waters. The March 1989 grounding of the EXXON VALDEZ, and the subsequent release of nearly 35,600 tons (11 million gallons) of oil in Prince William Sound, Alaska, renewed longstanding debate over the need for more pollution-resistant tank vessel designs.

In the aftermath of the Alaska spill, the largest ever in U.S. waters, the Coast Guard asked the National Research Council to assess whether alternative tank vessel designs would improve maritime safety and environmental protection. A committee was convened under the Marine Board to conduct a comprehensive review of the safety, economic, and environmental implications of alternative designs, and to determine how these designs might affect the overall consequences of accidents. The intent of the study was to assemble technical information and recommendations that could be used by the Coast Guard in determining design requirements for tank vessels operating in U.S. waters.

The committee was charged with:

- collecting and analyzing data to update what is known about tank vessel accidents, their environmental consequences, and the effectiveness of alternative designs in preventing oil pollution;
- identifying and assessing the technical concerns about alternative tank vessel designs, including concerns about the effects of alternative designs on salvage, safety, and ship operations;
- considering whether new materials or design approaches or any other developments merit research; and
- elucidating the safety, environmental, and economic costs and benefits of alternative new or retrofitted tank vessel designs.

Passage of the Oil Pollution Act of 1990 (P.L. 101-380), on August 18, further defined the committee's task. The Act states that ". . . the Secretary [of the Department of Transportation] shall determine, based on recommendations of the National Academy of Sciences or other qualified organizations, whether other structural and operational tank vessel requirements will provide protection to the marine environment equal to or greater than that provided by double hulls, and shall report to Congress that determination and recommendations for legislative action" (Title IV, Sec. 4115).

The committee was composed of members with expertise in tank vessel design and construction, tank vessel operation, maritime salvage, maritime safety, vehicle dynamics, structural engineering, economic analysis, risk assessment and management, environmental and maritime law, and maritime environmental protection. The principle guiding the constitution of the committee and its work, consistent with the policy of the National Research Council, was not to exclude members with potential biases that might accompany expertise vital to the study, but to seek balance and fair treatment. The committee was assisted by representatives of the Coast Guard, the Maritime Administration, the American Bureau of Shipping, and the U.S. Navy, who were designated as liaison representatives.

Reducing the risk of oil spills entails the ability to achieve the following fundamental goals: (1) an adequate understanding of spillage scenarios (e.g., groundings, collisions, etc.); (2) reduction of the likelihood of such scenarios; and (3) reduction of the extent or magnitude of spillage from such scenarios.

Risk reduction can be accomplished through various means, of which improved tank vessel design is only one. The relative importance of design, in relation to other factors influencing risk of accidents, is open to debate. In any case, vessel design is a significant influence on pollution risk, along with such other factors as vessel maintenance and operations practices, sea conditions, traffic density, vessel speed, and crew competence.

This study describes how alternative tank vessel designs might influence the safety of personnel, property, and the environment, and at what cost. Tank vessel design and operation, and risks worldwide, are addressed. To provide a context for its technical analysis, the committee addressed the feasibility and ramifications of implementing the various design options. In selecting designs to be considered, the committee included certain operational options that could minimize the amount of oil spilled in an accident, particularly if the procedure had structural implications. An example of such an operation is reducing the level of cargo carried in tanks. The committee also considered emergency mitigation options and associated equipment when related to ship design; cargo handling systems, for example, can help minimize oil spillage when placed at certain locations on the vessel. Among the options not addressed were: means of averting accidents, such as improved navigational aids; methods of altering the form of the cargo, such as mixing it with dispersants or jelling agents; and responses to oil spills, such as cleanup operations mounted from the vessel once cargo has been spilled.

Risk never can be eliminated entirely, but it can be reduced at some cost to a value acceptable to society. Society enjoys the benefits of oil transportation but now demands a substantial risk reduction, for which it presumably is prepared to pay reasonable costs.

This assessment addresses ocean-going oil tankers and barges over 10,000 DWT. Combination carriers "trading wet" (carrying oil) also are included. Chemical carriers and liquefied gas carriers were *not* addressed, because their cargoes, operations, and rules are vastly different from those of oil tankers.

The committee met seven times beginning in late 1989. The effort to gather and analyze information reached around the world, reflective of the breadth and complexity of the oil transportation system. The committee received more than a hundred contributions consisting of background material and suggested tank vessel designs, including some from Europe and Japan. Presentations were solicited from the tanker and barge industries, oil companies, the Coast Guard, ship classification societies, shipbuilders, university researchers, and environmental groups. In addition to reviewing existing studies and data on accidents and tank vessel design, the committee requested or conducted new investigations of the theoretical effects of collisions and groundings, the pollution-reduction potential and cost-effectiveness of alternative designs, and risk reduction potential. The committee also toured the EXXON VALDEZ during its repair in dry dock and the double-hulled CHEVRON OREGON, in port for maintenance.

The report is organized into four parts. Chapters 1 and 2 provide a basic overview of the tank vessel industry, covering traffic patterns, casualty history, design and operational principles, and the regulatory framework. Chapters 3, 4, and 5 comprise the technical core of the report. These chapters discuss the physical phenomena and engineering issues related to tank vessel design and present the committee's technical evaluation of 17 alternative tank vessel designs. Chapter 6 completes the committee's overall analysis with an economic assessment of the most promising designs. Chapter 7 examines the need for research.

The committee's conclusions and recommendations, contained in the Summary, are based on presentations made to the committee, the professional experience of committee members, reviews of past studies and new investigations developed by the committee, and deliberations by the committee. The report was reviewed according to the criteria established by the NRC's Report Review Committee.

The committee gratefully acknowledges the many individuals who contributed to the study. The broad scope of this report, and the difficulty of obtaining comprehensive information related to numerous topics, demanded an unusual effort on all fronts. In particular, the project could not have been accomplished without the help of the following people: Sean Connaughton of the American Petroleum Institute, who provided extensive background information; Thomas Allegretti of the American Waterways Operators, who provided information about the barge industry; Donald Liu of the American Bureau of Shipping, who helped explain the complex tank vessel design process; liaison representatives from the U.S. Coast Guard who offered continuing guidance, including Captain James MacDonald and his associates Stephen Shapiro, Daniel Sheehan, and Captain James C. Card; and other liaison representatives, including Frederick Seibold of the Maritime Administration and Lieutenant Commander Bert Marsh of the U.S. Navy.

In addition, the committee appreciates the assistance of those who provided data and information for analysis by the committee: David St. Amand of Booz-Allen & Hamilton, Inc., who prepared tanker activity graphics from data compiled by Lloyd's Maritime Information Services Ltd.; Ian White of The International Tanker Owners Pollution Federation Ltd., who provided ship and events data for major oil spills; Richard Golob of World Information Systems Inc. of Cambridge, Massachusetts, and the personnel of Cutter Information Corporation, of Arlington, Massachusetts, who provided quick response information about specific accidents; David Palk and his associates at Clarkson Research Studies Ltd. (London), who supplied world fleet statistics; Constantine Foltis, naval architect, Arlington, Virginia, who researched accident damage data from selected grounding and collision reports; and Arthur McKenzie, Director, Tanker Advisory, who provided much data about U.S. and world double-hull tank fleets. Cost information was obtained from those who made presentations to the committee as well as from several persons who provided written reports. The committee also appreciates the extensive cost analyses for VLCCs provided by John Parker, Chairman, Harland and Wolff Shipbuilding and Heavy Industries Ltd., Belfast,

Northern Ireland, and by Mitsubishi Heavy Industries Ltd. The committee also appreciates the comments received from many ship salvors who responded to the committee's survey concerning tanker salvage.

The committee acknowledges all who submitted proposals for alternative tank vessel designs and the many members of maritime industries and environmental community who contributed their insight on various issues. Special thanks is extended to those who made presentations at committee meetings, listed in Appendix J.

Finally, the Chairman wishes to thank all members of the committee for their time and expertise, especially those who made an extra effort to obtain or analyze information on specific topics, to attend extra meetings or work sessions, or to prepare drafts of particular sections of the report. These special contributions were essential to the interpretation of a difficult subject and the production of a well-informed report.

Contents

Executive Summary

This study, prompted by the March 1989 grounding of the EXXON VALDEZ in Prince William Sound, Alaska, focused on how alternative tank vessel (tanker and barge) designs might influence the safety of personnel, property, and the environment, and at what cost. In selecting designs to be considered, the committee included certain operational options that might minimize the oil spilled in an accident. The study did not consider means of averting accidents, altering the form of cargo, or responding to oil spills.

RISK OF SPILLS IN U.S. WATERS MAY BE INCREASING

The threat of pollution exists wherever tank vessels travel, and traffic in U.S. waters is increasing: Projections call for up to a 50 percent increase in imports of crude oil and petroleum products by the year 2000.

U.S. control over the condition of the tankers calling on U.S. ports is limited. More than 80 percent of these tankers are foreign flag. With the depletion of the present Alaskan oil fields, the proportion of U.S.-flag carriers is likely to decrease further during the coming decade.

One five-hundredth of 1 percent of the total amount of oil moving through U.S. waters is spilled. The 9,000 tons (on the average) of crude oil and petroleum products that is spilled annually in U.S. waters can be damaging from environmental, economic, and social perspectives. Large spills (30 tons, or about 10,000 U.S. gallons, and greater) comprise less than 3 percent of the events, but they cause nearly 95 percent of the accidental spillage in U.S. waters. Huge spills such as the EXXON VALDEZ can be devastating.

IMPROVED TANK VESSEL DESIGNS SHOULD REDUCE, BUT WILL NOT ELIMINATE, THE RISK OF OIL SPILLS

Improved tank vessel design is one way to reduce the risk of oil pollution, but the risk cannot be eliminated entirely. However, to assure some risk reduction, tank vessels of any design must be properly constructed, operated, inspected, and maintained. Whether adequate risk reduction can be achieved at reasonable cost through design changes is a controversial question; opinion varies widely.

Accidental oil spills, which constitute 20 percent of the total oil pollution from maritime operations and accidents, can result from collisions, groundings, structural failure, fires, explosions, or machinery failure. No single cause has dominated the historical record of accident-related oil spills, although groundings have been the dominant cause in U.S. waters, and the cause of the next large spill cannot be predicted. Therefore, all causes must be considered in tank vessel design.

Results of this study indicate that no single design is superior for all accident scenarios. Therefore, the Oil Pollution Act of 1990 (OPA 90), which mandated double hulls for tankers traveling in U.S. waters, should be viewed as only an interim step to reducing oil spills. More work remains to be done, as follows.

EXISTING TANK VESSEL DESIGN STANDARDS ARE NO LONGER ADEQUATE

Modern vessel design methods have produced such "efficient" structures that traditional allowances for errors, unknowns, and deterioration have been reduced to a significant degree. As a result, what were once secondary concerns can become critical; in short, modern tank vessels are less robust than their predecessors. Some ship classification societies, which set structural standards, are questioning publicly whether current strength criteria are adequate. The committee believes they are not adequate.

Existing design standards should be strengthened to ensure proper (1) corrosion protection (the surface area to be protected in double-hull vessels can be nearly three times that of single-hull ships); (2) dimensions of structural members; and (3) use of high-tensile steel. Concerted action should be taken by the U.S. Coast Guard, through the International Maritime Organization (IMO) and the classification societies, to effect these changes in a timely manner.

Furthermore, naval architects traditionally have not designed tank vessels, at the detail level, to withstand collisions and groundings. Design based on the possibility of accidents, a practice common in many industries, should be considered for tank vessels. Research should be conducted to confirm the promise and methodology of incorporating this approach into design practices, irrespective of whether vessels have double hulls or any other basic structural concept.

AVAILABLE INFORMATION IS INADEQUATE FOR DECISION MAKING

The paucity of data on tank vessel accidents and oil outflows, gaps in knowledge concerning vessel structural behavior in accidents (especially groundings, where characteristics of obstacles are critical yet undocumented), and uncertainties concerning the quantification of environmental benefits resulting from design improvements make any but the most primitive conclusions subject to conditional scenarios, assumptions, and judgments ranging from the informed to the intuitive.

The following conclusions are adequately supported with facts:

- Double hull vessels in low-energy (typically low-velocity) accidents should not pollute.
- Vessels that carry cargo in contact with a single skin (with the sea on the other side) will cause some pollution in any accident where a cargo tank is penetrated. However, certain design alternatives will minimize the amount of pollution in some specified scenarios.
- High-energy accidents nearly always result in pollution. The relative advantages of various design alternatives in reducing pollution from particular scenarios are highly dependent on the assumptions made in the scenarios.
- A comprehensive research effort can address the endemic lack of data and knowledge, and would result in placing decisions more nearly on a factual basis, rather than on the basis of informed opinion.

Until the gaps in understanding are filled, all other findings, by this committee or any other analyst, will be influenced by numerous variables including personal judgments. An example is the Det norske Veritas (DnV) report, which analyzes the pollution-prevention effectiveness of tank vessel design alternatives (detailed in Chapter 5 and Appendix F). In using DnV data to estimate oil outflow for design combinations (not calculated in the DnV report), the committee found that the results were driven by DnV assumptions regarding extent of damage and configurations of cargo tanks. In particular, outflows for small tankers appear inconsistent with results for large tankers, primarily because of the cargo tank configurations selected for the small tanker. Thus, outflow estimates should be viewed only as indicative examples rather than as absolutes.

In view of the above, the committee's judgment was by necessity a key element in every aspect of this report. However, the committee did reach a strong consensus on its findings, as follows.

DOUBLE HULLS SHOULD REDUCE POLLUTION FROM GROUNDINGS AND COLLISIONS

When OPA 90 requirements are fully implemented over the next 25 years, double hulls should save (in absence of other risk-reduction measures) an estimated 3,000 to 5,000 tons of oil spillage per year in U.S. waters from collisions and groundings, based on available historical data. The savings represent roughly half of the current average annual spillage from vessel accidents in U.S. waters. The added transport cost would be more than $700 million a year, or on the order of one cent per gallon transported. On the basis of cost-effectiveness, the double hull is among the best values of the designs evaluated by the committee.

Double hulls are particularly effective in low-energy (typically low velocity) groundings and collisions. Of the alternatives considered by the committee, this option is the only one that places two structural barriers between the entire cargo and the ocean; thus, the double hull can prevent outflow as long as the inner barrier is not breached.

In addition to providing sufficient between-hull spacing for grounding and collision protection as well as sufficient accessibility to all spaces for maintenance and inspection, double-hull designs must incorporate, to achieve full potential, adequate outer plate thickness, and cargo and ballast tank sizes and arrangements that assure adequate stability when the ship is damaged. Minimum thickness criteria for outer plates should be established promptly; in the interim, the minimum should be no less than the outer plate thickness required of new single-hull vessels.

Accessibility of void spaces for inspection and maintenance is a significant concern, which grows with the number of compartments (as in double hulls). A spacing of about 2 meters minimum between hulls in tank vessels over 20,000 deadweight tons (DWT) should allow room for inspection.

More research is needed to determine the spacing between hulls that best satisfies all concerns. To provide pollution protection, inter-hull spacing should be approximately the ship's beam divided by 15 (B/15) or 2 meters, whichever is greater. However, shipowners should not be required to make double-bottom heights (whether as part of a double hull or as an independent design) much greater than 3 meters; higher double bottoms should contain permanent fixed access to ensure adequate inspection of the underside of the inner bottom. Further study of void space dimensions in small tank vessels (including those under 20,000 DWT) should be conducted to determine the impact of the "B/15 or 2 meter" criterion.

The latest single-skin tank vessel designs exceed international criteria meant to ensure adequate stability following hull damage. The introduction of more ballast compartments subject to flooding (as with double hulls and related concepts), and larger cargo tanks, could result in designs that are less stable, while still meeting current criteria. The committee is persuaded that current criteria do not ensure adequate stability following high-energy (high-speed) accidents. The criteria can and should be revised to ensure that double-hull and other possible new designs are not less stable following damage than single-hull tank vessels.

MERITS OF OTHER DESIGN ALTERNATIVES FOR NEW VESSELS

A total of 17 design concepts, as well as 3 combination concepts, were evaluated by the committee. Alternatives and combinations that received all evaluations—technical, outflow performance, and cost—were:

- double bottom
- double sides
- double hull
- hydrostatically balanced loading (hydrostatic control)
- smaller cargo tanks
- intermediate oil-tight deck with double sides (IOTD w/DS)
- double sides with hydrostatic control
- double hull with hydrostatic control

The committee did not identify any design as superior to the double hull for all accident scenarios. Double bottoms and smaller tanks do not offer comparable pollution-reduction potential, although each is cost-effective. Double sides are not cost-effective.

The committee sought to establish whether there were design alternatives that would meet or exceed the pollution control and cost-effectiveness of state-of-the-art tankers built to current standards (codified in the International Convention for the Prevention of Pollution from Ships, or MARPOL). OPA 90 establishes a different standard, specifying that design alternatives "provide protection . . . equal to or greater than that provided by double hulls." However, there are no generally accepted criteria for evaluating the equivalency of two designs. The difficulty is that the results of any analysis of the performance of tank vessels depends on the assumptions and the particular accident scenario analyzed. Different designs may perform better in some situations and worse in others.

The committee considered adding an operational feature, hydrostatic control, to various structural alternatives for new tank vessels. This feature makes use of the laws of physics involving hydrostatic pressure to reduce or prevent oil outflow resulting from bottom damage. The combinations considered were double hulls with hydrostatic control, double sides with hydrostatic control, and a conventional single hull with hydrostatic control. These concepts are not considered as desirable as the double hull; the primary reasons are:

- the effectiveness of hydrostatic control depends on operators' strict adherence to rules, rather than on a permanent feature of vessel design and construction; and
- hydrostatic control does not provide complete protection against oil outflow, due to tidal variations or wave action following a grounding.

Therefore, the committee does not favor hydrostatically balanced design options for new tank vessels.

In theory, another design concept, which also employs hydrostatic principles, could perform better than the double hull in certain circumstances and scenarios. The IOTD w/DS could spill less oil in high-energy groundings (under circumstances that would have penetrated the inner hull of the double hull) and in some collisions (because it employs wider side tanks). This design is unique in that the pressure of seawater forced into the vessel (rather than oil flowing out) in the event of tank rupture would be significantly greater than the pressures available in all other design alternatives. Another favorable attribute of this design is that it employs less structure in void spaces than double hulls; this should result in less risk of corrosion and easier inspection.

However, significant factors may undermine the pollution-prevention benefits of this design, including the advantages of its hydrostatic pressure features. Because the cargo tanks have a single bottom, these tanks would be penetrated in groundings more often than would cargo tanks within a double hull; furthermore, when the bottom of an IOTD w/DS vessel is penetrated, at least some oil would be spilled (i.e., poorer performance than the double hull in low-energy groundings). Moreover, the spillage could be aggravated by tidal or wave action or ship motion subsequent to an accident. The cargo tanks of the IOTD w/DS would be a greater challenge to inspect than a single-hull vessel built under existing standards. An undetected structural failure of the middle deck of the IOTD w/DS would effectively transform the vessel into a conventional single-bottom tanker with equivalent outflow potential. Failure of mechanical features (i.e., valve actuators) that isolate upper and lower tanks, and operational mishaps (i.e., human error) that compromise the separation of upper and lower tanks, would eliminate the hydrostatic balance integrity of the ship. Finally, this design increases operational complexity with respect to loading, unloading, and preparation for docking.

Weighing the pros and cons, some committee members consider the IOTD w/DS a practical and innovative application of available technology, which, if treated as an equivalent to the double hull and allowed to trade in commerce to the United States, would reduce pollution in several classes of accidents, including high-energy groundings (which have caused some of the largest oil spills). Others of the committee hold the judgment that the gap between theoretical principles and practical application is wide, that the design and its implementation are unproven, and that, pending more extensive evaluation, the nation deserves designs employing physical barriers, such as double hulls, over those that continue to employ a single hull, such as the IOTD w/DS.

Some committee members recommend that the U.S. Coast Guard extend the committee's evaluation of the IOTD w/DS alternative to determine whether this design is equivalent or superior to the double hull standard of OPA 90. The results of this analysis should be presented to the Congress and the IMO for action, as appropriate. Committee members who are skeptical of this design recommend that a more extensive, critical evaluation be conducted, and the results evaluated by the IMO, before the Coast Guard takes action relative to OPA 90.

Any of the design options considered by the committee, if fully implemented in U.S. waters, would add roughly one or two cents per gallon to the cost of oil transported, which might be reflected in a similar increase in the cost of gasoline to consumers. This amounts to total annual costs of at least $340 million, and, at most, roughly $2 billion, depending on the design chosen.

Other design alternatives may be proposed in the course of future research. New proposals should be considered.

Regardless of designs chosen, certain features should be standard on all new tank vessels. Towing fittings of a design recommended by IMO would help in towing a vessel, if that should be necessary after an accident or machinery breakdown (conventional mooring fittings are not as sturdy). All new tankers and barges should be built with towing fittings mounted on the bow and stern.

In addition, all new tankers should have a reliable onboard system for transferring cargo from a damaged tank to an intact tank or another vessel. Practical concepts have been developed utilizing available equipment for this purpose. These features are likely to be less applicable to barges, as they are unmanned.

In addition, IMO and the Coast Guard should prohibit the placement of cargo piping in ballast tanks, to reduce the danger of fire and explosion due to hydrocarbon vapor leakage. Finally, the passive vacuum system (for use on fully loaded cargo tanks) deserves further research and development.

FEWER OPTIONS ARE AVAILABLE FOR BARGES

In general, fewer pollution-resistant design options are appropriate for barges. Barges are unmanned and their attendant tugs have minimal crew complements; thus, the dedicated presence of skilled personnel for operational control is limited.

Also, barges carry a wide range of petroleum products of varying densities and properties. The diversity in cargo-loading levels (depth of oil or product in tanks) required to implement hydrostatic control implies crew vigilance, knowledge, and presence to a degree not consistent with barge operations. At the same time, towed barges are less maneuverable than self-propelled ships and hence, more damage prone. Physical structural barriers for secondary containment, such as double hulls, double sides, or double bottoms, are more reliable for barge operations. The intermediate oil-tight deck might be adaptable for barges, but would require carefully controlled loading and discharge practices.

DOUBLE HULLS NEED NOT INCREASE INCIDENCE OF FIRES OR EXPLOSIONS, IMPAIR POST-ACCIDENT STABILITY, OR COMPLICATE SALVAGE

Various issues related to double bottoms, double sides, and double hulls (or any design incorporating large void spaces) have caused considerable controversy. Increased risk of fire and explosion, possible vessel instability after an accident, perceived salvage difficulties, and increased personnel hazards are cited by critics of these designs. The committee received various arguments along those lines but little conclusive evidence.

Void spaces involve risk of fires and explosions due to the potential accumulation of hydrocarbon vapors, which can enter through cracks or pits in bulkheads adjoining cargo tanks, or through a defect in the cargo piping system. Although double hulls have significantly increased void spaces compared to single-hull tankers (with up to three times the shell and bulkhead area exposed to corrosion), there is no reliable evidence of increased incidence of fires or explosions in existing double-bottom or double-hull ships.

However, the risk cannot be ignored; planned maintenance and thorough inspection are critical. Coast Guard inspections already are considered barely adequate, and owners must be relied upon to conduct complete inspections. Additional requirements for particular structural configurations, especially those increasing the need for proper maintenance and inspection, will add to the existing workload. Furthermore, most existing double-hull vessels carry petroleum products, which usually are less corrosive than crude oil. In double-hull crude carriers, diligence will be needed to monitor for corrosion from the cargo tank side as well as the ballast tank side, especially on horizontal surfaces. Means of detecting and combating the problem should be assessed as well.

Double-hull designs have large void spaces exposed to potential damage, flooding, and additional weight, which implies possible instability following an accident. However, tankers designed to international (MARPOL) standards tend to be exceptionally stable and generally exceed current damage stability requirements. Double-hull vessels can be designed not only to meet requirements for damage stability, but also to survive more severe accident damage than currently required.

The committee believes, however, that criteria for evaluating damage stability of double-hull tank vessels should be tightened to assure damage stability approaching that of conventional single-skin MARPOL tankers.

Salvage concerns are only partly supported. Piercing of void spaces of loaded ships (whether at bottom or sides) will cause sinkage and/or heeling depending on the position and size of the void. This may or may not be a problem, depending on whether the vessel remains firmly stranded following an initial grounding. Greater amounts of flooded void spaces would help keep a stranded ship firmly aground, thus reducing the possibility of pounding under wave action. However, this would increase the amount of cargo to be removed to refloat the ship, or the amount of water to be blown from flooded voids.

Most tankers grounding (or stranded as a result of grounding) at service speeds (14 to 16 knots) will remain firmly stranded; the possibility of losing the entire cargo from a stranded double-bottom tanker (where a single-bottom tanker would have survived) is remote. There are no salvage-related concerns that should limit the use of properly designed double hulls.

PERSONNEL HAZARDS DIFFER WITH VARIOUS DESIGNS, BUT ARE VERY DIFFICULT TO EVALUATE

Some design features may increase the hazard to the crew during normal operations. These include greater amounts of void spaces, which will have to be periodically inspected; void spaces lacking direct access to the weather deck (such as a double bottom or double hull); and greater numbers of cargo tanks, or cargo tanks located below other tanks (such as in the intermediate oil-tight deck design).

There is insufficient evidence to judge the degree of personnel risk associated with various designs. The committee believes these concerns do not represent unmanageable risks in any of the designs considered, but rather constitute important factors that will demand continuing vigilance on the part of ship operators.

EXISTING VESSELS WILL COMPRISE THE MAJORITY OF THE FLEET SERVING THE UNITED STATES FOR MANY YEARS

The phase-out of single-hull tank vessels trading to U.S. coastal ports will begin in 1995, and over the following 20 years they will be replaced with double-hull vessels. Therefore, means for improving pollution control on existing vessels should be considered.

Serious consideration should be given to requiring that all existing crude oil tankers promptly meet the latest IMO provisions for pollution prevention for new tankers (MARPOL). Most significant are the requirements for segregated ballast tanks (SBT carry only ballast, never cargo), crude oil washing (a system for rinsing out cargo tanks with oil), and possibly protective location of ballast tanks. The 1978 conference for the MARPOL convention recommended that IMO set a date by which new ship requirements would be applied to existing ships, thereby enhancing protection against operational outflow, but, in fact, this action never has been implemented. A multi-year phase-in period would be necessary, as "pre-MARPOL" vessels would require shipyard work.

Another possibility is a requirement for hydrostatic control, which could be implemented immediately without structural overhaul, although more research is needed to determine the optimal method of achieving hydrostatic control. Moreover, the design and condition of each vessel would have to be checked, and all bulkheads inspected, to ensure sufficient strength to withstand the added liquid cargo sloshing. Finally, leakage from a hydrostatically loaded tank vessel following a grounding may be exacerbated by wave action, current, falling tide, or vessel heel or trim (inclination). Implementation of this concept for the entire tank vessel fleet serving the United States would increase the number of vessels operating in U.S. waters, thereby potentially increasing the risk of accidents. Moreover, the cost of this measure could be substantial and warrants further study. The passive vacuum system might be another alternative, depending on the results of further testing.

Retrofitting double hulls on existing tank vessels is possible, but potential difficulties could result from combining new and old structural members. Furthermore, retrofitting would be, in general, much more expensive than hydrostatic control. Replacement of the entire ship forward of the machinery space would entail fewer technical difficulties and would require less time in a shipyard; however, it may be more expensive than retrofitting a double hull.

The committee urges prompt evaluation of a multi-faceted program to improve pollution control for existing vessels. One element would be the complete implementation of existing MARPOL requirements. The second element would be the consideration of implementing hydrostatic control. Additional emergency features, such as towing fittings and emergency cargo transfer systems that can help reduce the consequences of casualties, also should be considered.

While the potential costs and benefits of such a program are difficult to evaluate, it is reasonable to conclude that such requirements might expedite the construction of new vessels with double hulls. Implementing hydrostatic control on existing vessels would reduce cargo capacity on the order of 15 to 20 percent. If all existing vessels had been built at present costs, the limited data available to the committee indicates that reducing cargo capacity to accommodate hydrostatic loading (if implemented immediately) would cost about $1.1 billion per year, while saving 3,700 to 4,300 tons of oil spilled annually. Existing vessels were built over many years at a wide range of costs, so the total cost could be considerably less than the dollar amount stated. The actual impact on ship rates would be influenced by the relationship between vessel supply and demand. If existing vessels were insufficient to meet the cargo capacity shortfall, then costs would increase until new vessels were built.

This simplified model does not consider the additional person-hours necessary to monitor and enforce hydrostatic control, nor the increased vessel traffic resulting from carrying less cargo in each vessel. Furthermore, the benefits (and costs) of this program would diminish as new vessels replaced existing ships.

In sum, this type of program for existing vessels deserves consideration because of the potential environmental benefits. The program should guard against encouraging retention of older vessels.[1] It is outside the scope of this committee to judge whether similar benefits could be achieved at more or less cost through non-design measures (such as crew training, electronic display charts, or vessel traffic systems).

A COMPREHENSIVE RESEARCH PROGRAM SHOULD BE MOUNTED IN THE UNITED STATES

The state of knowledge regarding the precise circumstances and structural effects of actual tank vessel accidents is so inadequate that any assessment of design alternatives will produce results that are dependent on the chosen assumptions and accident scenarios—artificial rather than actual criteria. This is true even of this study.

Design criteria tend to "fix" technology at a point in time, thus inhibiting innovation and removing the incentive to advance ship technology and design. Performance standards are preferable, in that they tend to promote new development in terms of structural and operational innovations that would result in meeting or surpassing the standards. However, to achieve that goal, needs include: (1) an integrated micro-understanding of the dynamics of ship structural failure, and related factors; (2) long-range research in failure theory; (3) protocols leading to mandatory engineering documentation of casualties; and (4) computational models resulting in outflow predictions.

Major research programs directed to design of a "spill-free" tanker are being mounted in other countries. Japan plans a seven-year study on alternative tanker designs. Norway plans a five-year study on tanker design, operations, and oil spill cleanup. The U.S. federal government, industry, and academia should cooperate in a coordinated, substantive research effort directed to:

• performing a comprehensive risk-assessment study that would lead to establishment of future risk-based design goals for tank vessels with attendant compliance guidance;
• accomplishing the basic research needs noted;
• testing and evaluation of design concepts (including theoretical analyses, model tests, and field trials);
• advancing the capability to assess and value natural resource damages (better understanding of the environmental effects of oil spills and the feasibility and cost of restoration, and the development of accepted methodology for valuing environmental benefits would permit more reliable cost-benefit analyses of vessel design alternatives and other means of pollution control); and
• achieving optimal pollution control by integrating use of design alternatives with operational considerations.

OPA 90 authorizes such a research program. The program should be coordinated with foreign research centers, notably in Norway and Japan, and the efforts of the IMO. A comprehensive research effort is essential if important, far-reaching decisions concerning the future of oil transportation by tanker are to be made on the basis of fact, as contrasted with informed opinion.

NOTE

1. Failure to require pre-MARPOL vessels to retrofit SBT requirements before employing hydrostatic control would give these vessels an economic advantage compared to newer vessels meeting MARPOL standards.

TANKER SPILLS

1

Introduction

THE NATURE OF THE PROBLEM

Wherever tank vessels travel, laden with crude oil or petroleum products, the threat of pollution exists. The magnitude of the risk in U.S. waters is related to many factors, which this chapter will attempt to define.

Factors that warrant examination include tank vessel traffic patterns around the United States, the types of vessels involved, and local hazards. Also worth considering are historical patterns of accidents and oil spills, which reflect all of the factors listed above.

From the following discussion, a set of facts will emerge that should guide any attempt to reduce the pollution risk. These facts assisted the committee in determining which tank vessel design alternatives would be most effective.

Clearly, in the current popular view, too much oil is lost, particularly in the occasional very large spills. In other words, the risk of pollution is unacceptably high. The present study represents an attempt to identify, and recommend ways to reduce, that risk.

Tanker Traffic in U.S. Waters Is Heavy and Is Increasing

Along every U.S. coast, the low silhouette of a laden tank vessel—a tanker or barge—is a common sight. These vessels carry the leading commodity in international trade, crude oil, and its many refined products such as gasoline.

Since the end of World War II, tank vessels have become more numerous and more imposing. More than 3,000[1] are in service each day, some making voyages of more than a month, others putting to sea for only days at a time. They can be mammoth, up to roughly 400 meters in length. These giants are the largest moving objects made by man; a single cargo tank can hold more than twice as much oil as an entire World War II-era tanker.

Tank vessel traffic around the United States is heavy and likely to increase. Of the 1.7 billion tons[2] of crude oil and products transported by sea annually worldwide (including intercoastal trade), more than a third (0.6 billion tons) passes through U.S. waters.[3] On average, nearly five tankers (10,000 DWT and larger) a day call on Houston, and four a day on New York, the nation's first and second busiest ports for tanker traffic (10,000 DWT and greater).[4] In 1988,[5] more than 97 million tons of crude oil were shipped from Valdez, Alaska, the only major U.S. crude loading port (these "exports" go to other U.S. ports and to the Caribbean). The same year, more than 101 million tons of crude and product passed through New York, the port handling the greatest volume. Figure 1-1 reflects U.S. seaborne movements of crude and product for the leading 15 ports in the United States.

This activity is expected to intensify throughout the 1990s, due to the nation's growing thirst for oil coupled with its declining domestic production. In a 1988 report, the Department of Energy projected that daily crude oil

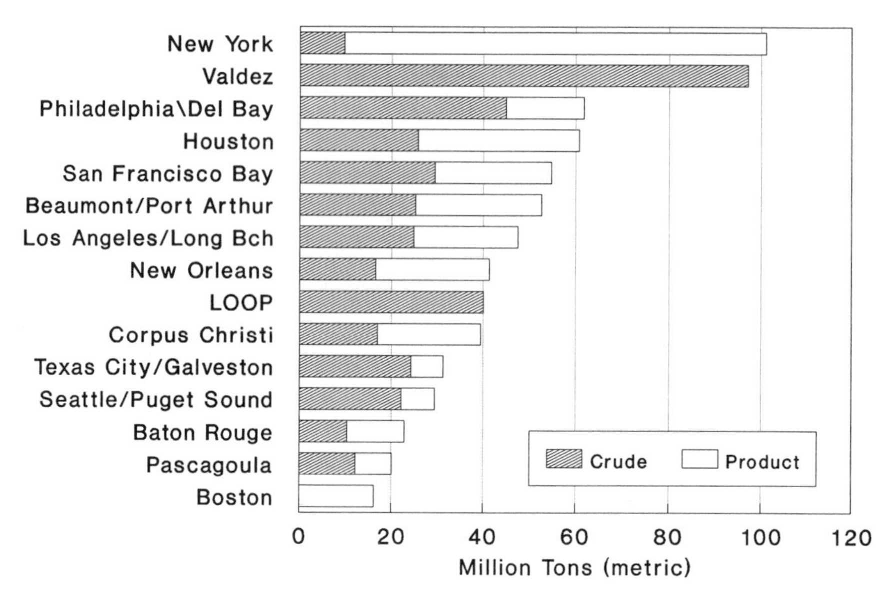

FIGURE 1-1 Petroleum volume by major port complexes: crude and product import/export—1988. Source: U.S. Army Corps of Engineers.

consumption would rise from roughly 17 million barrels per day (mbd) (2.3 million tons) to 19 mbd (2.5 million tons) by the end of the century, while domestic production is expected to fall from 10.5 mbd (1.4 million tons) per day to less than 8.6 mbd (1.2 million tons). (Energy Information Administration, 1989) In fact, this predicted increase in consumption may be conservative, preliminary 1990 data indicate. Although the precise effect on imports will be influenced by the price of oil, the EIA projects an increase in import volume of up to 50 percent by the end of the century.[6] Figure 1-2 reflects the range of projected imports for 1990, 2000, and 2010 and their relation to the oil price. These estimates preceded the current war in the Persian Gulf; while significant changes in supply, demand, and oil cost may be brought about by this conflict, any attempt to adjust projections at this stage would be highly speculative.

In 1988, more than 7 mbd (equivalent to 360 million tons per year) of crude oil and refined products were loaded or discharged through U.S. ports by tanker, in an estimated 14,000 port calls.[7] The average tanker size was 80,000 DWT. The projected increase in seaborne imports would lead to a 50 percent increase in tanker port calls to 21,000 by the year 2000, assuming no change in average tanker size.

Regardless of the extent of the increase, these data portend the nation's growing reliance on foreign oil and petroleum products—and on seaborne transport, which brings in most of the foreign supply. This year (1990), half of the nation's oil demand will be met by foreign suppliers. The continental United States will continue to receive oil shipments from Alaska by sea, but existing production levels in that northernmost state are expected to decline sharply during the decade. This supply would be replaced in the late 1990s either by development of present arctic oil reserves, or,

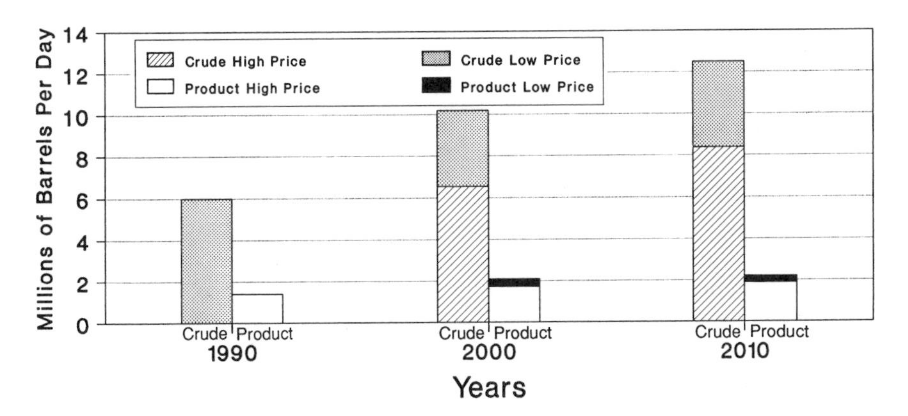

FIGURE 1-2 U.S. petroleum import projections: high and low oil price scenarios. Source: U.S. Department of Energy.

more likely, by foreign imports. In sum, the amount of crude and product imported to North America by tankship is expected to increase steadily in the current decade.

Traffic Is Heaviest from Alaska to the West Coast and Along the Gulf of Mexico and for Imports to the East Coast

Most of the tank vessel activity in U.S. coastal waters involves the importation of foreign crude oils and products. Thirty percent of the oil comes from the Middle East, 25 percent from the Caribbean, and 20 percent from West Africa. Smaller but still substantial amounts come from Northern Europe and Southeast Asia.

During 1989, seaborne movements of crude oil and product in U.S. waters consisted of the following major traffic patterns[8]:

• Traffic along the West Coast was mostly crude shipments from Alaska to the West Coast ports and Panama (97 million tons). This traffic, involving an average of more than 850 tanker loadings or trips per year, accounts for nearly one-sixth of the tanker traffic in U.S. waters, including intracoastal movement. Foreign imports and intracoastal domestic movements bring the total West Coast traffic to 131 million tons. By the end of the decade, much of this oil supply is likely to be replaced by foreign tanker-borne imports from Indonesia and the Caribbean region, as well as pipeline imports from Mexico and other parts of the United States. (Projections of imports from the Middle East are highly speculative at present.)

• Traffic along the U.S. Gulf of Mexico consists of crude oil imports to Gulf ports and product shipments along the Gulf and to the East Coast (approximately 310 million tons). Crude imports come primarily from the Persian Gulf and the Caribbean; the remainder comes from Africa and the North Sea, as well as from Alaska via Panamanian pipelines and transshipment by tanker.

• Traffic along the East Coast is primarily foreign imports of crude and refined products from the Persian Gulf, Africa, and the Caribbean. In addition, domestic movements from the Gulf to the East Coast brought total movements to an estimated 151 million tons. Tanker traffic (barges are not included) passing through the Florida Straits is reflected by the number of laden transits and the size of tankers shown in Figure 1-3. This is one of the most concentrated maritime shipping lanes along U.S. coasts.

Intercoastal and Intracoastal Activity Is a Large Part of U.S. Trade

Intercoastal and intracoastal trade by tankers or ocean-going barges[9] is a significant portion of the petroleum tank vessel activity in U.S. waters,

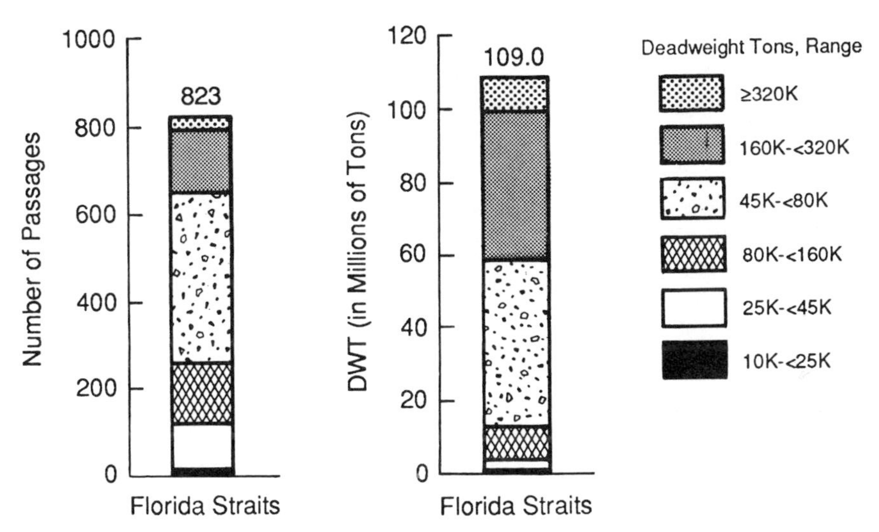

FIGURE 1-3 Number and tonnage of laden transits in the Florida Straits. Source: Lloyd's Maritime Information Services, Ltd.

comprising over a fifth of the total seaborne crude and product trade (125 million tons in 1988). Most coastal activity, for tank vessels over 10,000 DWT, consists of lightering[10] of crude oil from large tankers to refineries along the Gulf Coast (22 percent). Product distribution by tanker along the Gulf Coast and from there to the East Coast is also significant (38 percent of the total). Large ocean-going barges carry most of the product transported along the East Coast (17 percent of the total national coastal tonnage). Barges play a secondary role to tankers in Gulf product distribution; barges are not a major element on the West Coast.

Coastal trade in crude oil remained steady from 1983 to 1988. Over the same time period, coastal product shipments declined by about 17 percent to about 85 million tons; the difference was replaced by increased product imports.

A significant result of the expected need for seaborne imports,[11] added to possible increases in coastwise tanker traffic and cargo transfer from large to smaller tankers (lightering), will be a major increase in the number of tank vessels in U.S. intercoastal and intracoastal service. Because of port limitations, vessel size often is limited to about 80,000 DWT. The average size of vessels calling on U.S. ports may increase, however, as product increasingly is delivered directly from overseas in ships that are larger than those typically employed in domestic trade. In addition, offshore ports may be expanded, or new ports built to accommodate larger tankers, by the end of this decade or later.

The Type and Control of Tank Vessels in U.S. Trade Varies

In 1989, 1,500 different tankers visited (often more than once) the 15 busiest U.S. port complexes surveyed by the committee (see Figure 1-4). The vast majority—over 80 percent—were foreign flag.[12] The exceptional case is the Alaska trade, in which U.S.-flag ships command 96 percent of the activity due to legal requirements.

Overall, about 50 percent of the tanker calls in U.S. ports are made by ships under direct U.S. control, as indicated by the activity in 15 major U.S. ports shown in Figure 1-4. However, much of the Valdez, Seattle, and Los Angeles/Long Beach activity accounted for exports and imports of Alaskan oil; foreign-flag tankers dominate the remainder of U.S. import activity. Of the active worldwide fleet, in tonnage (approximately 250 million DWT for tankers over 10,000 DWT), 92 percent is owned and operated by foreign-flag companies and countries. Owners are primarily independent shipping companies (67 percent of total worldwide tonnage), and the remainder are oil companies (17 percent) and governments (16 percent), as shown in Figure 1-5. Of 3,048 tankers in the world fleet, 203 or 7 percent are U.S. owned (Maritime Administration, 1990), and 257 tankers or 8 percent are U.S. flag.[13]

The condition of these tankers, both foreign- and U.S.-flag, varies widely. Construction standards typically are implemented by legislation in each flag

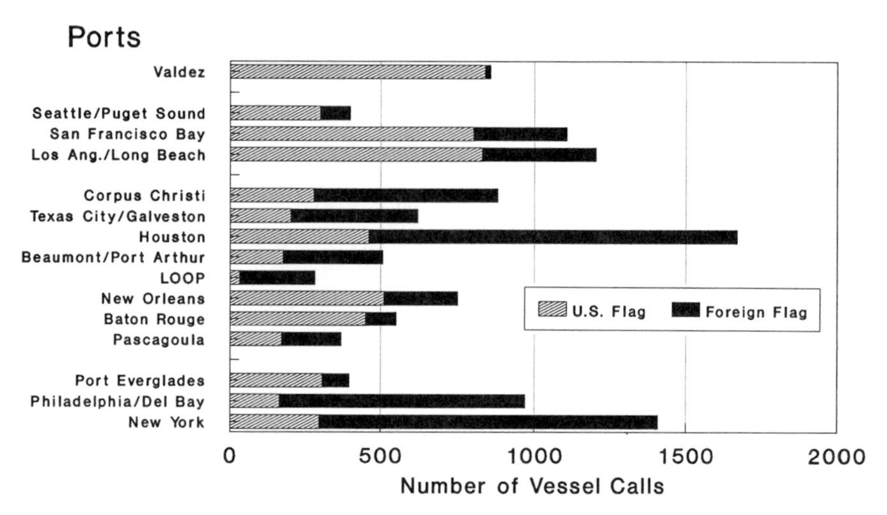

FIGURE 1-4 U.S.- and foreign-flag tanker calls* to major U.S. ports—1989. Source: Lloyd's Information Service, Ltd., and committee estimates from U.S. Army Corps of Engineers data. Note: Texas City/Galveston, New Orleans and Seattle/ Puget Sound port calls are committee estimates derived from U.S. Army Corps of Engineers 1988 data. * Calls are for tankers, 10,000 DWT and greater.

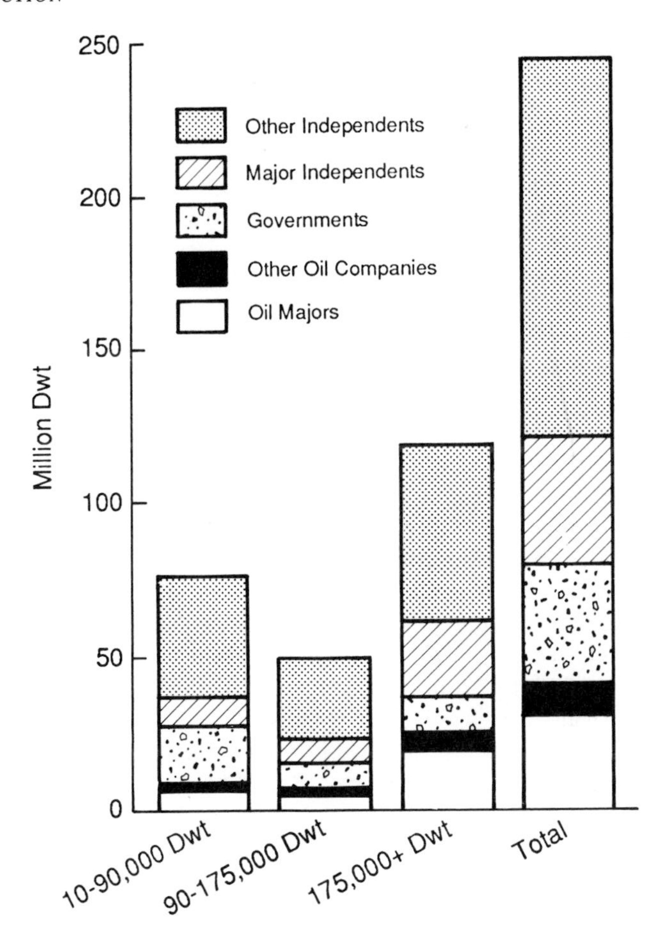

FIGURE 1-5 World tanker fleet ownership—1990. Source: Drewry Shipping Consultants, Ltd.

state and by the various ship classification societies; thus, standards differ. Moreover, even though ships must pass periodic surveys and inspections to remain "in class," ship owners adopt different approaches to vessel manning, maintenance, and equipment.

Construction practices also vary worldwide. Ships built for the U.S. ocean-going domestic trade (ships operating between U.S. ports) are required by law to be constructed in U.S. shipyards (so-called "Jones Act" ships). No such constraint applies to ships engaged in international trade with U.S. ports; therefore, ships are built at the yards offering lowest costs and are built to acceptable international design and construction standards. At present, no tankers over 10,000 DWT are being built in the United States.[14]

In recognition of varying flag-state regulations, the United States is among the nations that have adopted a practice of "port state control" to guard against the hazards posed by unsafe vessels. Thus, the Coast Guard may inspect foreign-flag vessels entering U.S. ports to ensure compliance with international standards for safety of life and structural condition. In addition, U.S. law requires that a foreign vessel's casualty and pollution history be evaluated prior to its entry into U.S. waters (U.S. Coast Guard, 1990b). Coast Guard practices regarding both domestic and foreign vessels are discussed in Chapter 2.

In all, the material condition of a tanker is influenced by a host of factors, including age, size, country of registry, classification society, and owner maintenance policies (U.S. Coast Guard, 1990b).

With the preceding uncertainties noted, some specifics can be ascertained regarding tank vessels serving the United States.

The size of vessel chosen for a particular trade route depends on the length of the voyage, the type of cargo, the storage capacity at the receiving end, and physical port restrictions (primarily water depth) at the ports visited. Because of port depth limitations, most voyages in U.S. waters are made by tankships in the midsize (50,000-100,000 DWT) category. In 1989, 301 such vessels—nearly half of the world total of 618 in this size range—made 1,369 trips, and possibly two to three times as many port calls, in U.S. waters.[15] The largest tankers (above 320,000 DWT) can call on only one U.S. port, the deepwater Louisiana Offshore Oil Port (LOOP), although larger tankers lighter off the U.S. Gulf Coast. A breakdown of the tanker sizes calling on eight leading U.S. ports is shown in Figure 1-6.

The vast majority of tankers calling on the United States—and most of those in the Valdez/West Coast crude oil trade[16]—have a single hull.[17] Indeed, 79 percent of tankers (10,000 tons and greater) worldwide have a single skin (Tanker Advisory Center, 1990). In the U.S. fleet, nearly one of every six tankers carrying crude oil or product (40 tankers) has a double bottom or full double hull. (These hull designs are described in Chapter 5.) U.S.-flag operational experience with double-hull and double-bottom tankers is extensive: more than 500 ship-years with 27 groundings reported (Tanker Advisory Center, 1990), none of which resulted in major damage or pollution. A similar finding was reported by the Coast Guard in an assessment of the pollution-control performance of tankships with double bottoms. In 54 groundings in U.S. waters between 1977 and 1987, no pollution resulted; foreign-flag tankers caused no pollution in 12 groundings (U.S. Coast Guard, 1990e). As a point of reference, overall casualty reports (for both U.S.- and foreign-flag tankers of single- or double-hull designs, involved in either groundings or collisions) indicated a pollution involvement (not defined) for about 9 percent of the casualties (U.S. Coast Guard, 1990e).

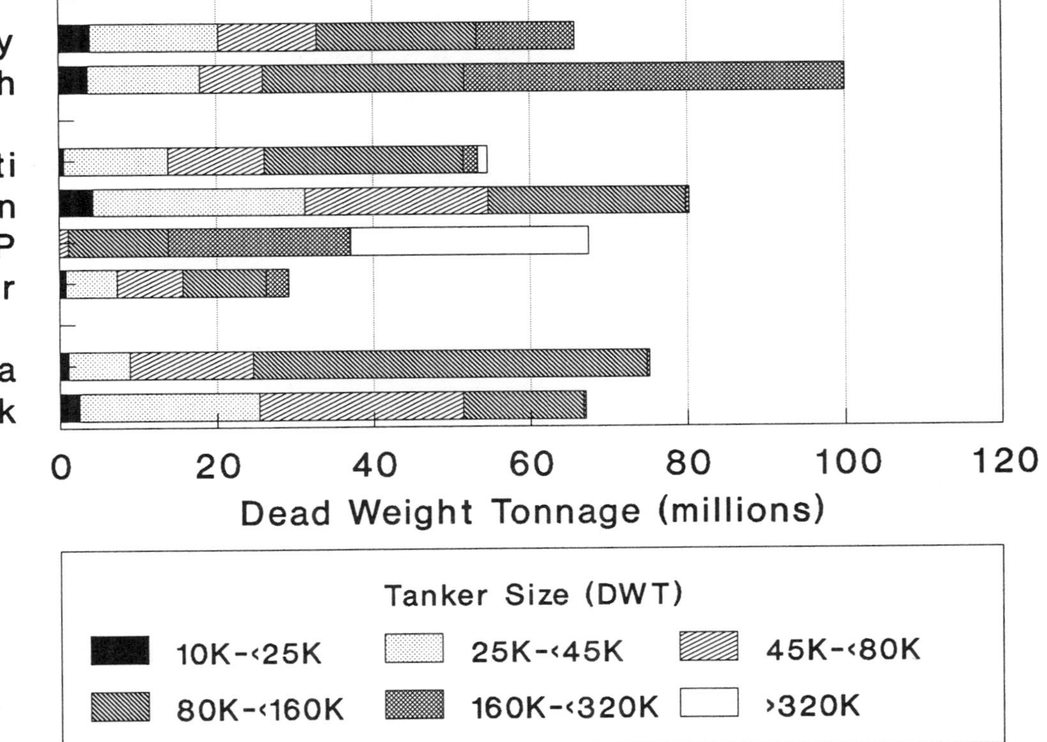

FIGURE 1-6 Deadweight tonnage: tanker size & port—selected ports—1989. Source: Lloyd's Maritime Information Services, Ltd.

The average U.S. double-bottom or double-hull tanker is approximately 70,000 DWT. About 12 percent of the combined crude and product tonnage, both U.S. and foreign, passing through the 15 largest U.S. ports is carried in these hull designs. Most double-hull vessels carry petroleum products, which are less corrosive than crude oil.

In general, smaller double-bottom tankers carry petroleum products or chemicals, and larger double-bottom tankers are crude oil or "combination carriers" for both dry and liquid cargo.

Ocean-Going Barges

Compared to the diverse tanker fleet, the ocean-going barge fleet serving the United States is relatively homogenous. Barges operating in U.S. waters are all engaged in the domestic trade, and are all U.S.-built and flagged. The number of seagoing barges over 5,000 GT (approximate 10,000 DWT) is small, only 72, all committed to handling petroleum products. Of these, nine have double bottoms and six have double hulls (American Waterways Operators, letter report to the committee, April 6, 1990).

Barges generally are smaller than tankers, rarely exceeding 30,000 DWT (see Figure 1-7). The cargo tanks are smaller as well, and they are more likely to contain product than crude oil. In addition, barges normally are unmanned and do not move under their own power; they are towed or pushed by tugboats. The Coast Guard routinely inspects all tank barges that carry combustible or flammable liquid cargo in bulk. Ordinarily, these vessels are inspected internally and drydocked at least every three years; they undergo operational inspections annually.

Tank Vessel Pollution Accidents Are Diverse

There is no "typical" tank vessel accident scenario that leads to pollution. Nonetheless, a general understanding of the causes, nature, and consequences of accidents should aid in selection of the measures most useful in preventing or reducing pollution.[18]

Accident-related spillage of crude oil and petroleum products from ocean-going tank vessels is not the major source of petroleum inputs to the sea, as shown in a 1985 report by the National Research Council. According to this analysis, tanker accidents accounted for approximately 13 percent (0.4 million tons) of the 3.3 million tons of petroleum hydrocarbons entering the marine environment each year from all sources.[19] (See Table 1-1.) Larger inputs came from tanker operations (non-accident) and municipal and industrial wastes and runoff, based on data estimated a decade ago. A more recent estimate of spillage from maritime transportation (both accidental and operational) indicates that accidents contribute 114,000 tons or 20 per-

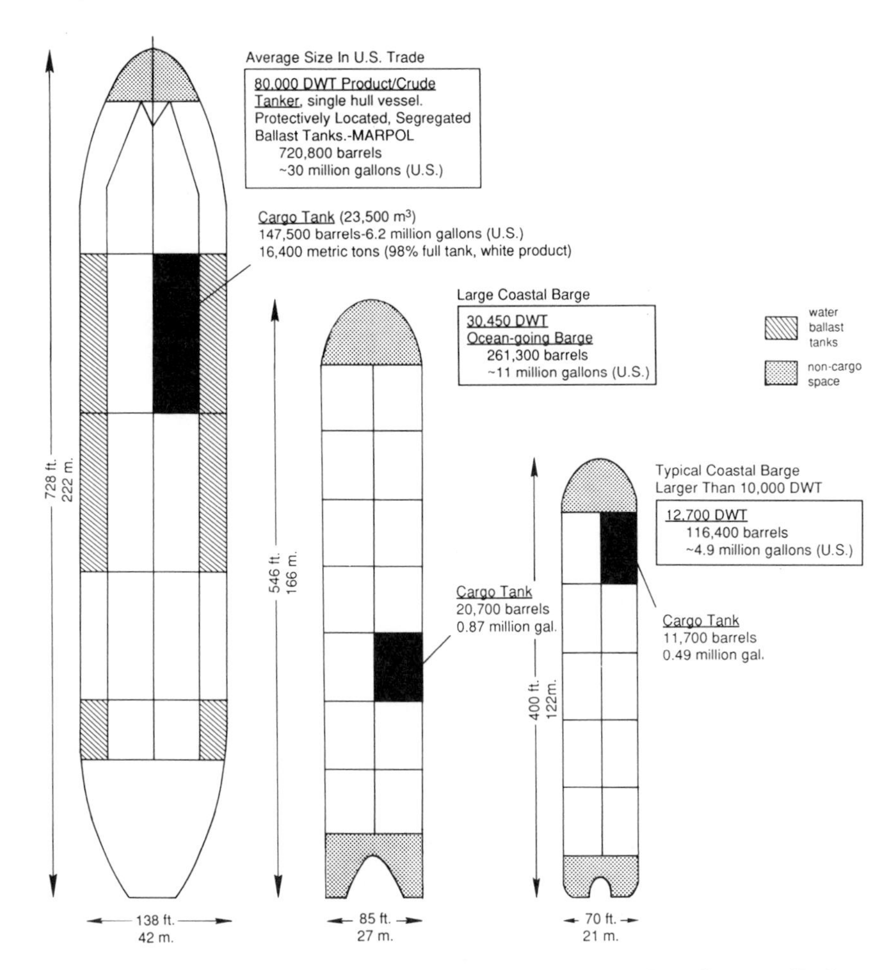

FIGURE 1-7 Comparison of tanker and ocean-going barge size. Sources: Tankers: Mobil Shipping and Transportation Company. Barges: MARITRANS and the American Waterways Operators.

cent of the 569,000 tons of oil entering the world's seas annually from maritime transportation (U.S. Coast Guard, 1990d). (See Table 1-2.) These total losses, while large, reflect a major reduction from the nearly 1.5 million tons (estimated) spilled annually worldwide from both tanker operations and accidents during the late 1970s and early 1980s (National Research Council, 1985). The decrease can be attributed to international cooperation in development and execution of rules for tanker design, clean ballasting, and vessel operations, and the supportive action of most major maritime states.

TABLE 1-1 Input of Petroleum Hydrocarbons into the Marine Environment—1985 Estimates

Source	"Best Estimated" Million Tons Annually (mta)	
Natural Sources		0.25
Offshore Production		0.05
Maritime Transportation		
Tanker Operation	0.70	
Tanker Accidents	**0.40****	
Other	0.40**	1.50**
Atmospheric Pollution Carried to the Sea		0.30
Municipal and Industrial Wastes and Runoff		1.18
Total		3.3**

SOURCE: National Research Council, 1985.
**Data rounded to the nearest 0.1 mta.

The number and volume of major accidental and operational spills from tankers also have decreased significantly during the last decade (see Figure 1-8). Tanker accident rates—collisions, groundings, fires and explosions, structural failures, and other incidents causing loss of life and/or pollution—were particularly high from 1976 to 1979 before returning to a lower level throughout the 1980s, although the volume of oil spilled continued to vary from year to year.

The rate of serious casualties to oil and chemical tankers (6,000 GRT

TABLE 1-2 Estimated World Maritime Operational and Accidental Sources of Oil Entering the Marine Environment (million tons annually)

	1990[1]	1981/85[2]	1973/75[3]
Bilge and Fuel Oil	0.25	0.31	*
Tanker Operational Losses	0.16	0.71	1.08
Accidental Spillage			
Tanker Accidents	**0.11**	**0.41**	**0.20**
Non-Tanker Accidents	0.01	—	0.10
Marine Terminal Operations	0.03	0.04	0.50
Drydocking	} 0.01	0.03	0.25
Scrapping of Ships		—	—
Total	0.57	1.50	2.13

SOURCES: [1]U.S. Coast Guard, 1990d; [2]National Research Council, 1985 (based on data from 1981); [3]National Research Council, 1975.
* Includes bilge and fuel oil.

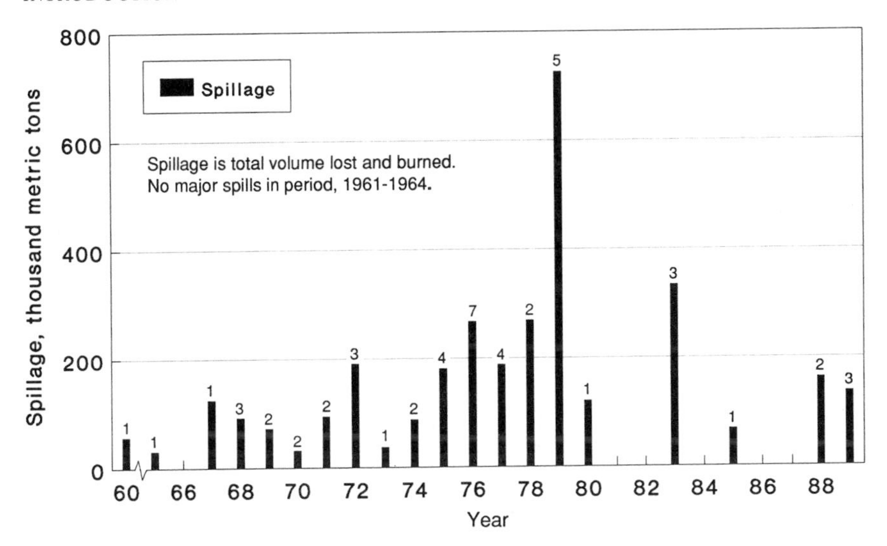

FIGURE 1-8 Tanker spillage and number of events: 50 major oil spills, 1960–1989. Source: The International Tanker Owners Pollution Federation Ltd.

and above, roughly equivalent to 10,000 DWT and above) reported by the International Maritime Organization was 1.92 incidents per 100 operating tankers in 1988. An analysis by the committee, comparing the estimated metric ton-miles of seaborne trade for crude oil and products to the number of serious casualties for the same tanker population, shows less change during the 1980s in serious casualties worldwide[20] (see Figure 1-9).

Most Tanker Casualties Do Not Cause Pollution

The vast majority of casualties do not result in pollution. Only 6 percent (518) of the accidents reported in an assessment of 9,276 accidents resulted in oil outflow (Lloyd's Register of Shipping, 1990). (In 18 percent of the cases—1701—information was insufficient to determine whether pollution had occurred.) At the request of the committee, Lloyd's analyzed its database of casualties resulting in outflow greater than 30 tons from tankers 10,000 DWT and larger. The results, covering 370 pollution-related accidents, are broken down by cause in Figure 1-10. These data show that, worldwide, groundings/strandings and collisions/rammings (contact) cause roughly equal numbers of major pollution incidents (those resulting in spillage of more than 30 tons, or approximately 10,000 gallons) for tanker sizes considered by the committee. This finding is also valid when considering volume spilled. The breakdown of attributed accident causes for the 50 largest tanker spills between 1960 and 1989 is shown in Table 1-3. This

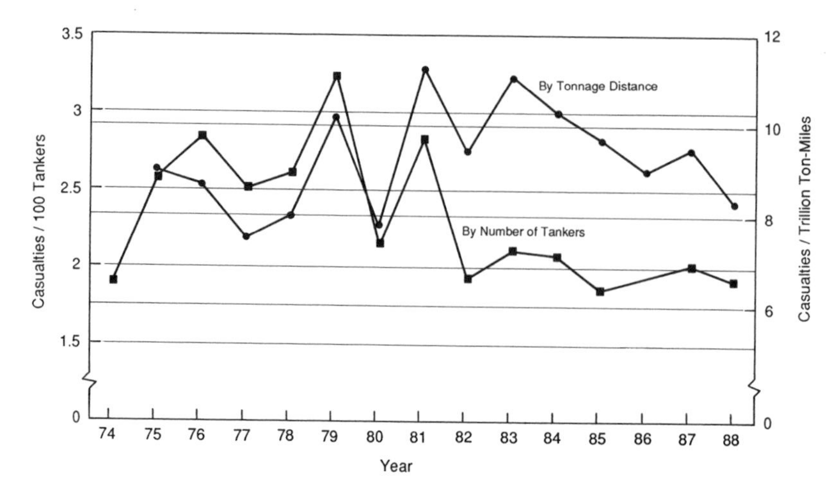

FIGURE 1-9 Worldwide rates of serious casualties to oil/chemical tankers, 1974-1988 (6,000 GT and above). Source: International Maritime Organization Analysis of Serious Casualties to Sea-Going Tankers, 1974-1988. Tonnage-distance analysis was prepared by the committee based on IMO casualty statistics and International Association of Independent Tanker Owners for miles data. *Casualties/trillion ton-miles data covers period 1975-1988.

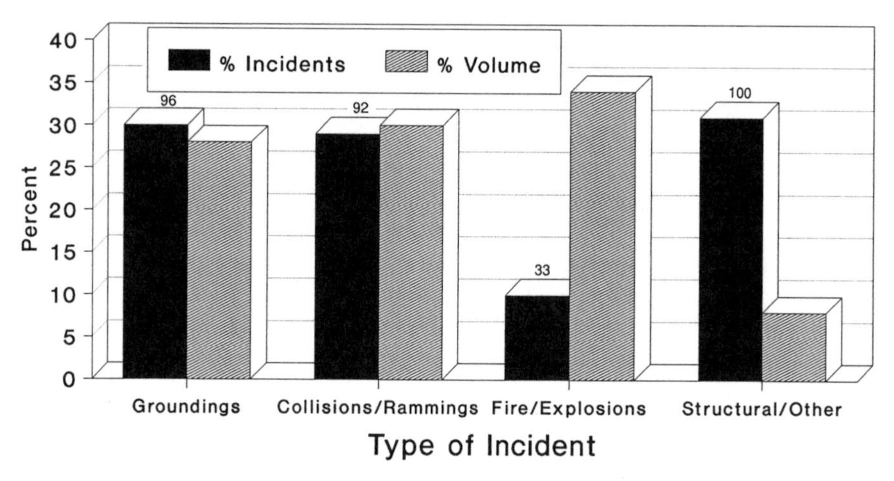

FIGURE 1-10 Major oil spills from tankers and causes: number of incidents and volume—world, 1976-1989. Source: Lloyd's Register of Shipping.

table shows the nearly equal prevalence of the various causes (grounding, collision, structural and machinery failure, and fires and explosions) in the 50 largest spills from tankers. Table 1-4 shows the years in which these spills occurred.

Particularly noteworthy in this table is the large volume of spillage resulting from a relatively small number of fires and explosions. As the initiating cause, fires and explosions produced the greatest outflow of the 50 largest spills. Two recent disasters were vivid reminders of this historical pattern: the KHARK 5, which sustained an explosion and fire leading to release of 76,000 tons of crude oil off Morocco in 1989 (the 11th largest spill since 1960), and the MEGA BORG, which caught fire in the Gulf of Mexico off Texas in 1990, losing 14,000 tons of crude. Fires and explosions also may cause pollution following a collision or other accident. Tanker spill databases are often unclear as to whether the "cause" refers to the casualty or the final event leading to loss of oil. This ambiguity can be misleading. For example, in the world's third-largest spill, the AMOCO CADIZ, oil outflow was due to grounding (stranding), while the casualty usually is attributed to failure of the steering system (machinery failure).

Groundings and Infrequent Large Spills Dominate U.S. Outflow

In U.S. waters, for tankers over 10,000 DWT, grounding events clearly dominate in terms of both numbers of accidents, and, particularly, volume spilled (see Figure 1-11). This should be no surprise in view of the shallow access to Gulf of Mexico and East Coast ports. Examples of shallow-water groundings include the ARGO MERCHANT in 1976 (approximately 25,000 tons outflow), the ALVENUS in 1984 (7,389 tons outflow), the ESSO PUERTO RICO in 1988 (4,050 tons), and the AMERICAN TRADER in 1990 (975 tons). These groundings account for a large portion of accidental spills from tankers that have occurred in U.S. waters since 1980.

Compared to worldwide spillage from major pollution incidents, major oil spills in U.S. waters are relatively low in volume, as shown in Figures 1-12a and 1-12b. The years 1982, 1984, and 1986, when particularly high or average U.S. spillage coincided with unusually low world spillage, are exceptions to this record. There is no evidence of a correlation between U.S. and world losses. The random nature of major spill events is reflected in the list of 50 major outflow incidents (shown in Table 1-3): The frequency, attributed causes, cargo losses, and locations all appear to lack any discernible pattern. The erratic nature of pollution is also evident in Figure 1-13: A few very large spills dominate the total outflow, comprising nearly 95 percent of spillage; but such spills comprise less than 3 percent of the events. In addition to the major groundings noted previously, the 13-year U.S. record is influenced strongly by the BURMAH AGATE (1979), the PUERTO

TABLE 1-3 50 Major Oil Spills from Tankers and Combined Carriers 1960 and 1965–1989

RANK/ RATING*	NAME	TANKER SIZE		SPILLAGE		TYPE	LOCATION	CAUSE				
		GRT	DWT	TONS $\times 10^3$ (Metric)	BARRELS $\times 10^3$			CO	GRN	F/E	S/H	N/K
1	ATLANTIC EMPRESS	128,398	292,666	257	1,890	CRUDE	WEST INDIES	x				
2	CASTILLO DE BELLVER	138,823	271,540	239	1,760	CRUDE	SOUTH AFRICA			x		
3	AMOCO CADIZ	109,700	237,439	221	1,628	CRUDE	FRANCE, ATLANTIC				x	
4	ODYSSEY	65,746	138,392	132	925	CRUDE	MID ATLANTIC			x		
5	TORREY CANYON	61,263	121,000	124	909	CRUDE	U.K., CHANNEL		x			
6	SEA STAR	63,989	120,300	123	902	CRUDE	GULF OF OMAN	x				
7	HAWAIIAN PATRIOT	45,246	101,038	101	742	CRUDE	HAWAIIAN ISLANDS				x	
8	INDEPENDENTA	88,690	152,408	95	696	CRUDE	TURKEY, BOSPORUS	x				
9	URQUIOLA	59,723	111,225	91	670	CRUDE	SPAIN, NORTH COAST		x			
10	IRENES SERENADE	50,903	105,460	82	600	CRUDE	GREECE			x		
11	KHARK 5	138,394	284,632	76	560	CRUDE	MOROCCO, MEDITER.			x		
12	NOVA	118,654	239,435	68	500	CRUDE	IRAN, GULF OF	x				
13	WAFRA	36,697	68,600	62	480	CRUDE	SOUTH AFRICA		x			
14	EPIC COLOCOTRONIS	37,469	64,000	58	427	CRUDE	WEST INDIES		x			
15	SINCLAIR PETROLORE	35,744	56,000	57	420	CRUDE	BRAZIL			x		
16	YUYO MARU NO 10	43,724	52,836	42	375	WH. PROD	JAPAN	x				
17	ASSIMI	33,847	59,032	50	370	CRUDE	OMAN			x		
18	ANDROS PATRIA	99,460	122,173	48	350	CRUDE	SPAIN, NORTH COAST			x		
19	WORLD GLORY	28,323	45,000	46	337	CRUDE	SOUTH AFRICA				x	
20	BRITISH AMBASSADOR	27,114	44,929	46	337	CRUDE	JAPAN				x	
21	METULA	104,379	206,719	45	330	CRUDE	CHILE		x			
22	PERICLES G.C.	38,915	59,096	44	324	CRUDE	QATAR			x		
23	MANDOIL II	25,313	45,000	41	300	CRUDE	USA, WEST COAST	x				
24	JAKOB MAERSK	48,252	88,000	41	300	CRUDE	PORTUGAL		x			
25	***BURMAH AGATE***	32,285	62,663	41	300	CRUDE	USA, GULF	x				

26	J. ANTONIO LAVALLEJA	68,931	131,663	38	280	CRUDE	ALGERIA		x			
27	NAPIER	23,690	38,561	37	270	CRUDE	CHILE		x			
28	***EXXON VALDEZ***	94,999	214,861	36	267	CRUDE	USA, ALASKA		x			
29	***CORINTHOS***	30,705	56,882	36	266	CRUDE	USA, DELAWARE R.	x				
30	TRADER	21,999	35,000	36	263	FUEL (CGO)	GREECE				x	
31	ST. PETER	20,678	34,730	33	246,700	CRUDE	ECUADOR			x		
32	GINO	26,167	48,760	32	240	CARBON BLK	FRANCE, ATLANTIC	x				
33	GOLDEN DRAKE	16,231	30,004	32	238	CRUDE	BERMUDA			x		
34	IONNIS ANGELICOUSSIS	35,269	68,106	32	236	CRUDE	ANGOLA			x		
35	CHRYSSI	19,183	29,653	32**	232	CRUDE	BERMUDA				x	
36	IRENES CHALLENGE	21,090	34,884	31	228	CRUDE	PACIFIC OCEAN				x	
37	***ARGO MERCHANT***	18,743	28,691	28	225	FUEL (CGO)	USA, EAST COAST		x			
38	HEIMVARD	35,335	55,000	31	225	CRUDE	JAPAN	x				
39	PEGASUS	11,089	—	25**	225	WH. PROD	USA, EAST COAST					x
40	PACOCEAN	17,328	30,016	31	225	CRUDE	NORTH WEST PACIFIC				x	
41	***TEXACO OKLAHOMA***	20,084	35,072	29	225	FUEL (CGO)	USA, EAST COAST				x	
42	SCORPIO	26,031	42,000	31	225	CRUDE	MEXICO, EAST COAST		x			
43	ELLEN CONWAY	27,931	47,566	31	225	CRUDE	ALGERIA		x			
44	CARIBBEAN SEA	18,589	30,661	30	225	CRUDE	EAST PACIFIC OCEAN				x	
45	CRETAN STAR	19,674	30,372	27	218	CRUDE	INDIA, WEST COAST					x
46	GRAND ZENITH	18,736	29,930	26	213	FUEL (CGO)	SOUTH AFRICA				x	
47	ATHENIAN VENTURE	18,251	31,016	26	200	WH. PROD	CANADA, NEWFOUND.			x		
48	VENOIL	152,328	330,954	26	191	CRUDE	SOUTH AFRICA	x				
49	ARAGON	122,583	238,959	24	175	CRUDE	MADEIRA				x	
50	***OCEAN EAGLE***	12,065	18,824	21	157	CRUDE	PUERTO RICO		x			
							Total, Number Accidents	11	13	12	12	
							Total, Tons ($\times 10^3$)	792	693	851	653	

Code: CO — Collision/Allision
GRN — Grounding/Stranding
F/E — Fire/Explosion
S/H — Structure, Hull, or Machinery Failure
NK — Not Known

SOURCE: The International Tanker Owners Pollution Federation, Ltd. (ITOPF) / Ship Tonnage from Lloyd's Register of Shipping (Technical Report No. STD R2-0590) and ITOPF.

*Ranking is based on barrels spilled. **Spillage data in doubt. Items italicized in bold occurred in U.S. waters.

TABLE 1-4 50 Major Oil Spills from Tankers and Combined Carriers 1960 and 1965–1989

Rank/ Rating*	Name	Spillage Tons ×10³ (metric)	1960	65	66	67	68	69	70	71	72	73	74	75	76	77	78	79	80	81	82	83	84	85	86	87	88	89	Rank/ Rating*
1	ATLANTIC EMPRESS	257																x											1
2	CASTILLO DE BELLVER	239																				x							2
3	AMOCO CADIZ	221															x												3
4	ODYSSEY	132																									x		4
5	TORREY CANYON	124				x																							5
6	SEA STAR	123									x																		6
7	HAWAIIAN PATRIOT	101														x													7
8	INDEPENDENTA	95																x											8
9	URQUIOLA	91													x														9
10	IRENES SERENADE	82																	x										10
11	KHARK 5	76																										x	11
12	NOVA	68																						x					12
13	WAFRA	62								x																			13
14	EPIC COLOCOTRONIS	58												x															14
15	SINCLAIR PETROLORE	57	x																										15
16	YUYU MARU NO 10	42											x																16
17	ASSIMI	50																				x							17
18	ANDROS PATRIA	48															x												18
19	WORLD GLORY	46					x																						19
20	BRITISH AMBASSADOR	46													x														20
21	METULA	45												x															21
22	PERICLES G.C.	44																				x							22
23	MANDOIL II	41						x																					23
24	JAKOB MAERSK	41												x															24
25	***BURMAH AGATE***	41																x											25

26	J. ANTONIO LAVALLEJA	38							x																			26
27	NAPIER	37										x																27
28	***EXXON VALDEZ***	36																									x	28
29	***CORINTHOS***	36												x														29
30	TRADER	36									x																	30
31	ST. PETER	33													x													31
32	GINO	32																x										32
33	GOLDEN DRAKE	32									x																	33
34	IONNIS ANGELICOUSSIS	32																x										34
35	CHRYSSI	32							x																			35
36	IRENES CHALLENGE	31														x												36
37	***ARGO MERCHANT***	28													x													37
38	HEIMVARD	31		x																								38
39	PEGASUS	25					x																					39
40	PACOCEAN	31						x																				40
41	***TEXACO OKLAHOMA***	29								x																		41
42	SCORPIO	31													x													42
43	ELLEN CONWAY	31												x														43
44	CARIBBEAN SEA	30													x													44
45	CRETAN STAR	27												x														45
46	GRAND ZENITH	26												x														46
47	ATHENIAN VENTURE	26																								x		47
48	VENOIL	26														x												48
49	ARAGON	24																									x	49
50	***OCEAN EAGLE***	21					x																					50
TOTAL (TONS × 10^3)			57	31	-	124	92	72	32	96	191	37	87	181	267	188	269	726	120	–	–	333	—	68	—	—	162	136	

SOURCE: The International Tanker Owners Pollution Federation, Ltd.

*Ranking is based on barrels spilled, crude or product.

Items italicized in bold occurred in U.S. waters.

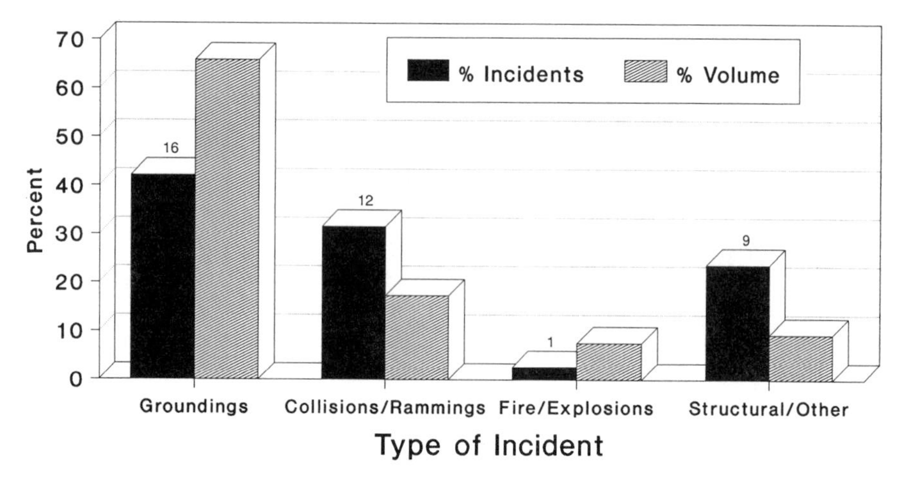

FIGURE 1-11 Major oil spills from tankers and causes: number of incidents and volume—U.S. waters. Source: Temple Barker & Sloane, Inc.

RICAN (1984), the GEORGIA, and the OLYMPIC GLORY (both in 1980). The record of outflow from smaller spills (under 30 tons or 10,000 gallons) in U.S. waters shows significant improvement since the 1970s and has remained relatively constant, between 0.1 and 0.2 thousand tons, since 1983.

Since 1984, the total annual volume of small spills has been less than 5 percent of the major spills. However, the reverse has been true regarding the number of incidents: On average, there have been over 50 times as many small spills as major spills.

The few but calamitous large spills, such as from the EXXON VALDEZ, emerge clearly from the statistics, but the smaller spills that are not among the top 50 also may have significant environmental consequences. In 1989 and early 1990, four tanker spills over 200,000 gallons (equivalent to 650 tons of crude, or 600 tons of fuel oil) occurred near the continental U.S. coast; they received much public attention and were noted by Congress prior to passage of the Oil Pollution Act of 1990. They were:

- The WORLD PRODIGY, which grounded and spilled 900 tons (also reported as high as 1,365 tons) of light fuel oil off southern Rhode Island in 1989. Fortunately, most of the oil evaporated quickly, but some local environmental damage was sustained.
- The PRESIDENTE RIVERA, which grounded in the Delaware River near Marcus Hook, Pennsylvania, in 1989 and spilled nearly 900 tons of fuel oil. Local environmental damage was sustained.
- The RACHEL B, which collided with a coastal towed barge and lost more than 800 tons of partially refined crude in the intersection of the Houston Ship Channel and Bayport Ship Channel.

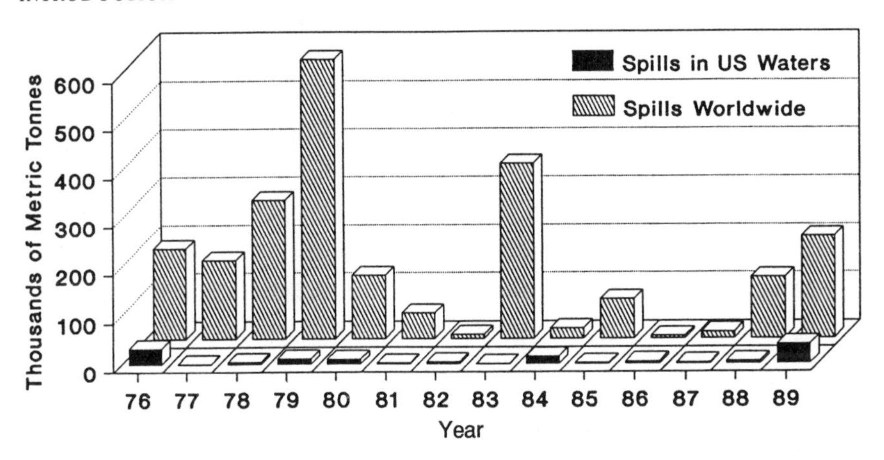

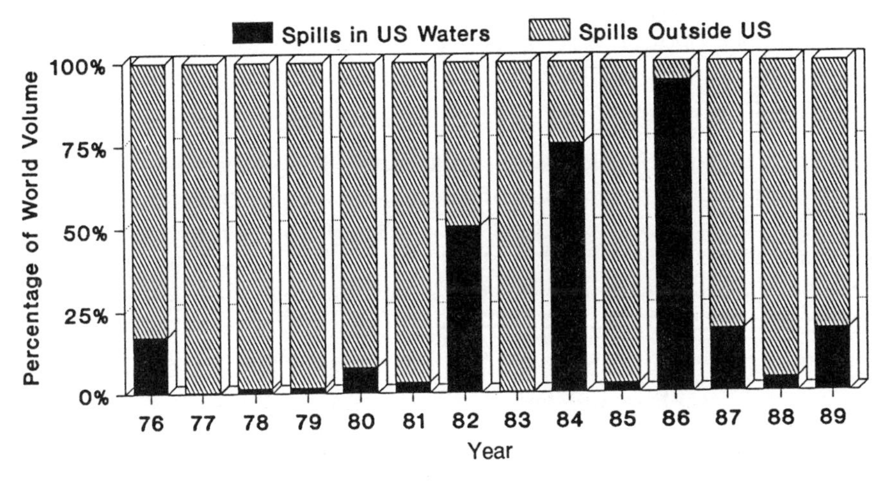

FIGURE 1-12 Comparison of major spills—U.S. versus worldwide.
*Over 30 tons, or approximately 10,000 gallons (U.S.).

- The AMERICAN TRADER, which grounded one mile off Huntington Beach, California in 1990 and spilled an estimated 1,200 tons (395,000 gallons) of light crude. While over a third of the light Alaskan crude oil was recovered, local pollution and beach damage was sustained.
- The B.T. NAUTILUS, which grounded in the Kill van Kull waterway, New York Harbor, in 1990, and spilled over 700 tons of fuel oil.

Another accident, in 1988, demonstrated how a spill smaller than those listed above (220 tons), which occurred in an ecologically sensitive area, can be damaging and can capture the public attention, causing international repercussions. The barge NESTUCCA was struck while under tow on Puget Sound, and the hull was breached. The released oil, which floated inshore and affected beaches and shorelines in Washington, reached well along the

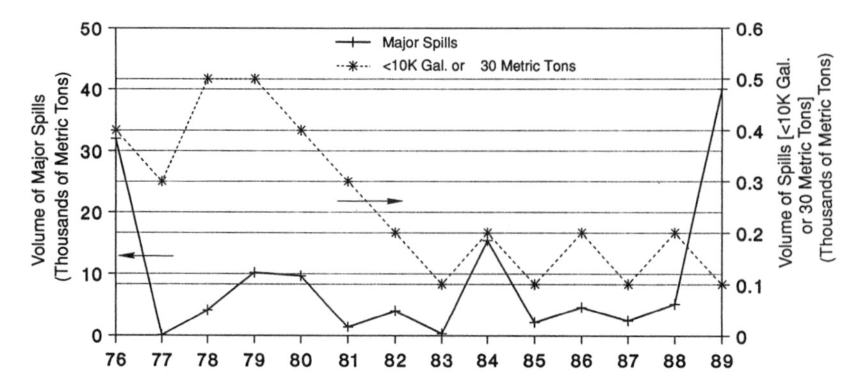

FIGURE 1-13 Tanker oil spills in U.S. waters—volume of spills.*>10,000 gallons (U.S.) or approximately 30 tons metric (crude oil). Source: U.S. Coast Guard/ Temple, Barker & Sloane, Inc.

coast of Vancouver Island, Canada, causing significant short-term shoreline damage over much of the island and adjacent Puget Sound areas.

These accidents underscore the need to consider, when developing new design and operational practices, the types, sizes, and condition of vessels (and their areas of operation) engaged in the brisk import and coastal petroleum trade.

Vessel Size and Age May Play a Role in Casualties

The IMO database for the world tanker fleet (IMO, 1989) provides a large sample of incidents, both polluting and non-polluting, in most accident categories to offer some valid insight about accidents and tanker characteristics over a 15-year period ending in 1988.

Size of Tanker

The size of tanker[21] most common in U.S. waters—50,000 to 100,000 DWT—is within size range with the worst overall worldwide casualty rate of all size categories based on IMO data.[22] The casualty rate for collisions and groundings for this size range is about equal to that for all tankers, but the rate for fires/explosions exceeds the average by 33 percent (a rate of 0.76 per 100 ships per year compared to 0.57). Unfortunately, data related to many fire and explosion causes either are lacking or are inconclusive.

Age

There is some evidence of a link between vessel age and serious casualties: Older vessels have more accidents, particularly fires/explosions and structural/machinery damage (Ponce, 1990). However, the rate for collisions is relatively constant for all age groups. There is no clear evidence of a link between the rate for cargo-related explosions (range 0.18-0.25) and ship age between 5 and 25 years (Lloyd's Register of Shipping, 1990).

Safety of Life

Fires and explosions are the major accidental cause of injuries and death aboard tankers. Any consideration of tank vessel hull design configurations and operations must take into account the influence on risk to the crew of fire and explosion. Unfortunately, there is little accident data that might relate crew casualties to the cause of the accident or the environment, or that might allow detailed analysis of how ship structure, compartment arrangements, ventilation, and safety systems relate to fatalities and injuries. The committee recognized, however, that steps to reduce pollution from tank vessels should not be taken without regard to the safety of the ship's crew.

Again, the 1989 analysis conducted by IMO provides the only comprehensive guidance available to the committee. Over the 15-year period 1974-1988, an average of six major cargo fire incidents occurred per year (IMO, 1990). There was a noticeable reduction in cargo fires and explosions during the years 1986 through 1988; this suggests that the international requirement for inert gas systems (IGS, described in Chapter 3) is having a positive effect. Cargo fires represent about a third of all tanker fires (others occur in the pump room or machinery spaces) and pose the greatest hazard to the crew.

The IMO analysis in this area covers not only oil and chemical carriers, but also combination and gas carriers. Over the 15-year period, 1,209 lives were lost—an average of 81 persons per year. Of these deaths, 67 percent (829) resulted from fires and explosions, and over half of these (480) were due to cargo fires and explosions; 167 lives were lost because of fires and explosions on ships under repair, including 83 in a single casualty in 1978.

The fatality record, in some cases, does not clearly reflect the effect of secondary causes such as a fire or foundering following a collision. Secondary causes account for the loss of many of the 149 persons linked to collisions over a 15-year period.

REDUCING THE RISK OF POLLUTION FROM TANK VESSELS

Background

The world at large has enjoyed the benefits of industrial technology for almost two centuries. Historically, owners and managers of industry and transportation, as well as the public, have focused on the benefits and typically have underestimated or even ignored the inherent risks in deployment of these technologies. Society has reaped the benefits first and has delayed paying the price of risk until later.

In the last 20 years or so, affluent societies have begun to acknowledge that there are risks associated with these benefits. Risk and safety issues have moved to the forefront of public discussion. Almost daily, newspapers and television news provide coverage of risk in modern society. Regulatory agencies are playing a growing role in risk assessment and its management. Society is more risk conscious than ever before.

Major accidents, with attendant media coverage, underscore the risks: The nuclear accident at Three Mile Island; the chemical accident at Bhopal, India; the nuclear accident at Chernobyl in the U.S.S.R.; the explosion of the space shuttle Challenger; and the wreck of the EXXON VALDEZ. Economic factors and litigation are also compelling industrial owners and managers to consider risk along with the benefits.

Defining Risk

The term "risk" can be defined as the possibility of suffering harm from a hazard. A hazard is a source of risk and refers to a substance (such as crude oil), an event (e.g., an oil spill) that harms the environment, or a natural hazard (e.g., a hurricane).

In contemplating alternative designs and operations for tank vessels, one way to define risk is to ask the following fundamental questions:

- What can go wrong?
- What is the likelihood of that happening?
- If it does happen, what are the consequences?

The answers to these questions constitute a risk analysis or assessment. For purposes of this report, the answers might be arranged in three columns, with headings such as (1) tank vessel accident scenario, (2) the likelihood or frequency of that scenario, and (3) magnitude or volume of oil spilled in that scenario.

Is the Existing Risk of Pollution Acceptable?

Risk cannot be eliminated but it can be reduced to a level acceptable to society. There are no universally acceptable risks, so decision makers typically identify levels of risk that are tolerable.

In the case of the EXXON VALDEZ, society enjoys the benefits of oil transportation but demands, and subsequently requires action to achieve, a substantial risk reduction.

Figure 1-14 illustrates in a simplified manner the concept of acceptable versus unacceptable risks. Prior to risk-reduction efforts, the risk coordinates (F1, C1) typically fall into the "unacceptable" region. In industry and transportation, changes typically come in the form of design modifications, operational enhancements, administrative features, and changes affecting the management's (e.g., tank vessel owner's) safety/environmental culture. Such changes, if properly executed, can result in substantial risk reduction.

Before deciding how much risk reduction may be desirable, the magnitude of existing risks must be assessed. In the case of tank vessels, the ratio of oil gallons spilled per gallon transported may be useful. Based on Coast Guard statistics for 1974 through 1978, the ratio for tankers is 3.32×10^{-5}, (i.e., 3.3 gallons lost per 100,000 gallons transported) and for barges (both ocean-going and inland) 1.73×10^{-5} (National Research Council, 1981).

A perspective on the 1980s can be gleaned from statistics gathered by the committee (presented in Chapter 6, in Table 6-1). In the average year, 9,000 tons of oil are spilled in U.S. waters, and 100,000 worldwide; in the worst year, those numbers rise to 40,000 and 360,000, respectively. For comparative purposes, the volume of oil moved annually is roughly 600 million tons in U.S. waters and 1,500 million tons worldwide.

From these figures,[23] the ratio of oil spilled to oil transported is:

Location	Average Year	Worst Year
U.S.	1.5×10^{-5}	6.7×10^{-5}
World	6.7×10^{-5}	24.0×10^{-5}

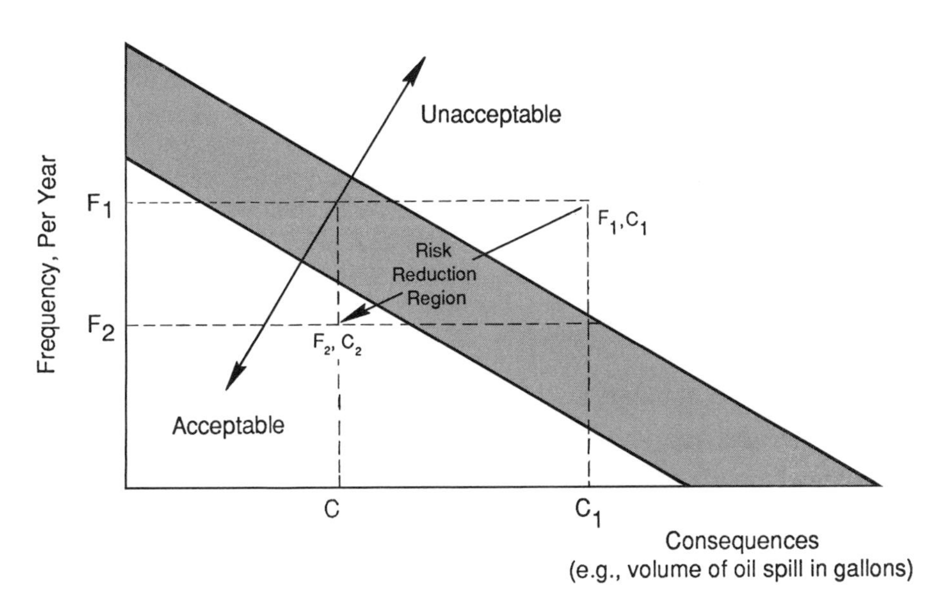

FIGURE 1-14 Unacceptable/acceptable risks for oil spills—frequency versus consequence diagram.

The record suggests that the average risk in U.S. waters has decreased somewhat since the 1970s, but the reluctance of society to accept that risk has increased.

However, to put these numbers in perspective, risk ratios from other industries are useful. For example, recent statistics for the offshore oil pipeline industry indicate that the ratio is 1×10^{-5} (Ronald E. Prehoda/Department of the Interior, personal communication to D. Perkins, National Research Council, November 1990). In the airline industry, the probability of a landing being aborted is less than 1×10^{-7} per landing (or one in 10 million, or 1/100 of the pipeline example). The latter example may not be entirely comparable, as landing accidents are a direct threat to human life, while tank vessel accidents often are not. Nevertheless, there is certainly reason to attempt to make significant reductions in risk; a comprehensive risk-assessment study could be undertaken, as the basis for establishing risk-based design goals for tank vessels.

SUMMARY

The preceding discussion makes several points that will influence the analysis in later chapters. First, a threat of pollution exists wherever tank vessels travel, and traffic around the United States appears to be increasing. In addition, the type and condition of tank vessels traveling in U.S. waters varies, and this affects safety and risk; control over these factors is limited. Finally, pollution incidents are so diverse in nature that it may be difficult to select one pollution control measure that would be effective in all, or even most, cases. However, groundings have been the dominant cause of accidental oil pollution in U.S. waters.

The historical record shows that most spills are small, and there is no discernible pattern by which to predict the less frequent large spills. But vessels in the size range most common in U.S. waters appear to have the worst casualty record, based on worldwide data.

Finally, it appears that the risks of transporting oil by tanker may be greater than in related industries, such as the offshore oil pipeline system. A comprehensive risk-assessment study could be undertaken, as the basis for establishing future risk-based design goals for tank vessels, with attendant compliance guidance.

NOTES

1. Refers to ocean-going tank vessels, defined for purposes of this report as those over 10,000 deadweight tons (DWT). DWT refers to the weight of cargo, consumable liquids such as the oil and water, stores, the crew and its effects, and excludes the weight of the ship. This report focuses on ocean-going vessels.
2. For purposes of this report, "tons" refers to metric tons unless otherwise noted. One metric ton equals 2,205 pounds, or 7.33 barrels, or 308 gallons (based on average Arabian Light 33.5° API gravity). Some data bases (ship tonnage) are maintained in long tons (2240

pounds), short tons (2000 pounds), or barrels (42 U.S. gallons). U.S. imports and exports originally were expressed in short tons in U.S. Army Corps of Engineers data; these have been converted to metric tons for this report. (One billion is one thousand million, or 10^9.)

3. Includes import/export, inter- and intracoastal shipping statistics. Sources: Temple, Barker and Sloane, Inc., Energy Information Administration (EIA, 1989), British Petroleum Statistical Review of the World Oil Industry, and statistics provided to the committee by the U.S. Army Corps of Engineers.
4. The Port of Houston handled 1,670 port calls in 1989 by tankers over 10,000 DWT, the largest number of any U.S. port; the Port of New York and New Jersey followed with 1,409 port calls that year. (Data provided to the committee by Lloyd's Maritime Information Services Ltd.)
5. Data presented here refer to 1988 for statistics concerning petroleum import projections and volume in specific ports (source: EIA), and to 1989 for tanker calls in major U.S. ports (source: Lloyd's Maritime Information Services Ltd.) and exports and imports (source: EIA).
6. The "base case" EIA estimate is that imports would increase from 6.6 mbd (in 1988) to 10.0 mbd in the year 2000. By then, the forecast calls for consumption to range from 17.5 to 19.9 mbd, domestic production to range from 7.8 to 9.4 mbd, and net imports to range from 8.1 to 12.2 mbd (EIA, 1989).
7. A tanker or barge may make several port calls before its entire cargo is unloaded. As an example, a tanker out of Valdez, Alaska, might discharge at Puget Sound and San Francisco Bay.
8. Import data are from the Energy Information Administration (EIA, 1989); Alaskan shipments are based on Lloyd's data; intercoastal and intracoastal non-Alaska data are derived from Corps of Engineers statistics for 1988 but are indicative of present activity for tankers and barges over 10,000 DWT. Data are stated in metric tons.
9. Refers to barges over 5,000 gross registered tons (GRT) or about 10,000 DWT. GRT are a measurement of the volumetric capacity of a vessel.
10. Lightering, a cargo transfer practice, is described in Chapter 2.
11. Beyond the additional imports required to meet higher demand, a significant increase in the importation of refined petroleum products and resulting product tanker traffic may result from limitations on: (1) the growth of U.S. refinery capacity and (2) the availability of deep-water offloading and storage facilities for crude oil transported in Very Large Crude Carriers (VLCCs). These factors are not reflected in Department of Energy projections for crude and product imports shown in Figure 1-2.
12. Lloyd's and committee estimates.
13. Includes 117 tankers owned by independent operators (comprising 45 percent of the total), 66 owned by major oil companies (26 percent), and 74 owned by the U.S. government (29 percent). The government total includes five re-flagged Kuwaiti tankers (Arthur McKenzie/Tanker Advisory Center, Inc., personal communication to D. Perkins, National Research Council).
14. Frederick Siebold/U.S. Maritime Administration, personal communication to D. Perkins, National Research Council, November 27, 1990.
15. Lloyd's data provided to the committee.
16. Several Alaska-trade tankers in service have double bottoms.
17. Lloyd's data provided to the committee.
18. Definition of casualty terms is essential to understanding statistics as obtained from the U.S. Coast Guard, Lloyd's Register of Shipping, and the International Maritime Organization. The following terms are used in this report:

    Groundings. The tank vessel is reported in contact with the sea bottom or a bottom obstacle, struck object on the sea floor, or struck or touched the bottom; includes ships reported "hard and fast" for an appreciable period of time (wrecked/stranded).

    Collision/Ramming. The tank vessel struck or was struck by another vessel on the water surface, or struck a stationary object, not another ship (an allision). In this category, Lloyd's data on collisions and contact are combined.

Fire and Explosion. The fire and/or explosion is the initiating event reported, except where the first event is a hull/machinery failure leading to the fire/explosion. Therefore, casualties involving fires or explosions after collisions or groundings are categorized under "collision" or "grounding," respectively.

Structural/Machinery/Other. Hull/machinery damage, missing, and miscellaneous non-classified reasons. This category also combines ships sunk due to either weather or break-up related to causes not covered by other casualty categories (foundered).

19. The NRC has not updated these comprehensive estimates of petroleum hydrocarbons entering the sea. However, the maritime transportation component was updated by the Coast Guard in 1990.
20. The rate averaged about 0.009 per billion ton-miles over the 10 years through 1988; the rate ranged from a minimum of 0.008 in 1977 to 0.011 in 1983 and was slightly over 0.008 in 1988.
21. Oil and chemical carriers, not combination carriers.
22. Normalized data have been provided in the IMO databases; these data have been adjusted for numbers of vessels at risk over the time period. These IMO data for casualties were grouped in a range from 45,000 to 150,000 DWT.
23. The time periods covered are 1980 to 1989 (U.S.) and 1976 to 1989 (world).

REFERENCES

American Waterways Operators. 1990. Letter report to the Committee on Tank Vessel Design, NRC, April 6, 1990.

Energy Information Administration. 1989. Petroleum Supply Annual. Washington, D.C.: Government Printing Office.

International Maritime Organization. 1989. Analysis of Serious Casualties to Sea-going Tankers, 6,000 gross tonnage and above, 1974-1988. London: IMO.

Lloyd's Register of Shipping. 1990. Statistical Study of Oil Outflow from Oil and Chemical Tanker Casualties. Report conducted for the American Petroleum Institute, Washington, D.C. Technical Report STD R2-0590.

National Research Council. 1975. Petroleum in the Marine Environment. Report based on a workshop held by the Ocean Affairs Board, Airlie, Virginia, May 21-25, 1973.

National Research Council. 1981. Reducing Tankbarge Pollution. Washington, D.C.: National Academy Press.

National Research Council. 1985. Oil in the Sea. Washington, D.C.: National Academy Press.

Ponce, P. 1990. An Analysis of Total Losses Worldwide and for Selected Flags. Marine Technology 27(2):114-116.

Tanker Advisory Center. 1990. Guide for the Selection of Tankers. New York: TAC.

U.S. Coast Guard. 1990b. Report of the Tanker Safety Study Group, Chairman H. H. Bell (rear admiral USCG, retired). Washington, D.C.: U.S. Department of Transportation.

U.S. Coast Guard. 1990d. Update of Inputs of Petroleum Hydrocarbons Into the Oceans Due to Marine Transportation Activities. Paper submitted to IMO Marine Environment Protection Committee 30, London, September 17, 1990.

U.S. Coast Guard. 1990e. Assessment of Success of Tankships with Double Bottoms and PL/SBT in Mitigating Pollution Due to Casualties. Internal analysis by the USCG, Washington, D.C., March 12, 1990.

U.S. Maritime Administration. 1990. Foreign Flag Merchant Ships Owned by U.S. Parent Companies. Report prepared by Office of Trade Analysis and Insurance, Washington, D.C., January 1, 1990.

2
Tank Vessel Design, Operation, and Regulation

Before embarking on a technical analysis of alternative tank vessel designs, standard design and operational practices related to pollution prevention should be understood. This chapter will discuss the evolution of these practices, as well as the legal and regulatory framework governing the tank vessel industry. The discussion applies primarily to tankers, although some sections, as noted, apply to barges.

TANK VESSEL DESIGN AND OPERATION

Tanker Design

Crude oil and petroleum products have been carried in ships for more than 100 years. The practice of carrying the oil directly inside the single hull of a ship has been common since this type of ship was first built in 1886. The hull provided far better security for the cargo than barrels, or casks, which could split and spill oil, creating fire and explosion hazards.

Tanker designs established in the late 1880s remained virtually unchanged until shortly after World War II. Tankers commonly were of 10,000 to 15,000 DWT, with a single skin, the engine to the stern, and multiple compartmentation with either two or three tanks across. Cargoes were usually refined products, most often light or "white" oils, which were not considered polluting as they rapidly evaporated if spilled. The non-polluting cargo meant that tanks could be rinsed out with water (which then was dumped at sea), and the same tanks could be used for ballast (sea water).

Separate ballast tanks, other than the peak tanks (at the ends of the ship), were virtually unheard-of until after World War II.

After the war, the world economy expanded with a resulting huge increase in demand for energy in the form of oil. At the same time, a new shipping pattern evolved: Crude oil often was transported from distant sources, such as the Persian Gulf, to major marketing areas, notably North America, Northern Europe, and Japan, where the crude was refined and redistributed as product. These long voyages set the stage for a dramatic increase in ship size, which started about 1950. Between 1950 and 1975, the largest tanker in the world grew from about 25,000 DWT to over 500,000 DWT. (See Figure 2-1.) The numbers of tankers in the world fleet also multiplied many times over.

Meanwhile, significant technical developments were afoot, including the following:

- Welding replaced riveting, a major benefit to the tanker industry in assuring tightness of tanks. The practice initially led to some cracking, and ships breaking in half, but these problems were solved with better materials, welding, and design.
- The empirical, or rule-of-thumb, design approach was augmented and partially supplanted by theoretical techniques. This trend was facilitated by the introduction of computers in the 1950s and 1960s, and, in fact, was necessitated by the growth in ship size from vessels of around 500 feet to over 1,400 feet, with an increase in deadweight of over twenty-fold in less than 20 years.
- While the basic types of static and dynamic forces acting on ship structure had been known in general for years, it was not until the 1960s that naval architects were able to quantify the loads precisely and to carry out the stress analysis needed to design ships on a theoretically sophisticated basis. By the 1970s, reliable theoretical quantification of loads and structural response was common for tankers; however, practical service experience remains vital to verify structural integrity and detail design.
- As newer design techniques were introduced, "safety factors" (design allowances for unknown factors) were reduced, in the desire to keep costs down and to get maximum deadweight for minimum draft (the depth of water a vessel draws). The effect is shown in Table 2-1.[1] The significant reduction in ratio of lightweight (ship weight without cargo, crew, fuel, or stores) to deadweight directly reduces the cost of a ship per ton of cargo; this means a ship can carry more cargo for a given draft. It also implies more efficient structure, and, in general, less margin to tolerate construction or maintenance errors or unusual operational events.
- Structural weight reductions were accompanied by a reduction in the number (and resulting increase in size) of compartments; the intent was to lower construction cost and simplify operations. While this change has

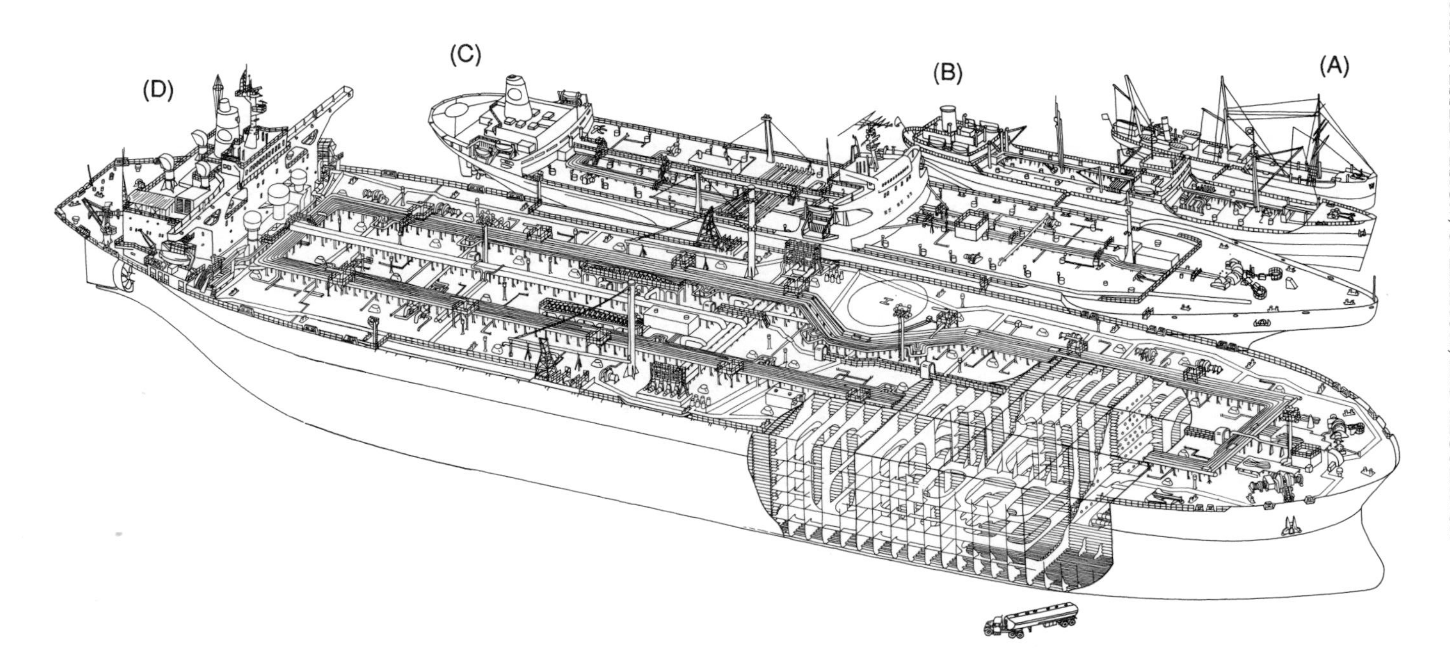

FIGURE 2-1 Evolution of the tanker. Sources: National Geographic Magazine, July 1978, and Tanker Advisory Center. (A) 1886, GLUCKAUF—First prototype tanker, 3,000 DWT. (B) 1945, T-2, World War II workhorse, 16,500 DWT, 525 built. (C) 1962, MANHATTAN—115,000 DWT (after conversion to an ice-breaker in 1969), the largest U.S.-flag ship at time of building. (D) 1977, KAPETAN GIANNIS—(formerly ESSO ATLANTIC) 517,000 DWT, length: 1,334 ft., third largest tanker in the world.

TABLE 2-1 Reduction in Tanker Lightweight to Deadweight Over Time

Years	Deadweight	Approximate lightweight	LW/DW
1940s	16,500 (T-Z)	6,000	.36
1950s	50,000	12,000	.24
1960s	100,000	27,000	.27
1960s	200,000 (VLCC)	30,000	.15
1970s	300,000	40,000	.13
1970s	500,000	65,000	.13

been criticized on safety grounds, large tankers (with two or three longitudinal bulkheads, multiple transverse bulkheads, and a nearly continuous upper deck) seldom have stability problems experienced by other types of ships with much larger (relative to the size of the ship) open spaces. A valid concern with larger compartments is the increased amount of oil that could be spilled if the tank were breached.

As tanker design practices evolved, problems, of course, periodically surfaced. Among the more significant problems was buckling of internal structures, encountered in larger tankers in the late 1960s and early 1970s. The solution was use of more precise finite element and more sophisticated frame analysis techniques. The most dramatic problem from the industry standpoint was explosions, especially after three VLCCs exploded (two were total losses) in one week in 1969. The solution was inert gas systems, described in Chapter 3, which were mandated by international agreement for progressively smaller ships during the 1970s.

In sum, there are two key features of modern structural design of tankers. First, introduction of new stress analysis techniques (employing finite element analysis and three-dimensional frame analysis) have permitted reductions in the structural weight. This in turn has led to a substantial reduction in cost (steel, measured by weight, is a major component in ship cost), and a modest increase in cargo-carrying capability. Second, improved welding and steel-making techniques have led to increased use of high-strength steel in tanker hulls, with attendant economic benefits. Even with these more sophisticated methods, however, ship design still must be conservative because loads never can be precisely predicted for all environments.

The exact design of a particular tank vessel depends on many factors. There are 10 basic ship characteristics that must be considered:

- ship dimensions
- hull form
- machinery size, type, and location
- speed and endurance
- cargo capacity and deadweight

- accommodations arrangements
- cargo/ballast tanks arrangements
- subdivision and stability accommodations
- relative amounts of mild or high-tensile steel
- basic scantling and structural arrangement

Irrespective of any specific design chosen, however, the technical advances in the design process have fostered a number of difficulties. These concerns, involving corrosion resistance, design margins, and fatigue resistance, are detailed in Chapter 4. Suffice to say at this point, advancements in design techniques and analyses unquestionably have made modern tankers more vulnerable to failure under conditions of unusual stress, or less-than-diligent maintenance. This matter has been of considerable interest to the committee.

The committee also has noted that prevention of damage or rupture of structure due to collisions or groundings heretofore has *not* been a design consideration for merchant ships, except in rare cases (e.g., barriers to nuclear reactors and, to a lesser degree, chemical carriers and liquified natural gas carriers). In attempting to reduce pollution risk, incorporation of these additional criteria into tanker design practices warrants serious attention.

Finally, it should be noted that design practices, while technically oriented, have a direct bearing on cost, and this is a factor in design decisions. The process is essentially circular, for the following reasons.

Because the structural rules, which are developed by classification societies, determine the weight and thus a major component of the cost of the ship, "class" decisions to a large degree control cost. The differences among classification societies—non-profit groups in competition—are factors that attract clients (ship owners who pay fees to "class" their ships). At the same time, classification societies are managed fundamentally through boards of directors composed mainly of ship owners but with some representation from shipbuilders, insurers, and government. This situation offers the potential for conflict of interest, in that it makes this aspect of the industry essentially self-regulating.

The owners' interest in maintaining reliable ship structure has kept class rules, in the majority view, to a high technical standard. Nonetheless, it must be acknowledged that competition among classification societies and among shipyards has produced strong pressure to produce a minimum cost ship that will perform to an adequate structural standard.

Barge Design

Ocean-going barges have been used in the shipping industry for many years. In the United States especially, barges have become extremely important in coastal transportation, particularly since World War II. There are two principal reasons. First, under U.S. manning regulations, a non-propelled cargo section pushed or pulled by a tugboat requires a much smaller crew than a tanker, thus offering a major economic savings. Second, structural and safety requirements have been less stringent for barges. Until recently, unmanned barges could be built to lesser scantling (dimensions of structural members) requirements than ships with cargo sections of the same size. In addition, the absence of a crew meant that unmanned barges could escape many of the safety requirements (related to fire fighting, life-saving, anchoring, etc.) imposed on tankers. Similarly, barges had more liberal (lower) freeboard assignments[2] than ships and usually could be operated without ballast, thus providing substantial economic advantages.

In recent years, many of the differences in technical standards between tankers and barges have been eliminated, and the U.S. Coast Guard has applied structural provisions of international tanker conventions to larger barges. However, the manning requirements remain quite different, and this—besides encouraging the continued use of barges—influences vessel design, including the choice of appropriate alternatives for pollution control.

The basic design process for offshore barges carrying petroleum products is similar to that for tankers. Structurally, barges are somewhat different in that they tend to have heavier side structure (to accommodate loads from contact with piers, locks, and tugboats), and they have greater breadth than tankers of the same length.

Tank Vessel Operations

Tankers generally operate between single or multiple loading and discharge ports. When tankers were smaller in the 1950s and early 1960s, they often loaded and discharged alongside a pier, usually in a harbor. As ships got larger, requiring deeper ports, new port complexes were constructed, principally in the Middle East, Europe, and Japan. While many of these facilities were essentially conventional, with a sheltered port and a permanent fixed pier, the tanker industry also developed offshore multi-buoy moorings (MBMs) and, eventually, single-point moorings (SPMs).

Tankers are loaded near the midship point by shore pumps, or gravity if storage tanks are elevated, through hoses or "hard-arms" (hard pipe structures with swivel connections) connected to the ship's piping system. Tankers commonly have pipes on deck and running down to the ship bottom. When a tanker is discharging, the ship's power drives cargo pumps, usually located in a pump room between the engine room at the stern and the cargo tanks. The cargo is pumped up to the deck, and then ashore through the hoses or hard-arms. The process is essentially identical for the handling of crude oil, fuel oil, or refined products. Most tankers have two, three, or more separate piping and pumping systems, so as not to intermingle cargoes of different grades or characteristics. These systems are powerful enough to transfer cargo in an emergency, provided they remain intact; however, setting up the system to drain cargo from a damaged tank quickly is difficult, because the installed transfer piping suction is on the tank bottom and the oil remaining normally will float above the suction. A more detailed discussion of cargo systems can be found in Chapter 4.

Tankers operate on many different trade routes and serve thousands of delivery points. Most tankers load cargo in one area, take it to another area for discharge, and then return empty to a loading area. Thus, tankers typically are loaded roughly half the time, and otherwise are "in ballast." To deal with buildup of sludge, and to clean tanks between switches of grades in product trades, cargo tanks are cleaned periodically. For many years this was done with either cold or hot water. And, until the late 1960s, sea water often was placed in some empty cargo tanks for the ballast voyage. These practices, of course, led to some mixing of oil and water, and discharge of rinsing or ballast water inevitably caused some pollution of the seas. Concern about this "operational pollution" led to a series of new practices mandated by international conventions in the 1970s; these are among the provisions described later in this chapter.[3]

Lightering

Another measure adopted widely by the tanker industry in the post-war years was lightering (or lightening), the process of transferring cargo from one floating vessel directly to another. Lightering is used principally to remove cargo from larger vessels to make them lighter (and to reduce their draft), to allow them either to enter a harbor or to approach a pier to discharge remaining cargo. Sometimes the entire cargo is removed offshore ("lightering to extinction"). Lightering is relevant to the present study because its use may be encouraged by the Oil Pollution Act of 1990, which prohibits some tank vessels from approaching U.S. shores.

Nearly all types of dry and liquid cargo, in bulk or in containers, can be transferred by lightering. Crude oil often is moved in this manner, at least in the following areas:

- the English Channel;
- Argentina;
- the Mediterranean, Middle East, and Far East;
- the lower Delaware Bay (where barges are commonly used to take some cargo out of larger arriving tankers, to allow them to proceed to refinery terminals farther up the Delaware River);
- the U.S. Gulf of Mexico; and
- San Francisco Bay.

Lightering became common in the U.S. Gulf of Mexico in the early to mid-1970s, when crude oil imports increased rapidly along with the size of ships. With no resolution of the debate on the wisdom of deep-water ports, lightering was the industry solution for accommodating the largest tankers, which could not be berthed in shallow U.S. harbors even in lightened condition. Lighters in the 50,000 to 80,000 DWT size range were employed; at times, probably six to eight of these vessels were engaged full-time in the Gulf of Mexico and the Caribbean. Then, as oil imports declined and the first—and still only—U.S. deep-water port (LOOP, 18 miles off the coast of Louisiana) came onstream in early 1982, use of lightering dwindled.

Now, with crude oil imports increasing again, imports of refined products rising, and the LOOP operating at full capacity, tanker activity in the Gulf—including lightering—will intensify once more.

Another offshore facility, TexPort, is under consideration by an oil company consortium; it would be located 27 miles off Freeport, Texas, where there are no depth constraints to even the largest tankers. At its greatest proposed capacity, TexPort could handle nearly a third of the present U.S. oil import volume (about 100 million tons per year). The consortium estimates that this facility will be operable five years after local, state, and federal regulatory hurdles are overcome.

Reliability and Safety of Lightering. The industry took several steps in the early 1970s to assure the reliability and safety of lightering. Special techniques involving large fenders, special mooring arrangements, and ship handling techniques were explored at both model and full-scale levels, to develop the best methods for mooring and maneuvering two ships together. Lightering with both ships underway was found to be, in many instances, the preferred method, as the best course could be determined with regard to wind and waves, thus simplifying ship handling. Also, with both ships underway the mooring of the smaller vessel alongside the VLCC could be accomplished with full control over the lighter. The procedures for maneuvering vessels, mooring, and handling cargo were codified in a detailed Ship To Ship Transfer Guide (International Chamber of Shipping, 1978).

The committee is unaware of specific records kept on the safety of lightering, but, based on inquiries to concerned industry organizations, there is no evidence of any major accident or pollution incident stemming from lightering of tankers.[4]

Economic Considerations

Tank vessel design is an important factor in the economics of tanker transportation, and this should be taken into account when considering future design changes. The basic principles involved are described here; a fuller explanation, including the economic impact of specific design alternatives, can be found in Chapter 6.

There are three basic categories of costs that comprise the total cost of carrying oil by tanker. These are:

- Capital cost, the predominant component of overall cost for tankers of nearly all sizes. It may be expressed either as the vessel price or as a down payment plus loan payments, just as with a house. Tankers are expected to last 15 to 30 years, so operators must think in terms of depreciation or capital recovery.
- Operating costs, covering the crew, maintenance and repair, stores, supplies, insurance, overhead, and administration. This is the second largest component of overall cost and is heavily influenced by vessel cost (which affects insurance rates), the complexity of the vessel (which affects maintenance and repair), and the size and nationality of the crew.
- Voyage costs, including fuel, port charges, piloting, and berthing tugs.

There are economies of scale in the tanker industry, as shown in Figure 2-2. In other words, a large tanker nearly always can carry oil at a substantially cheaper rate than a smaller tanker, assuming there is adequate cargo to fill the ship and that the ports can accommodate large vessels.

There are three main reasons for the economy of scale: (1) The amount of steel needed to contain a given quantity of cargo does not increase in proportion to the deadweight of the ship; (2) The horsepower needed to propel the ship at service speed does not increase nearly in proportion to the increase in ship size (offering economies in both vessel price and subsequent voyage cost for bunkers); and (3) The size of tanker crews seldom is related to vessel size. Thus, a crew of 20 to 25 will suffice for both a small tanker (20,000 to 30,000 DWT) and a ship 5 to 10 times larger.

Of course, tanker economics depends on more than cost structure, because tankers in international trade are essentially regarded as commodities. This is because tankers of similar size and characteristics are essentially interchangeable. Thus, market factors driven by supply and demand will play a dominant role in the actual shipping rates ultimately earned by tanker owners.

In summary, when considering possible design changes, the impact on capital costs, operating expenses, and the marketplace should be borne in mind.

Safety Considerations

As economic concerns drove the tanker world to increasingly larger ships, the size increase resulted in two important safety improvements. First, the number of ships needed to carry the world's ever-growing demand for petroleum was far less than it would have been with smaller ships. The increase in traffic and congestion would have been enormous if maximum ship size had remained at the World War II level (16,500 DWT); about three times the

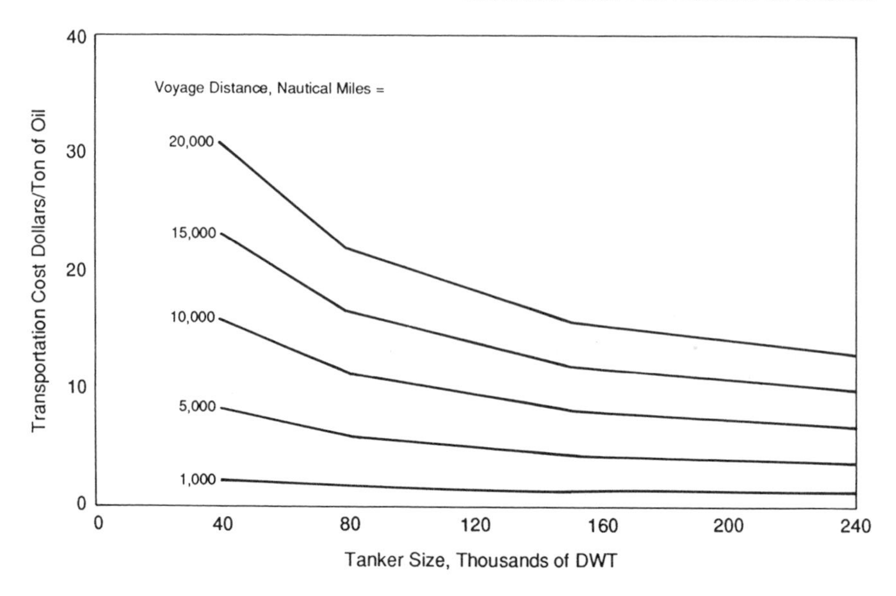

FIGURE 2-2 Tanker transportation economics. Source: Temple, Barker & Sloane, Inc.

roughly 6,000 ocean tankers (over 5,000 DWT) now sailing the world seas would be needed.

Second, larger ships generally discharge cargo at remote locations. This has kept them at distance from some, though not all, of the world's busiest harbors. However, the United States was very slow to adopt deep-water ports, offshore SBMs, and other new facilities to accommodate the largest tankers. Thus, it remains common in the United States to see tanker drafts limited to roughly 40 feet, which restricts tanker size to a maximum of roughly 60,000 to 80,000 DWT. By contrast, most of the rest of the world deals with much larger ships at more remote locations.

These changes may have helped control the incidence and impact of pollution, albeit indirectly. The industry also took other steps in its continuing concern for the consequences of groundings and collisions. Structural solutions (i.e., "crash-proofing") were studied, although they never achieved prominence or widespread adoption. Instead, the industry concentrated on:

- radar and electronic navigation, which became important for ships of all types but particularly tankers;
- harbor and coastal traffic control systems, which became prominent in much of Europe and were adopted to a lesser degree in the United States;
- ship handling and bridge team simulator training, which were promoted by some tanker operators and were adopted in a number of prominent fleets, though by no means universally;

• research on ship maneuverability, to the point that prediction of ship maneuvers became commonplace in tanker design and operation; and
• an international agreement covering crew training and licensing, ratified by many nations though not yet by the United States.

The effectiveness of these measures, either individually or collectively, is difficult to assess. Data on serious tanker casualties do seem to show that tanker accidents have decreased over the last 10 years, based on the ratio of accidents to the number of tank ships in service worldwide. However, a comparison of accidents to a general measure of exposure—ton-miles per year of oil and products shipped—while demonstrating a decrease in casualties since the mid-1980s, does not show such a clear safety improvement. (The data were presented in Figure 1-9.)

LEGAL REQUIREMENTS FOR VESSEL DESIGN AND POLLUTION PREVENTION

This section discusses the evolution of international conventions and laws related to tank vessel design and pollution prevention, as well as the current status of enforcement activities related to tankers operating in U.S. waters.

Tankers must satisfy a substantial number of design requirements when initially constructed, for purposes of safety and pollution prevention. These requirements fall into three broad categories: international legal requirements, domestic legal requirements, and classification society requirements.

International Legal Requirements

The International Maritime Organization (IMO) is the United Nations agency responsible for maritime safety and environmental protection of the oceans. All of the world's major shipping nations are members of IMO. Each member nation is encouraged to accept the international agreements adopted by IMO. These include 22 full conventions or treaties and 17 codes (as of December 31, 1989) as well as numerous resolutions containing recommendations and guidelines.

Regulation of ship design for safety and pollution prevention is achieved primarily through three international conventions:

• The International Convention on Load Lines (1966), or ICLL;
• The International Convention for the Safety of Life at Sea (1974) and its 1978 Protocol (SOLAS); and
• The International Convention for the Prevention of Pollution from Ships (1973) and its 1978 Protocol (MARPOL).

The ICLL establishes the deepest draft to which a ship can be safely loaded. These "loadlines" are commonly seen as the "Plimsoll Mark" line on a ship's side. Historically, the objective of the loadline was to assure that ships were not so overloaded as to run undue risk of sinking, or to create unsafe working conditions.

The overall objective of SOLAS is to assure safety of the crew, ports, passengers, ships, and cargo, and, indirectly, the environment. Among the more important provisions are: (1) subdivision and stability requirements, to prevent ships from capsizing, and to ensure survival under specified collision and grounding damage circumstances; (2) general construction principles, to ensure the ship is strong enough for its intended trade; (3) safety equipment requirements, to assure the carriage of sufficient lifeboats and other safety equipment; (4) fire protection requirements, to ensure that ships could withstand certain fire damage and fight fires effectively; and (5) radio telegraphy requirements, specifying the communications and navigation equipment ships must carry.

Both ICLL and SOLAS have the indirect effect of preventing oil spills and consequent marine pollution. The MARPOL convention seeks to prevent pollution directly, both from normal operational discharges and accidents. MARPOL specifies design, equipment, and procedural requirements to prevent pollution of the seas from oil, chemicals carried in bulk, harmful substances carried in packages, sewage, and garbage. Each of these five potential sources of pollution is addressed in regulations set out in an annex to MARPOL.

Annex I of MARPOL addressed the prevention of pollution from oil, and it and certain SOLAS provisions include these major requirements:

- **Segregated Ballast Tanks (SBT).** Oil carriers over 20,000 DWT built after dates specified in MARPOL '78, and tankers over 70,000 DWT built after dates specified in MARPOL '73 (see Figure 2-3), are required to carry ballast in SBT. Only in severe weather can additional ballast be carried in cargo tanks. In such cases, this water must be processed and discharged in accordance with specific regulations.
- **Protective location of SBT.** The required SBT must be arranged to cover a specified percentage of the side and bottom shell of the cargo section.[5] Thus, the protectively located segregated ballast tanks (PL/SBT) are intended to provide a measure of protection against oil outflow in a grounding or collision. To be credited, each wing tank or double-bottom tank must meet certain minimum width or depth requirements, respectively (generally 2 meters).
- **Draft and trim requirements.** To assure safe operation of the vessel in ballast condition, the SBT must be of sufficient capacity to permit full submergence of the propeller. They are to provide a molded draft (d) amidships of not less than $d = 2.0 + 0.02 L$, and a trim (horizontal tilt) by stern not greater than 0.015L, where L is the ship length in meters.
- **Tank size limitations.**[6] To minimize pollution in case of side or bottom damage, the maximum length of cargo tanks is limited to values between 10 meters and 0.2 L, depending on tank location and longitudinal bulkhead

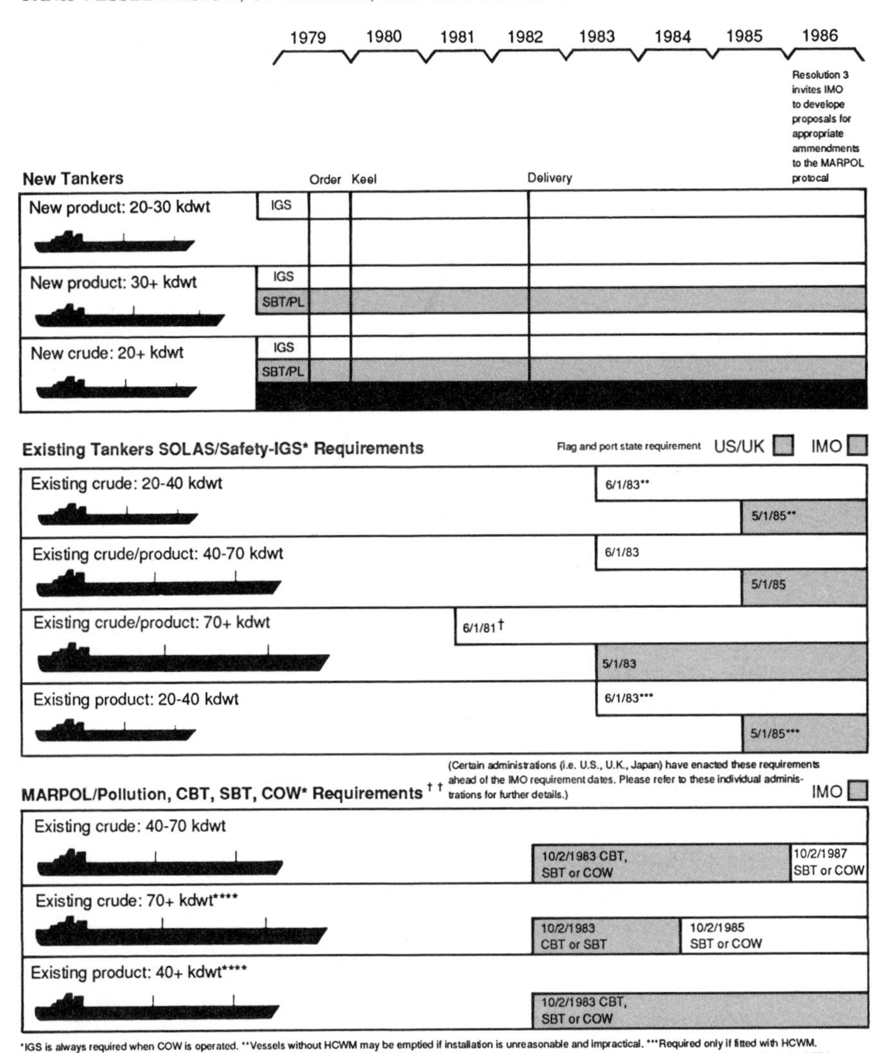

*IGS is always required when COW is operated. **Vessels without HCWM may be emptied if installation is unreasonable and impractical. ***Required only if fitted with HCWM. ****Existing crude or product tankers over 70 kdwt for which the order was placed after 12/31/73 or the keel laid after 6/30/76 or delivered after 12/31/79 must be provided with SBT in accordance with Regulation 13 (1) MARPOL Protocol 1978. † Japan enforced the same requirements on 11/1/81. †† Oil discharge monitoring and control system required late 1985.

FIGURE 2-3 Time frame for implementation of MARPOL and SOLAS. Sources: Exxon Marine and Clarkson Research Studies Ltd.

arrangement. The maximum volume of each cargo tank may vary up to 22,500 m^3 for side tanks and up to 50,000 m^3 for center tanks, depending on tank arrangement and location.

• **Hypothetical outflow of oil.**[7] Formulas establish the maximum allowable hypothetical outflow of oil in case a cargo tank is breached at any location on the ship. For the purposes of these calculations, the regulations specify certain assumed longitudinal, transverse, and vertical damage. The key damage assumptions, where B is the ship's beam or breadth, are:

Side transverse extent: B/5 or 11.5 meters, whichever is less
Bottom vertical extent: B/15 or 6 meters, whichever is less

- **Subdivision and stability.** For a specified assumed shell damage, the regulations require that tank subdivision and ship features be such that, under certain specific damage conditions: The final water line is below any opening leading to progressive down-flooding, and the heeling angle (tilt to one side) does not exceed 25 degrees (or 30 degrees if the deck edge is not submerged).
- **Crude Oil Washing (COW).** New crude oil tankers must be fitted with an effective tank cleaning procedure that uses cargo oil as the washing medium. COW is a superior system of cleaning cargo tanks using the dissolving action of crude oil to reduce clingage and sludge. Furthermore, elimination or reduction of water washing has helped reduce operational oil pollution of the seas.
- **Inert Gas System (IGS).** This system supplies the cargo tanks with an atmosphere lacking in oxygen so that combustion cannot take place. Treated flue gas from main or auxiliary boilers, inert gas generators, or other sources may be used for that purpose. This important safety system is described in more detail in Chapter 3.
- **Slop tanks.** Tankers must be fitted with slop tanks of specified capacity to retain on board all slop, cargo drainage, sludge, washings, and other oil residues. Their discharge then is monitored in accordance with the regulations.

The Significance of MARPOL

The IMO Conferences of 1973 and 1978 together produced fundamental changes in the way tankers are designed and operated. The significance of these changes in relation to the present study will be summarized here. Most important is the fact that the "MARPOL vessel" represents the current standard, against which any further design changes should be measured. MARPOL also set a precedent by establishing major retrofitting requirements for tankers, applying new equipment requirements (IGS and either SBT or COW) to existing tankers for the first time. The following other points may be less obvious.

Historically, control of pollution from operations had been accomplished through the "load on top," or LOT, system.[8] This method was highly dependent on the vigilance of the crew and was difficult to monitor; MARPOL introduced structural means of achieving the same goal, clearly an advantage. However, these provisions involved inherent difficulties. The introduction of segregated ballast basically changed tankers from deadweight-limited carriers to cubic-limited carriers, and this in turn tended to increase the amount of oil outflow in groundings.[9] This drawback was noted by the drafters of MARPOL; the result was increased pollution risk in some accidents for newly built SBT crude carriers, and even most "black fuel" carriers.

The time frame for implementation of MARPOL and SOLAS requirements, including SBT/PL, is shown in Figure 2-3. Even now, the world fleet remains a mix of carriers, many of which are exempt from these SBT and SBT/PL regulations due to age or year of construction. About 35 percent of the world tanker fleet over 10,000 DWT has SBT, and about half of these ships have SBT protectively located (Clarkson Research Studies, Ltd., 1990), thus meeting full MARPOL requirements.

In new vessels, the attempt to satisfy MARPOL requirements in the most economical way has led to two changes in ship proportions. First, to make up for cubic lost to SBT, tankers became deeper in relation to their length; the length to depth ratio (L over D) is now lower, and the ratio of draft to depth (H/D) has decreased, because freeboard is increased. Second, in the interests of economy and with improving hull form design, ships generally were made broader and shorter, as breadth is a cheaper dimension to increase than length. Thus the ratio of length to beam (L/D) generally has decreased.

In comparing pre-MARPOL to SBT ship designs, four general observations can be made (U.S. Coast Guard, 1973) that are relevant to the present study: (1) Depth must be increased in SBT ships to obtain sufficient volume for ballast; (2) for a given cargo volume, the ballast volume increases a great deal in SBT ships—in the range of 234 to 334 percent, which is indicative of the additional area that must be protected from corrosion; (3) expected oil outflow in groundings increases by up to 90 percent in many SBT designs; and (4) the greater depth (for a given draft and deadweight) in SBT designs means that deck and bottom plate thickness can be reduced significantly.

Implementation of International Law

Implementation of IMO conventions is not straightforward, because the procedures—and their effectiveness—can vary. Requirements are imposed on a vessel through its flag state. (The flag usually is carried at the stern with the city of registry and ship's name.) Each tankship therefore is governed in design, arrangements, and construction by the international agreements ratified by its flag state.

Nations that have formally ratified or approved IMO conventions usually implement the requirements through legislation. When a vessel is judged to have been designed and built to international standards, the flag state issues a certificate for each convention to which the vessel complies. Each certificate is valid for five years, provided an annual inspection—afloat—demonstrates that the ship has been maintained in accordance with convention requirements. After five years, the ship undergoes a major inspection and, as deemed necessary, renovation, prior to renewal of the certificate.

In traditional maritime nations, the inspection of vessels for compliance with both international and domestic requirements usually is carried out by government agencies, such as the Department of Transport in the United Kingdom, the Coast Guard in the United States, and the Coast Guard in Canada. Increasingly, with open registry or "flag of convenience"[10] ships, however, enforcement and inspection is conducted on a contract basis, in which the flag state contracts for all these services to be handled by a classification society.

Domestic Legal Requirements

In addition to complying with international convention requirements, ships must adhere to any additional requirements imposed by the flag state. Compliance becomes further complicated when nations, as port states, impose unilateral requirements.

There are very few unilateral port state requirements related to basic ship design and construction that represent a variance or extension of the international standards. Unilateral regulations imposed by port states usually deal with matters such as employment of pilots, hours in which ships can operate particular channels, use of tugs, and other issues peculiar to a certain locale. The United States, however, has imposed several requirements that vary significantly from international standards, as described in the following pages.

Each flag state may require its own vessels to meet any set of regulations deemed appropriate. The regulations may apply as vessels travel anywhere in the world. However, each port state may require foreign-flag vessels entering its territorial waters to meet its own set of regulations. Foreign-flag ships have the option of either abiding by port state requirements or not traveling in those waters.

Flag state requirements are subject to continual monitoring. Basically, this is handled by yearly inspections, which are fairly routine, and with more thorough inspections occurring at five-year intervals. No extension can be granted for five-year surveys.

IMO conventions do not specify penalties for noncompliance, other than removal of the current certificate. They direct that penalties (by indictment, warning, fine, or imprisonment of the person(s) responsible for the violation) be imposed by the flag state.

Classification Requirements

Classification societies establish standards, guidelines, and rules for the design, construction, and survey of ships. There are eleven leading classification societies, as represented by the membership of the International Association of Classification Societies (IACS).[11] A ship that has been constructed in accordance with the rules of a society is issued a classification certificate. To maintain their classification, vessels must be presented for survey at regular intervals.

Class requirements essentially are concerned with the structural integrity of the ship and its propulsion and steering systems; they do not address safety equipment or crew qualifications. Class requirements embrace: (1) materials for hull and key machinery components; (2) structural design requirements including scantlings (dimensions of structural elements) and details of all structure and key machinery components (i.e., main engine, shafting, propeller, etc.); and (3) supervision, inspection and certification of manufacture of steel, welding, machinery components, hull structure, etc.

These requirements must be met for the ship to comply fully with international convention requirements and to obtain more favorable insurance rates.[12] The requirements are not statutory in nature; however, under the SOLAS Convention, each ocean-going ship[13] must have a Safety Construction Certificate attesting to the adequacy of the construction. Being "in class" does not, in itself, satisfy SOLAS requirements, but, when authorized, a classification society may issue a SOLAS certification on behalf of a flag state.

Survey Procedures

Once a ship has been delivered, it can maintain its "in class" status only by meeting continuing survey requirements of the classification society. Essentially, the ship has to satisfy the society of its suitability to continue trading by passing both hull and machinery surveys at various intervals. Surveys can be of either a continuous or periodic nature, or a combination thereof.

Historically, periodic surveys were set up based on a system of annual and special surveys. Annual surveys do not include in-depth inspection of the ship's machinery and structure, unless there is cause for concern. Special surveys of hull and machinery are spaced at four-year intervals, although the society often grants a "year of grace" unless there is a compelling reason to deny it. Therefore, special surveys tend to fall at age 5, 10, 15, etc. The basic purpose of the special survey system is to assure the vessel's ability to trade successfully until the next scheduled special survey.

Special surveys become increasingly rigorous, at least in theory, as a ship gets older. The first special hull surveys will include, for example, a general examination of the most critical parts of the ship structure. By the second special survey, a more thorough examination is conducted, including measurement of the thickness of certain key structural members. The third special survey includes a comprehensive examination of the ship's structure and measurement of thickness of structure.

Continuous surveys, in lieu of special surveys, are applied particularly to machinery, and sometimes to the hull. Under this system, various parts of the ship are inspected by classification surveyors during port calls, while the vessel remains in service.

One approach to conducting structural surveys on large tankers is described in Appendix C. The inspection of the vast structural areas of a modern VLCC (see Table 4-2) requires more time than is usually available while a tanker is in port, making operator inspection while the ship is underway and in ballast an attractive and economically efficient option.

When significant damage has occurred, or is suspected, which might affect the vessel's seaworthiness, the ship's owner is required to call in classification surveyors to inspect the damaged area. The surveyor then has three options: To allow the ship to continue trading "in class," to specify temporary repairs, or to require permanent repairs before the vessel can again be regarded "in class."

How Requirements Are Implemented in the United States[14]

The U.S. implementing legislation for Annex I of MARPOL applies not only to seagoing ships of U.S. registry, but also to foreign-flag seagoing ships while in U.S. waters. The law authorizes the inspection of such ships, while in a port or terminal under U.S. jurisdiction, for compliance with the requirements of MARPOL. If a violation is found, the ship may be detained until authorities determine that it can proceed without undue threat of harm to the environment.

In some respects, U.S. law exceeds the requirements of MARPOL. For example, in U.S. waters, crude tankers between 20,000 and 40,000 DWT were required to have SBT or COW by 1986, or upon reaching 15 years of age, whichever occurred later. Product tankers in the same range were required to retrofit to SBT or operate dedicated clean ballast tanks by 1986, or upon reaching 15 years of age. Neither of these measures were required under MARPOL. Only one example of a flag-state requirement going beyond international requirements related to tanker design (other than for ice-navigation capability) has been identified outside of the United States: Finland's imposition of a large surcharge in imports of crude oil carried in single-hull ships.

The Oil Pollution Act of 1990, enacted on August 18, requires that all ships trading to U.S. waters meet standards that exceed the construction and design requirements of MARPOL in compliance with a phase-in schedule. Specifically, all new tank vessels (contracted after June 30, 1990 or delivered after January 1, 1994) operating in U.S. waters or the Exclusive Economic Zone[15] must be fitted with double hulls. Existing single hull tank vessels are permitted to operate until the time limits set forth in the Act; the timetable ends January 1, 2010. Existing tank vessels with a double bottom or double sides meet a separate schedule that ends in 2015.

A number of exceptions are provided to the requirements for double hulls. Tankers used exclusively for responding to oil spills, and tank vessels under 5,000 gross registered tons (about 10,000 DWT) fitted with a double containment system, are exempt. Also exempt until January 1, 2015, are tank vessels unloading or discharging at a deepwater port, off-loading in a lightering operation more than 60 miles from U.S. coasts, and under 5,000 GRT.

Coast Guard Responsibilities

In the United States, the Coast Guard is responsible for regulations and enforcement related to tank vessel design, construction, and safety. Specifically, the Coast Guard is responsible for safety of life and property at sea and protection of the marine environment under provisions of Title 46 of the U.S. Code, Part B—Inspection and Regulation of Vessels, and 33 U.S. Code, Chapter 33—Prevention of Pollution from Ships, respectively, and other laws. Regulations prescribed by the Coast Guard incorporate American Bureau of Shipping (ABS) rules. In carrying out inspections and vessel design plan reviews, the Coast Guard may rely on ABS reports, documents, and certificates (46 USC 3316).

New Construction

Vessel design plans are subject to Coast Guard approval, and inspection and review of plans and construction is a Coast Guard responsibility. The Coast Guard has delegated selected parts of its responsibility in this regard to the ABS, except for major safety aspects such as stability and fire fighting. The Coast Guard accepts ABS plan review and inspection as part of the certification process for new vessels, or vessels undergoing a major modification without review or attendance by Coast Guard personnel. The Coast Guard maintains an oversight program, overseeing approximately 20 percent of all ABS plan approval and inspection activities (U.S. Coast Guard, 1989b).

Vessel Maintenance

The Coast Guard is also responsible for ensuring that vessels are maintained to the appropriate standards. For this purpose, the Coast Guard requires inspection of U.S.-flag tank vessels every two years. Tankers, however, are often on a five-year operating and drydock cycle, conforming to most classification society survey intervals. In practice, this can mean that in a five-year period, the Coast Guard may have to conduct more than two biennial inspections on the vessel.

The Coast Guard is also required to determine whether foreign-flag tank vessels can operate safely in U.S. waters. Regulations require that each foreign-flag tankship be inspected or examined at least once a year, with detailed inspections for vessels over 10 years of age. In practice, Coast Guard inspections of foreign-flag tank vessels do not routinely include internal inspection of the cargo or ballast tanks. The Coast Guard relies on the flag state or classification societies to conduct internal tank inspections.

A 1989 study conducted by Coast Guard staff found that hull structural examinations of foreign-flag vessels "are at best minimal" (U.S. Coast Guard, 1990b). The study team also found that, in general, neither U.S.-flag nor foreign-flag vessels are prepared for Coast Guard inspections. Moreover, a shortage of trained or experienced inspectors results in short-cutting, deferred discretionary inspections, and, overall, barely adequate performance. A technical problem with important consequences is the difficulty of conducting internal inspections of large tank vessels while they are in service (because of difficult access to high or remote areas). The majority of defects in tank or hull structure, especially in vessels over 80,000 DWT, are found by the owner or classification society (considerably more man-hours are expended in inspections by these parties than by the Coast Guard).

The Coast Guard's field inspectors generally felt that the Coast Guard needs to do an independent, thorough internal investigation of large tank vessels at pre-set intervals. "The burden of safety assurance," the report states, "is falling more and more on the Coast Guard." According to the Coast Guard, economic pressures of the last decade, which have forced cost-cutting in the maritime industry, have resulted in major reductions in industry engineering staffs. As a result, many organizations have become reactive, rather than proactive, in handling inspection and maintenance.

Coast Guard inspection efforts are not sufficient to ensure structural safety of oil tankers. The Coast Guard deployed about 250 hands-on inspectors on more than 36,000 vessel inspections requiring 380,000 staff hours. In FY 1988 and FY 1989 combined, the Coast Guard performed more than 3,300 inspections of tankers, totaling more than 25,000 staff hours in hull inspections alone (see Table 2-2).

The Coast Guard spends between 11 and 36 person-hours for each inspection of hull structure related to a hull examination, inspection for certification, or reinspection. This effort is only a small fraction of the time needed to conduct a thorough examination of a tank vessel (see Chapter 4).

TABLE 2-2 Coast Guard Resource Hours for the Inspection of 46 CFR Subchapter D Tank Ships FY88 and FY89

Inspection Type	Number	Hull Hours	Mach Hours	Travel Hours	Extra Hours	Admin Hours	Trainee Hours	Trainee Travel	Trainee Extra	Total Hours
Insp for Cert	344	8,313	7,930	4,356	2,783	4,536	2,920	579	79	31,495
Reinspection	267	2,977	2,842	1,586	2,310	1,421	1,675	390	348	13,549
Hull Exam	233	8,451	1,142	1,921	5,461	2,259	1,530	434	1,401	22,600
Other	905	2,041	1,477	1,497	733	1,620	651	262	65	8,345
Repairs	215	1,448	758	493	87	463	350	195	25	3,819
Deficiency Ck	680	867	728	931	347	1,203	189	117	8	4,390
Damage Survey	122	617	110	435	592	383	67	45	1	2,249
Dry Dock Ext	8	52		9		21	6	4	2	93
Excursion Permit	5	9		5		15	6	2		36
Oth Agency Oversight	6	16	3	2		3				24
Hotline Invest	2	1	3	5		3	4	5		20
Const. Oversight	1		1	2		2	1	2		6
Initial Cert.	3	281	478	55		142	275	70		1,301
MARPOL Survey	42	38	8	40		74	4	1		165
MARPOL Test	4	27	12	7		13	27	8		92
Plan Review	1	24				8				32
Permit to Proceed	5	56	18	31		15	24	2		145
Reflag	14	189	115	304	601	799	81	46	220	2,352
Administration	502	76	40	49	209	1,141	10	3	1	1,485
		25,479	15,663	11,726	13,122	14,222	7,817	2,163	2,148	92,341

CODE	EXPLANATION
ADMINISTRATION	Administration type cases not covered by other areas, affecting the status of the COI without showing inspection activity.
ADMIN HOURS	Administrative time expended by all inspectors and trainees. It is all time expended by the inspector preparing to conduct an inspection and reporting the results. It includes: researching files, regulations, marine safety manual,

TABLE 2-2 *Continued*

CODE	EXPLANATION
	NVCs, etc.; communications with other units and making arrangements with vessel owners/operators; entering msis data and generating COIs and other documents; writing inspection books and reports, discussions with supervisors or colleagues regarding the inspection; making travel arrangements including inoculations, passports, visas; and preparing travel claims. "Parent commands should enter admin hours associated with the review and validation of detachment cases."
CONST OVERSIGHT	Initial construction oversight of third party associations during assembly of components into complete vessel systems.
DAMAGE SURVEY	Damage survey not involving a credit drydock exam.
DEFICIENCY CK	Deficiency check—follow up on outstanding CG-835 or response to reported deficiency (except hotline responses).
DRY DOCK EXT	Exam conducted to support decision to extend drydock interval.
EXCURSION PERMIT	Inspections associated with issuance of an excursion permit.
EXTRA HOURS	Total time expended for extraordinary delays by all qualified inspectors on the inspection. Extra time includes time consumed by unusual delays or otherwise lost, usually associated with TAD travel; e.g., time lost in a foreign yard because the vessel was not ready and the inspector could not leave. This includes all time between departure on and return from TAD, as stated on the travel claim, less all time accounted elsewhere.
HOTLINE INVEST	Inspection work performed pursuant to complaints or notifications received via the HQ 800 hotline. If any other kind of inspection results from the inspection done in immediate response to a hotline notification, a hull for example, enter the time expended on the immediate as hot and enter the subsequent inspection under the appropriate category. Hot is a special case of def.
HULL EXAM	Credit drydock examination—includes all alternative forms of credit drydockings such as underwater surveys or alternate intervals in lieu of hull. A hull entry is required for a credit drydock exam conducted in conjunction with a COI, in addition to the entry for COI.
HULL HOURS	All onboard time expended for the inspection by hull inspectors. Also, all T-Boat, barge, platform, and other inspections which are not identified as strictly hull or machinery work. This includes time for unsupervised trainees doing these inspections.
INITIAL CERT	Initial certificate of inspection—excluding re-flaggings and oversight of third parties. This includes certifications associated with new construction and conversions, i.e., any inspection leading to the issuance of a certificate to a vessel for the first time. Do not include hours reported under the heading of initial construction.

INSP FOR CERT	Inspection done prior to issuance of a COI to a previously certificated vessel.
MACH HOURS	All onboard time expended for the inspection by boiler inspectors on propulsion and auxiliary machinery, pressure vessels, piping and electric systems, etc. This includes time for unsupervised boiler trainees.
MARPOL SURV	MARPOL Survey
MARPOL TEST	MARPOL Test
OTH AGENCY OVERSIGHT	Oversight of other agencies not involving new construction.
PERMIT TO PROCEED	Special inspection type used to reflect issuance of a permit to proceed. Nullifies inspection status on VFLD, MISS, and MICP to reflect permit to proceed issuance.
PLAN REVIEW	Time associated with plan review of a subchapter T vessel, directly linked to a specific vessel.
REFLAG	Reflagging
REINSPECTION	Hours associated with periodic reinspection of a vessel, e.g., mid-period.
REPAIRS	Examination of repairs.
TRAINEE HOURS	The total onboard time expended for the inspection by any supervised trainees.
TRAVEL HOURS	Total travel time expended for this inspection by all qualified inspectors. It is the time spend en route to and from the inspection site, by whatever mode. When travel time supports both CVs and non-CVs missions, the inspector must allocate (approximately) the total travel time into CVS and non-CVS proportions. The CVS portion should be entered on the MIAR. The remainder should be entered on the appropriate activity report(s) for the non-CVS missions. When several CVS inspections are done consecutively, travelling from site to site, or at the same site, average the total time for all the jobs and assign the average to each inspection. Travel time to and from work, either at the office or for shipyard residents, should be reported only when it exceeds one round trip per normal work day; i.e., report all local travel beyond one normal commuting round trip per day. For TAD, add time expended awaiting change of mode, flight, or carrier at intermediate stops.

SOURCE: U.S. Coast Guard MSIS Extract Data compiled for the committee by G-MIM-2, 31 August 1990 inspections.

SUMMARY

The preceding discussion makes a number of important points that are relevant to the present study.

First, the prevention of oil outflow from groundings and collisions has not been a primary consideration in tank vessel design practices to date. Furthermore, advances in design have made modern tankers more vulnerable to failure under conditions of unusual structural stress, fatigue, or less-than-diligent maintenance. These facts are worth considering in taking steps to reduce pollution resulting from accidents.

In addition, the assessment and selection of alternative vessel designs should consider the impact on tanker economics, including capital costs, operating costs, and the marketplace. Enforcement and inspection capabilities are additional considerations. As noted, Coast Guard inspections already are considered barely adequate. Additional requirements for particular structural configurations, especially those increasing the need for proper maintenance and inspections, will add to the existing workload. A more detailed discussion of inspection concerns can be found in Chapter 4.

Finally, the committee was particularly impressed by two related facts concerning ship structural design and survey standards. First, responsibility for establishing and verifying adequacy of construction standards seems to be divided among multiple parties, including a ship's flag state administration, its classification society, and, to a degree, the International Association of Classification Societies (IACS). Second, the sophisticated computer design techniques pursued by the most advanced classification societies, while producing very efficient structures in terms of weight and cost, also have eroded traditional margins.

Uniform criteria for more robust structure could be established through joint review of structural design standards by IACS, the IMO, and the Coast Guard. Further, in light of the tangle of responsibilities regarding construction standards, and the fact that many classification societies do not belong to IACS (and perhaps do not possess the requisite technical capability to implement the more sophisticated design techniques), the Coast Guard and IMO, together with IACS, could undertake a comprehensive study aimed at aligning standards. The objective would be to develop linkage between statutory construction standards, as specified in SOLAS, and the actual standards adopted by classification societies. Such a study should recognize the capabilities of various societies and how each functions in relation to IACS, IMO, and to its government.

NOTES

1. This discussion of Table 2-1, comparing vessels of widely ranging sizes, tends to disguise the fact that an increase in the size of ships, all designed to the same standards, in itself would result in a significant reduction in the lightweight to deadweight ratios.

2. Less distance between the waterline and deck is permitted for barges, in comparison to tankers.
3. International conventions (MARPOL 73) that eliminate mixing of oil and water by requiring tanks dedicated to ballast only, have been applied to new vessels (see Figure 2-3); in pre-MARPOL vessels, ballast still is placed in empty cargo tanks, and the "load on top" (LOT) method is used for pollution control (explained later, in the discussion of the Significance of MARPOL).
4. Initiation of the fire on the tanker MEGA BORG off the Texas coast, June 9, 1990, occurred in the pump room and was unrelated to the lightering operation underway at the time (Cutter Information Corp., 1990).
5. To be considered protectively located, segregated ballast tanks are to be arranged so that the side shell and the bottom shell area of the tanks are from 30 to 45 percent (depending on the ship's size) of the total side and bottom area of the entire hull (MARPOL regulation 13E).
6. The full requirements are contained in Appendix A, which describes damage assumptions, hypothetical oil outflow limitations, and tank size and arrangement limitations. These criteria, initially adopted by IMO in 1973, are based on comprehensive review of the damage and outflow experience of many nations in the years prior to 1973. The term "hypothetical" reflects the fact that this body of regulations as a whole does not purport to represent precise determination of amounts of oil that will flow out of a damaged tanker, even under the specified conditions. Rather, the procedure, taken as a whole, provides a relative index of the outflow potential for various designs and tanker arrangements, given a consistent set of damage assumptions for collisions and groundings (U.S. Coast Guard, 1973).
7. See previous footnote.
8. LOT relies on the principle that when oil and water mixtures are left standing, the oil separates and rises to the top. The heavier clean water at the bottom can be drawn off and returned to the sea; oil and water mixtures that remain are transferred to a slop tank. At the next loading port, new cargo is loaded on top of the oil retained in these tanks.
9. A deadweight-limited carrier is one in which the cargo is sufficiently dense that it does not require all the available volume, or cubic, to load the ship down to its deepest allowable draft with minimum freeboard (the distance between the waterline and the deck). Conversely, a cubic-limited carrier is one in which there is insufficient volume for the density of the cargo to enable the ship to be loaded to its deepest permissible draft with minimum freeboard. Ships carrying very dense iron ore are typically deadweight carriers, whereas ships carrying LNG, or light petroleum products, such as gasoline, are typically cubic carriers. Cubic carriers can lose substantially more oil than deadweight carriers in groundings, due to the greater freeboard and higher static head above the ship's laden draft. (The concept of hydrostatic balance, and the effect on oil outflow, is explained in Chapter 3.)
10. Flag state selected (by non-resident ship owners) on the basis of favorable conditions for non-residents in terms of commercial flexibility and tax treatment.
11. IACS has 11 member societies and one associate member. The members are: the American Bureau of Shipping, Bureau Veritas (France), China Classification Society, Det norske Veritas (Norway), Germanischer Lloyd (Federal Republic of Germany), Korean Register of Shipping, Lloyd's Register of Shipping (United Kingdom), Nippon Kaiji Kyokai (Japan), Polski Rejestr Statkow (Poland), Registro Italiano Navale, and USSR Register of Shipping. The associate member is Jugoslavenski Registar Brodova (Yugoslavia).
12. Many ships operate, legally, while not "in class" with a classification society. Flag states may issue load-line certificates; however, these also conform to applicable international convention requirements. Ships sustaining damage to an extent that cancels valid class status still may obtain insurance coverage for salvage purposes at exorbitant rates.
13. Does not apply to barges.
14. MARPOL and its Protocols are implemented in the United States through the Act to

Prevent Pollution from Ships, P.L. 96-478, Oct. 21, 1980, 94 Stat. 2297, 33 USC 1901 *et seq.* and regulations promulgated thereunder, 33 CFR Parts 155, 157.

The International Convention on Load Lines is implemented in the United States under P.L. 99-509, Title V, Subtitle B, Section 5101(2) in part, 100 Stat. 1914, October 21, 1986, 46 USCG 5101 *et seq.*, and regulations promulgated thereunder, 46 CFR Parts 1 *et seq.*, 172.

SOLAS legislation and regulations for tank vessels are implemented under 46 USC 3301 *et seq.* and 46 CFR Subpart D.

15. The Exclusive Economic Zone generally is considered to extend to 200 nautical miles from shore.

REFERENCES

Clarkson Research Studies Ltd. 1990. FAX to D. Perkins, National Research Council, Washington, D.C., August 31, 1990.

Cutter Information Corp. 1990. Oil Spill Intelligence Report. Newsletter published by Cutter, Arlington, Massachusetts, June 14, 1990.

International Chamber of Shipping. 1978. Ship to Ship Transfer Guide. London: Witherby.

U.S. Coast Guard. 1973. Note by the United States—Report on Study I Segregated Ballast Tanker. Report prepared for IMCO International Conference on Marine Pollution, London, October 8-November 2, 1973.

U.S. Coast Guard. 1989a. Navigation and Inspection Circular 10-82. Published by the Coast Guard, Washington, D.C., September 18, 1989.

U.S. Coast Guard. 1989b. Marine Safety Center internal memorandum, May 25, 1989.

U.S. Coast Guard. 1990. Report of the Tanker Safety Study Group, Chairman H. H. Bell (rear admiral, USCG, retired). Washington, D.C.: U.S. Department of Transportation.

3
Physical Bases of Phenomena Active in Casualties

The preceding chapters have drawn a general profile of tank vessel activity worldwide, with particular regard to the United States. The report has presented, within an historical framework, the most up-to-date information available on tank vessel traffic, casualty patterns, basic design principles, and the regulatory system.

The report now turns to scientific and technical aspects of tank vessel design. This chapter describes the physical phenomena governing tankship and barge behavior with respect to the casualties described in Chapter 1. The intent is to provide basic insight into the laws of nature that must be considered in tank vessel design. Later chapters will expand upon the significance of these phenomena in designs aimed at preventing and/or ameliorating the oil spillage resulting from casualties. It is, perhaps, obvious that these discussions can be neither complete nor comprehensive.

Topics to be discussed in this chapter include: momentum exchange and energy dissipation in collisions and groundings; residual strength following damage; fires and explosions; corrosion and fatigue; hydrostatic pressures in opposition; and diffusion and dynamics of fluid/vessel motion.

STRUCTURAL AND DYNAMIC ASPECTS

Momentum Exchange and Energy Dissipation in Collisions and Groundings

In a collision or grounding, forces are generated that tend to bring the impacting bodies to the same velocity; or, in other words, to bring them to rest relative to one another. When this occurs with forces so small that ship structure isn't damaged, then the kinetic energies involved, for the most part, are not dissipated. Instead, they are momentarily "stored" in the potential energy of elastically deformed structure, and the momentum exchange is completed when the deformed but intact structures "spring back" to their original shapes.

Unfortunately, the kinetic energies involved in most collisions and groundings are so large, compared to the potential energy that can be "stored" in elastic deflection of structure, that virtually all of this energy must be dissipated rather than exchanged. In groundings, there are four basic mechanisms of energy dissipation:

- lifting of the ship against gravity forces;
- frictional forces generated by rubbing of the hull against the grounding bottom;
- forces involved in the plastic, i.e., permanent, shape distortion of the hull girder; and
- forces involved in the fracture of structural material.

Although substantial friction forces often are generated as the structure of a penetrating vessel rubs against the ruptured structure of the stricken vessel, only the last two mechanisms listed above are available as major dissipation mechanisms in collisions.

The dynamics of the grounding process are highly interactive. That is, the amount of energy dissipated through any of the mechanisms listed depends on the characteristics of the grounding bottom (whether hard or soft, level or inclined, etc.), local strength of the hull girder, and "global" parameters such as ship length, displacement, and velocity.

To comprehend the magnitude of the forces and distances involved in dissipating the kinetic energy of tankers, consider that the kinetic energy of a Boeing 747 (the largest operational commercial transport airplane) at landing touch-down speed is roughly 3×10^8 pound-feet. This is about twice the kinetic energy of a 40,000 DWT barge at normal tow speeds of 6½ kts. A 200,000 DWT tankship, however, at a cruising speed of 12 kts, has more than eight times the kinetic energy of the landing 747.

Despite the large energies involved, only a small percentage of grounding incidents lead to oil outflow.[1] Many minor groundings entail minimal hull damage and are unlikely to be detected until subsequent drydocking. This emphasizes the importance of the first three energy dissipation mechanisms in groundings. Straightforward analysis of an idealized ship running aground at a speed, V, on a flat sea bottom with coefficient of friction f, and angle of inclination with respect to the horizontal, α, shows that the stopping distance, x, along the incline is

$$\bar{x} = V\sqrt{\frac{k\Delta}{gTPF\sin\alpha(f\cos\alpha + \sin\alpha)}}$$

where g is the acceleration of gravity, Δ is the ship's displacement, K is added mass coefficient, and TPF is tons per foot immersion. This formula results when the ship's bottom is oriented initially at exactly the incline of the sea bottom, or when the energy dissipation does not include that reorientation. When typical values of pertinent parameters for MARPOL tankers are inserted, the following approximate expression is obtained:

$$\frac{\bar{x}}{L} = \frac{0.0753\text{V (in knots)}}{L\sqrt{(f\cos\alpha + \sin\alpha)\sin\alpha}}$$

This formula determines the stopping distance as a fraction of ship's length (in feet), L. Applying it in two illustrative examples shows that a vessel 950 feet long, grounding at 12 knots:

- on a bottom of rocky nature (f = 1.2) inclined about 0.6°, will come to rest in slightly more than 1/4 of its length; and
- on a bottom of muddy and sandy nature (f = 0.4) inclined about 0.3°, will come to rest in slightly less than 7/10 of its length.

Similar approximations show that ground reactions will vary between somewhat more than 5 percent to 15 percent of displacement—apparently without bottom rupture. This explains how even groundings with high kinetic energy may not result in bottom damage. These considerations suggest categorizing grounding and collision incidents in terms of the severity and extent of damage. From the pollution prevention point of view, the severity and extent of damage incurred in an accident is perhaps best measured by the fraction of the cargo volume exposed in a hull breach. A related definition (Wierzbicki et al., 1990) measures damage severity by the number of breached tanks. Such definitions avoid categorizing collisions or groundings simply by the amount of kinetic energy to be dissipated, and also recognize that damage termed "severe" in one grounding may not be in another. In the literature, grounding incidents commonly are characterized in terms of the speed of the impacting ship. This is acceptable if scenarios are restricted to a specific class of ground obstacles, such as narrow, hard rocks. Such an approach was adopted in the pollution-effectiveness analysis discussed in Chapter 5.

The sequence of events involved in dissipating kinetic energy when hull material *is* distorted and fractured is similar in groundings and collisions. It involves the successive plastic deflection, puncture, and rupture of various structural members. In collisions, these members are in the bow and side shell of the striking and the stricken ship, respectively. In a grounding they are in the ship's bottom structure. In either case, outer shell plating and attached longitudinal stiffeners deflect, and, through bending and membrane tension, they provide a resistive force to penetration. As the indenting body advances deeper into the structure, the shell deflection and tension increase. Both of these factors increase the resolved lateral forces acting on the adjacent web frames or transverse bulkheads. (A description of structural elements can be found in Appendix B.)

These forces eventually may reach values required to buckle and initiate crushing of the nearest frame or frames. If still larger deflection results, then the lengths of side shell from the flanking inter-web spacings begin to stretch, applying forces on the next pair of frames. This process continues until either sufficient energy is absorbed to prevent any further indenting, or the outer shell reaches its rupture strain. If the latter occurs, then the penetrating body will continue advancing, resisted only by the remaining intact structure, with some energy dissipation through friction of the penetrating and resisting structures rubbing together.

Beyond these similarities, there are important differences between collision and grounding energy dissipation mechanisms, as follows.

In a grounding, the work done when (and if) the ship's unsupported weight is lifted by running aground, as well as friction forces, can be significant contributors to dissipation of kinetic energy. The ship's roll or pitch will not contribute; the weight unsupported by buoyant forces must be lifted. In addition, grounding usually involves a damage process lasting for a relatively long time, often extending along some considerable portion of the ship's length. The width of a penetrating body can vary, resulting in either a knife-edge type of slitting action involving a minimum of metal distortion, the ripping of a wide swath from the bottom with substantial metal stretching and distortion, or wider or narrower indenting when fracture does not occur. Collision damage, on the other hand, tends to be more restricted to a specified location in a ship. It may involve a continually widening hole, determined by either the shape of the bow of the striking ship or, more often, the angle of impact.

Structural members that may be affected directly by collision or grounding damage are:

- hull plating;
- longitudinal or transverse stiffeners;
- web frames;
- transverse bulkheads;
- longitudinal bulkheads; and
- horizontal decks.

Transverse bulkheads usually are sufficiently stiff and are spaced closely enough along the ship's length that they limit, in collisions, the extent of damage to hull plating in the longitudinal direction. The same is true of longitudinal bulkheads in limiting the lateral extent of damage in groundings. Although bulkheads can be strained by hull plating distortion, they usually will not incur damage unless impacted directly. The distance between bulkheads, however, is an important parameter in total energy dissipation.

The crushing strength of the bow of a striking ship in a collision derives from the substructure behind the bow—usually intersecting, deep horizontal and vertical web frames. In a collision, these members take on compressive loads. The resisting force and the friction in a steady-state grounding (to be explained shortly) derive primarily from shell straining and rupture and the continuous bending and rebending of longitudinal stiffeners, the latter being particularly significant in longitudinally stiffened ships. The transverse bulkheads and web frames are less important in resisting bow crushing in a striking vessel, or the progressive shell opening in a grounding vessel, except as side-supports to keep transverse and longitudinal members from buckling prematurely. On the other hand, the energy absorbed by a ship *struck* in a collision is related strongly to the thickness of transverse web frames and bulkheads. For all three cases—striking and stricken ships in collisions and ships involved in groundings—the work done by forces generated in stretching hull plating beyond the elastic limit, the so-called "membrane" energy, is a very significant contributor to the total energy absorbed.

Damage to ship structure incurred in collisions and groundings can be classified in two broad categories: Sharp and localized cuts of the bottom or side plating caused by a hard wedge, or more uniform crushing and rupture resulting from excessive straining at some distance from load applications sites. Separate studies of these damage scenarios (Minorsky, 1959, and Vaughan, 1977, respectively) have led to development of techniques for predicting the extent of damage.

For convenience, groundings can be thought of as occurring in the following four stages, and collisions in the first three:

- outer dynamics;
- initiation of local damage (denting/rupture of the hull girder);
- interaction between overall ship motion and localized damage; and
- steady-state deformation.

"Outer dynamics" describes the rigid body motion of a ship with respect to the impacting body, without regard to any permanent damage that may occur. This motion largely determines the reaction forces between the ships, or between ship and bottom, at any point in the process.

"Initiation of local damage" occurs as the impacting body plows progressively into the ship's hull at a given location, with forces largely determined by the "outer dynamics." The subsequent damage propagation can

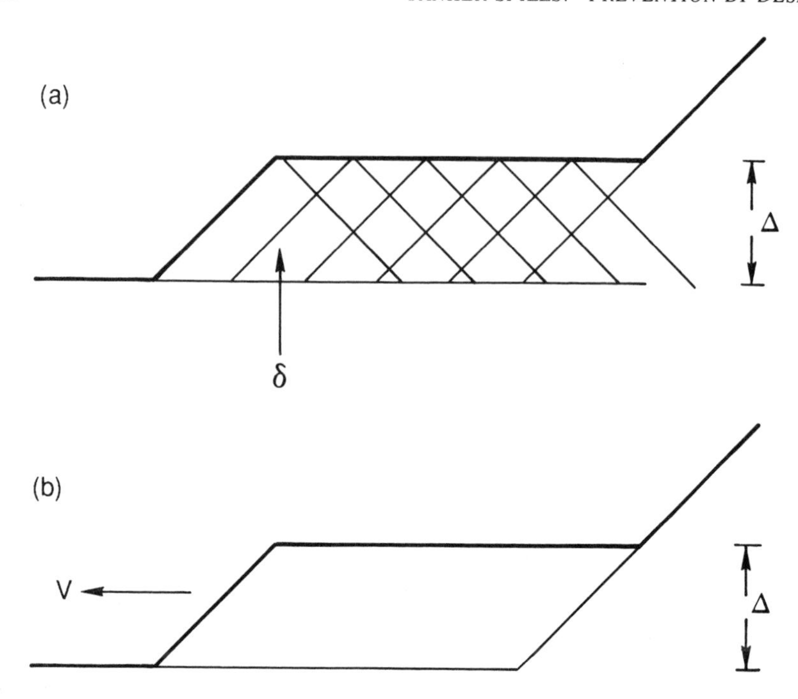

FIGURE 3-1 Two structural models of ship grounding: (a) sequence of re-initiation and (b) steady-state propagation.

be modeled either as a sequence of re-initiations of the idealized dentings, or as a continuing damage process (see Figure 3-1b) with increasing depth, either at the point of collision or at different points along the ship's length in a grounding. This is shown diagrammatically, for groundings, in the idealization of Figure 3-1a. Local force-deformation characteristics of the hull girder, under the indenting force, determine the lateral and longitudinal extent of damage and the length of transition zones between damaged and as-yet-undamaged parts of the hull.

The "interaction" phase is characterized by contact forces in the rigid body motion of the ship and local crushing/rupture modes, which determine the relative magnitudes of ship motion (mostly lateral and rolling in collisions, mostly vertical in groundings) and the local penetration depth. Light, stiff ships with hull plating resistive to rupture may ride over obstacles in groundings with minimal permanent damage; fully loaded VLCCs of current design are likely to incur damage almost equal to the height of the rock above the vessel's bottom, and with little vertical ship motion. These situations are contrasted in a highly idealized manner in Figures 3-2a and 3-2b.

In groundings, one of the most dangerous failure modes is the rupture of

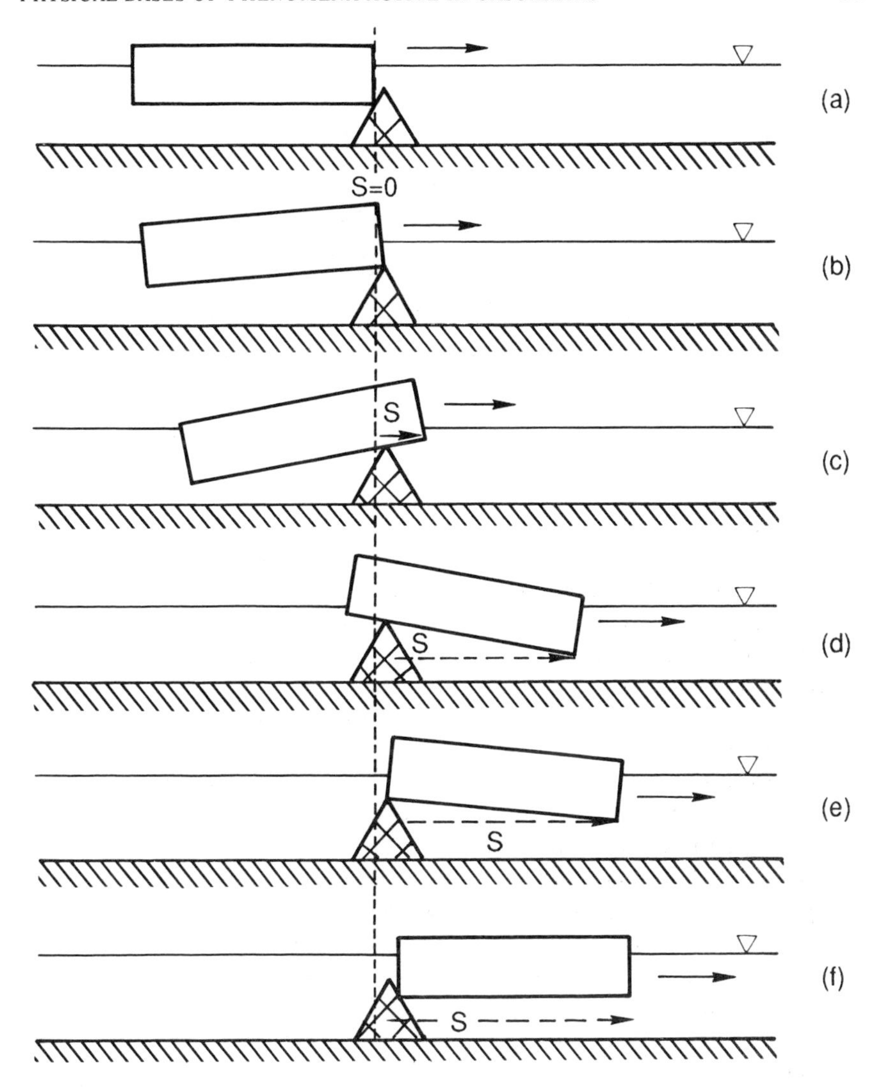

FIGURE 3-2a Idealized ship-rock configuration in a ride-over mode.

plating and the detachment of longitudinal stiffeners from the bottom plating. Both of these modes were observed during the committee's visit to the damaged EXXON VALDEZ. The rupture of plating, the more disastrous of two possible conclusions to the "initiation of local damage" phase, occurs when hull material's strain limit is exceeded. If this happens, and rupture occurs before the full kinetic energy is dissipated, then rupture progresses in the "steady-state" mode. Detachment of stiffeners usually follows; this

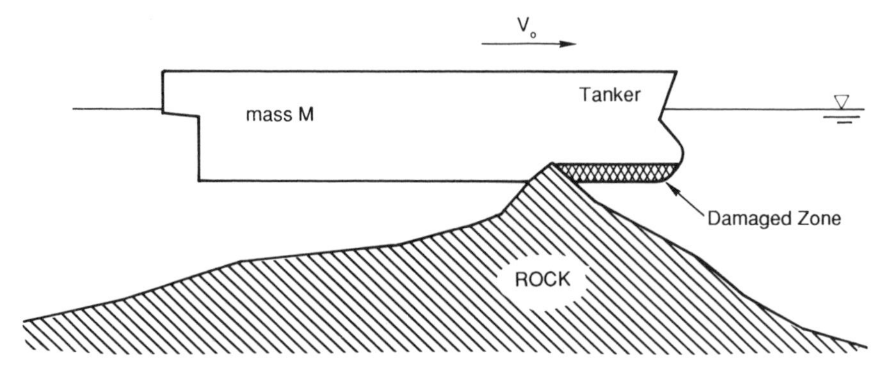

FIGURE 3-2b Idealization of a typical high-energy grounding scenario.

occurs when the shear stress in the welds attaching stiffeners to hull plating exceeds the allowable yield stress of the weld material. A critical penetration depth can be defined as the depth at which material separations occur, as determined by failure in either of the two modes.

The time scale of most collisions is too short to view the process between initiation of local damage, and either the disengagement of the ships or their attainment of equal velocities, as a steady-state mode.

A simple mechanics-of-materials model explaining how the crushing-tearing mechanism dissipates energy (in the failure initiation phase) is shown in Figure 3-3. As the component of indenting force in the plane of the material is resisted in pushing through the distance δ, energy is dissipated by plastic deformation and failure resulting in the membrane tension gap labeled "G-1." A similar model representing the steady-state bottom tearing process is shown in Figure 3-4. Energy dissipation also occurs in a membrane tension failure, evidenced by the gap labeled "G-2." Smaller amounts of energy are dissipated through the "moving plastic hinge," generated in the plate curling action shown in Figure 3-5.

If the energy absorption of all structural members of the hull girder participating in the damage process are collected, energy dissipation beyond that of bottom friction and weight lifting can be estimated. Examination of the associated forces also can be related to global ship parameters. This method, which represents the dissipative action of specific structural arrangements and scantlings, promises to predict the extent of damage to the hull girder in groundings and collisions (Wierzbicki et al., 1990). An earlier concept related the resisting contact forces between colliding ships only to the volume of distorted materials in the side of a stricken ship (Minorsky, 1959). This approach, which has the advantage of simplicity, "averages out" details of structural arrangements and scantlings, and requires that proportional constraints be determined on an empirical basis. Because struc-

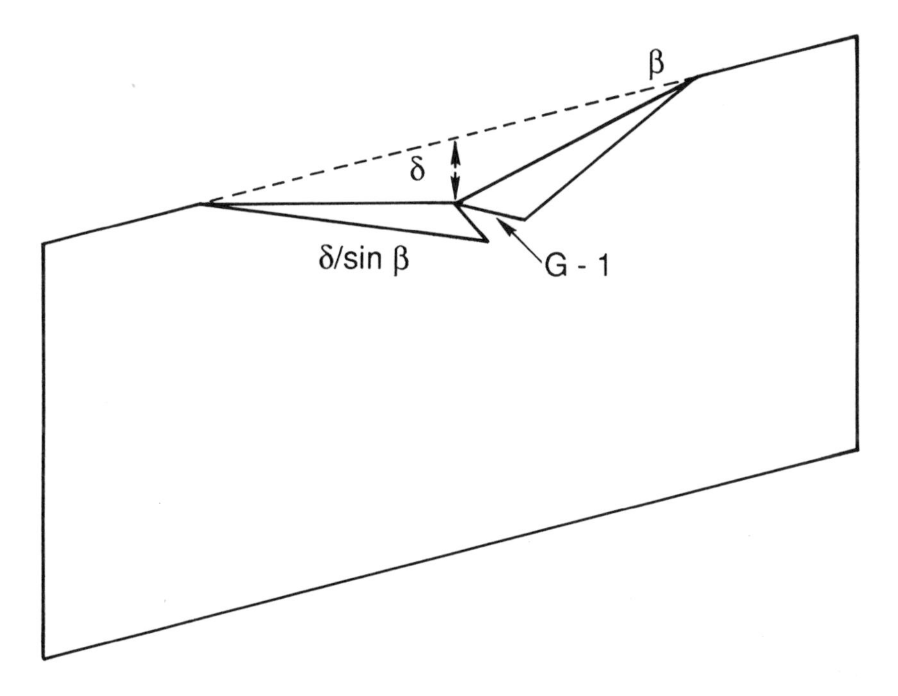

FIGURE 3-3 Simple computational model of the crushing of plates under inplane concentrated loading. (The gap, G-1, illustrates considerable membrane tension in the plate.)

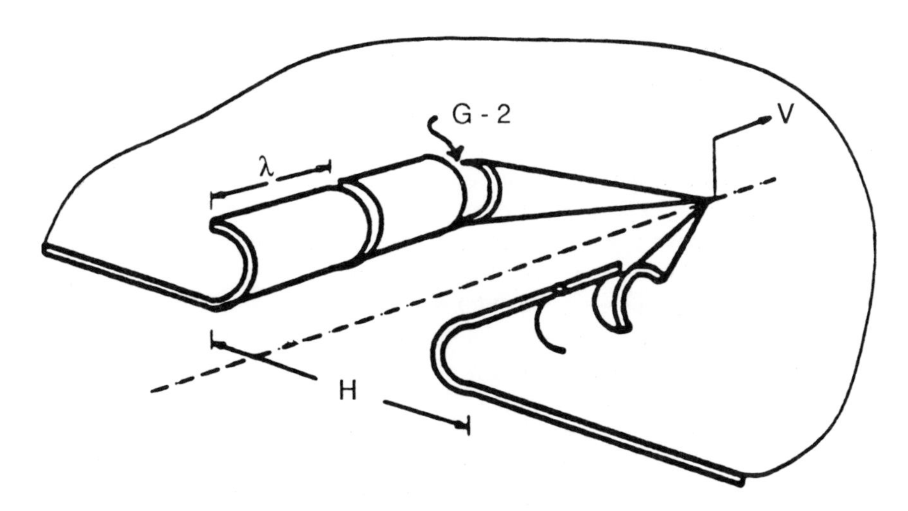

FIGURE 3-4 Computational model of steady-state bottom tearing.

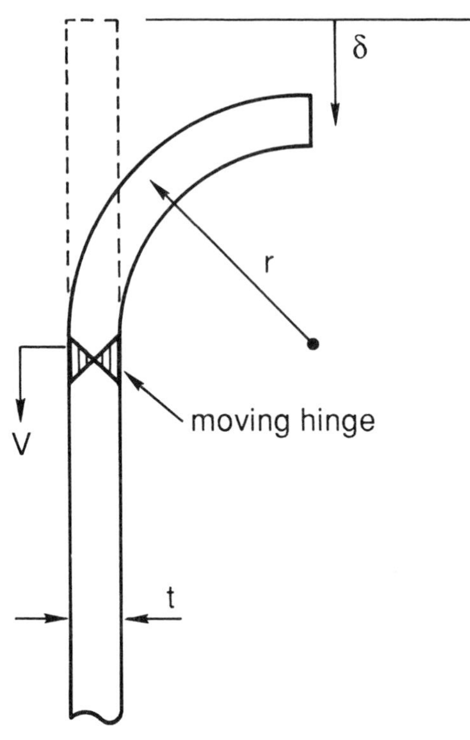

FIGURE 3-5 Progressive out-of-plane bending with a moving plastic "hinge."

tural members are subjected to compressive or crush loading in groundings, like the bow and side structures in a collision, the "volume of distorted material" approach can be extended to cover grounding of ships as well (Vaughan, 1977). This method[2] accounts for fracture of bottom plating by modifying the original empirical collision formula, through use of a term related to the area of the hull plating involved. The proportionality constant is determined from experimental data.

From results now available, it appears that the most efficient way to improve energy-absorbing capacity for ships involved in collisions or groundings may be (1) to increase the thickness of shell plating, (2) to employ material with increased capacity to withstand strain, and (3) to provide substructural (internal framing) arrangements that distribute strains as uniformly as possible, and thus delay rupture. This leads to two further conclusions. First, collision protection for a stricken ship with double sides, and grounding protection for a ship with a double bottom, would both be best served by distributing a given amount of double-shell material such that outer skins are as thick as possible and inner skins are as thin as allowable (Jones,

1990). Second, limiting the propagation of disruption through appropriate substructure design would serve to limit oil spillage, by preserving the integrity of tanks not in the immediate vicinity of hull penetration. Methods for predicting the propagation of damage through such structure, however, are neither well developed nor substantiated. Improvements in hull girder design, therefore, will be limited by appropriate developments in predictive analysis.

Residual Strength Following Damage

A ship whose basic structure (i.e., the "hull girder" formed by the deck, bottom, side plating, transverse and longitudinal bulkheads, web frames, and associated stiffeners) has been damaged severely is imperiled by loads that the intact structure could carry easily. Such damage might result from corrosion, fatigue-related structural failure, collision, grounding, explosion, or fire. If there are no basic changes in applied loads, then the determinative phenomena are, at the macro level, new load paths and, at the micro level, stress redistributions. The ability of the damaged structure to carry such unchanged loads can be called "residual strength." Virtually the same finite element analysis techniques used to predict load paths and stress distributions for the original structural design also can be used to estimate the residual strength of the damaged structure, with appropriate accounting for stress concentrators such as sharp cracks associated with the failure.

By focusing solely on achieving reductions in oil spillage, it is possible to conceive of designs that, once damaged, could place the entire vessel, crew, and cargo at unacceptable risk. Applied loads are unlikely to remain the same following such damage. Among other hazards, breached voids may fill with sea water, cargo may drain into void spaces, and, perhaps most foreboding, a grounded vessel may lose supportive buoyant forces if left impaled on a pinnacle by receding tides. In such cases, residual strength must be provided in the face of both redistributed and/or aggravated loadings *and* damaged structure. Tank vessel redesign with the intent of reducing pollution risk must consider residual strength very carefully. Aside from the potential for loss of life in the circumstances listed above, *all* of the cargo is likely to be spilled in the event of total vessel loss.

FIRE AND EXPLOSIONS

All hydrocarbon cargoes and their residues involve danger of fire and explosion. Such cargoes include crude oils, home heating oils and diesel fuel, and such products as gasoline, kerosene, and naphtha. The most volatile are, generally, the most dangerous, but virtually all are flammable at ambient conditions. Oil vapors accumulating in void spaces, or in ullage spaces (the air space above the cargo) in oil cargo tanks, can be highly explosive.

"Heavy end" fuels, such as those used for ship's bunkers, or power plant fuel, generally will not burn at ambient conditions. Such cargoes, having a flash point of at least 60°C, do not have explosive vapors at ambient conditions, and thus are not as hazardous as crudes or clean products. At elevated temperatures, however, these fuels exhibit the same flammability and vaporizing characteristics as crude and light products.

Asphalt is a specialty petroleum product that must be carried at elevated temperatures even to be liquid. It may be the only petroleum product which does not, of itself, pose some hazard of fire or explosion, unless water is present. In that case, because the high temperature of asphalt cargo can turn water quickly into steam (which expands and foams), a hazardous condition quickly arises.

Conditions for Igniting/Sustaining Combustion

For a fire or explosion to occur, three elements must be present concurrently: semi-solid, liquid, or gaseous hydrocarbons; oxygen in specific proportion relative to the hydrocarbons; and an ignition source. If the hydrocarbons are mainly in gaseous form, combustion, when it occurs, will be in the form of an explosion; otherwise, there will be a fire.

Prevention of fires and explosions requires removal of at least one of the three elements. In practice, neither hydrocarbons nor ignition sources can be removed reliably, for the following reasons.

Hydrocarbons always are present during normal operations, of course, because they constitute the tanker's cargo and its engine's fuel. Total removal of hydrocarbons to the extent that combustion cannot be supported is only possible when a ship's tanks are "gas free" to permit "hot work" in a particular tank; this may occur during either voyage or shipyard repairs. Therefore, unless tanks have been certified "gas free" by a marine chemist, it must be assumed that hydrocarbons sufficient to support combustion are present in the cargo tanks.

Ignition sources exist in several forms aboard tankers. Because of the multiplicity of potential sources, the tanker industry has adopted the position that all sources of ignition never will be fully removed from a tanker. Sources identified, often through costly experience, include the following:

- Static electricity, generated by washing of cargo tanks with water or, most often, crude oil. Sparks of electrical potential sufficient to initiate combustion can be released through potentials generated between the tank washing medium and the surrounding tank walls.
- Sparks from tank gauging tapes or other instruments, if not properly grounded when lowered into cargo tanks.

• Magnesium or aluminum sacrificial anodes (placed in cargo tanks to prevent corrosion) can become ignition sources if they drop to a tank bottom. Only zinc sacrificial anodes are now permitted in tankers; zinc does not produce an incendiary spark if it drops on steel from a height.

• Steel-on-steel contact, as occurs if a steel object drops to the bottom of a tank, or if improper tools are used by a crew in cargo tanks. Wrenches, buckets, shovels, and even personal items such as pocket knives and keys, can be dangerous in this respect. Another possible ignition source is "working" between steel members in a ship's structure following material failure or collision.

• External sources, such as lightning, striking a tanker's venting system, or the result of a collision. In the case of fires following a collisions, large quantities of oil almost always are burned.

• Ship machinery or electrical installations.

Given the inevitable presence of hydrocarbons and ignition sources, oxygen is the element that industry and governments have sought to eliminate during normal operations. This is done by "inerting" cargo tanks—filling the ullage space with gases that do not support combustion. The source of such gases can be either cleaned ("scrubbed") flue gas from the ship's boilers or, as is now common, a separate inert gas generator. The principal ingredients of the cleaned residue (derived from thorough combustion of hydrocarbon and oxygen) are nitrogen (roughly about 80 percent), and carbon dioxide (roughly 12-14 percent).

The incorporation of inert gas systems (IGS) in tankers over 100,000 DWT was mandated by the international SOLAS convention in 1974. This regulation was extended in 1978 to include all new or existing crude oil tankers over 20,000 DWT, and product tankers over 40,000 DWT.[3] This does not apply to barges.

Figure 3-6 shows, schematically, the major components of an IGS (Gray, 1979). Note that an IGS connects cargo tanks, which contain potentially explosive hydrocarbon vapors, to the ship's engine room, with its multiplicity of ignition sources. Cargo tanks are isolated from the ignition sources in the inert gas generator by a non-return valve and a water seal. The effectiveness of an IGS depends on maintaining a positive pressure in the cargo tanks so that oxygen will not enter. Tanks are vented through pressure vacuum valves that open whenever a predetermined pressure or vacuum in the tank is exceeded.

Fires and Explosions Outside of Cargo Tanks

Numerous potential sources of ignition exist in non-cargo spaces, such as the engine room, the ship's quarters, and the galley. In these spaces, just as on other types of ships, the main fire/explosion preventatives are simply good practices around hot machinery; e.g., removing sources of oil leaks,

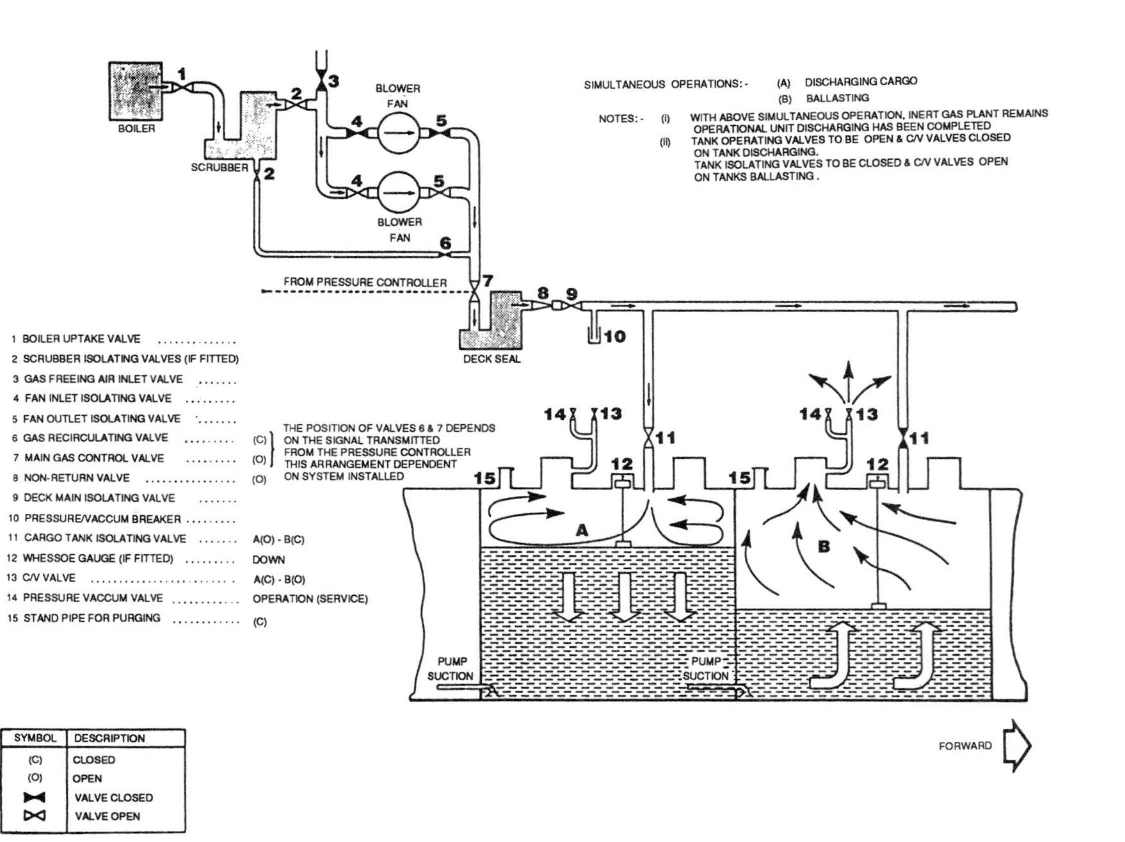

FIGURE 3-6 Inert gas system.

grease, and other flammables such as dirty rags. Tankers do, however, differ in one important respect from other ships in this regard, as they are fitted with a pump room.

Pump rooms usually are located centrally between the cargo tank block and the ship's stern, just forward of the engine room. The pump room houses the pumps and valves that control operation of the cargo system and ballasting system. The pump drivers, being potential ignition sources, generally are not in the pump room, but in the engine room; they are connected to the pumps by a shaft through a bulkhead. These drivers can be steam turbine, reciprocating steam engine, electric or hydraulic motor, or diesel engine power sources.

In the pump room itself, with its many pipes, valves, and pumps, the main defenses against fire or explosion are rapid ventilation to remove hydrocarbon vapors as soon as they are generated, and appropriate fire extinguishing systems. Because crew members are often in the pump room during cargo handling, an IGS system isn't possible there (the crew could not breathe). On modern tankers, much of the work could be handled from a cargo control room, but there is always a possibility of malfunction or operational difficulty, requiring crew members to enter the pump room. The pump room generally is considered the second most dangerous area in a tanker, next to the cargo tanks.

CORROSION AND FATIGUE

Corrosion of metal occurs through a spontaneous electrochemical mechanism in which a portion of the metal (the anode) is oxidized, and a substance in the environment (the cathode) is reduced. These two aspects of the process, oxidation and reduction, take place at different, though usually not widely separated, areas of the metal, with an electrical current flowing between the two.

If an alternate source of electrical flow is provided, then corrosion can be inhibited. This source can be provided by either active (powered) electrical systems or by passive "sacrificial" material chosen to act as an anode.

Adherent and nonporous films can protect surfaces from corrosive forces by increasing resistance in the path between cathode and anode, and by decreasing diffusion rates of reactants and products of electrode reactions to or from the metal surface. Naturally occurring oxidation films can inhibit further corrosion, if the oxide has (1) a higher specific volume than the metal, (2) forms under compression as a result, and (3) covers the entire surface under attack when the film has grown to only a few molecular layers in thickness. Many substances that act as anions, such as those encountered in seawater and salt-spray, break down oxide film and accelerate corrosion. Mechanical strains that cause cracking, of course, also can deprive films of their protective powers.

Oxygen concentrations in either fresh or sea water droplets can be non-uniform, and this so-called differential aeration causes areas of wet metal surfaces to be anodic where oxygen concentration is low and cathodic where it is high. Since the edges of droplets have good access to oxygen and the centers do not, anodic attack on the metal occurs near the center of droplets, with rust rings building up at the edges. This is among the phenomena leading to "pitting."

Internal stresses due to applied loads, or residual stresses from fabrication processes, generally make metals more anodic and tend to increase the rate of corrosion. Furthermore, once corrosion takes place, stress concentrations are aggravated particularly at points of attack, due to the sharp crack tips that usually occur there combined with the reduced metal thicknesses brought about by corrosion. This is one form of the phenomenon called stress corrosion and is by nature a progressive problem, as such cracks will continue to propagate.

Repeated cycles of stress ultimately cause structural materials to fail (fatigue) at stress levels considerably lower than could be tolerated under static conditions. Many structural materials have a protective endurance limit, an alternating stress level below which the material will not fail, even with an infinite number of applied stress cycles. But in the presence of corrosion, this limit falls precipitously. If corrosion continues long enough, almost any stress level will be sufficient to cause structural failure.

From this limited discussion, it follows that segregated ballast tanks, located between the outer hull and in-board cargo tanks, can be particularly susceptible to corrosion. In pre-MARPOL ships, seawater ballast was placed mostly in tanks that recently had held cargo; the oily film clinging to tank sides provided some measure of corrosion protection. But in a segregated ballast tank, only salt water and air may be held, so any metal plating without a protective coating is subject to a highly corrosive environment. In most new ships, ballast tanks are coated to prevent corrosion. Maintenance of these coatings is an important element in ship management. In all ships, the tank walls are part of the ship's primary structure (the hull girder), and as such are subject to stress, which contributes to deterioration of hull coatings.

In a ballast tank, the outer hull has thicker metal, but is subject to highly corrosive elements on both inner and outer surfaces. Moreover, the outer hull is usually the most highly stressed material in the ship's structure. Inner plating, while usually thinner, has the protection of cargo residue on one or more sides. Still, even very small, corrosion/pitting holes in the inner plating can have the most serious consequences; they can permit hydrocarbon vapors to enter empty ballast spaces, where they pose explosion hazards. Water-crude oil combinations, which tend to collect on cargo tank bottoms, are particularly troublesome. Resulting corrosion-related phenomena on tank bottoms make tanks susceptible to corrosion from the cargo side as well as the ballast/void side.

Protective coatings often are used to inhibit corrosion. Painted external hulls, of course, are in this category. Because many events can remove paint during normal operation, this protection usually is complemented by active systems, below the water line, which provide a flow of electrons to the cathodic material. For reasons explained previously, this alternate current source serves to inhibit corrosion. Most classification societies require that water ballast tank plating receive protective surface treatment, often complemented by "sacrificial" anodic materials or systems to act as alternate current sources. But such systems only function when the ballast tanks are filled with seawater, and corrosion can proceed when these tanks are empty. Sacrificial anodes, therefore, have limited use, and limited effectiveness.

HYDROSTATIC PRESSURES IN OPPOSITION

When a loaded tanker's hull is breached, seldom does the entire cargo spew into the sea. It can happen, but only in rare cases where the entire ship is lost (such as the ARGO MERCHANT and AMOCO CADIZ, listed in Table 1-3). In groundings, where the bottom of a cargo tank may be breached, usually only part of that tank's cargo is spilled. No cargo escapes, obviously, from cargo tanks that remain intact, or from breached voids, ballast tanks, or empty cargo tanks. The EXXON VALDEZ spill, great as it was, represented only about 20 percent of the total cargo. While an ecological calamity, the fact remains that 80 percent of the cargo was saved. This section seeks to explain how this is possible, by addressing the physical mechanisms governing oil outflow.

Hydrostatic pressure is an isotropic phenomenon. That is, at any given point in a fluid the pressure will be the same, regardless of the direction in which it is measured. The hull plating on a ship can experience a net pressure in or out, depending on the depth of the oil and water at that point. Oil will run out of a submerged tank opening only if the (interior) hydrostatic pressure of the oil at the opening is greater than the (exterior) hydrostatic pressure of the sea at the same point.

The hydrostatic pressure, or "head," of a column of liquid, h, is defined as its height times its specific weight, $g\rho$, where ρ is the mass density of the liquid. When comparing the heads of two columns of different fluids, then, it is convenient to think in terms of their specific gravities, ρ/ρ_{water}, where ρ_{water} is the density of fresh water. The specific gravity of crude oil varies between 0.83-0.90, a common value being 0.86. The specific gravity of sea water is 1.025. The relative heights of columns of oil and sea water in equilibrium (i.e., having equal pressure) at a given point will be inversely proportional to their specific gravities. Thus, the height of an oil column, relative to that of a seawater column with which it is in equilibrium, can be expressed as:

$$h_o = h_w \cdot sg_w/sg_o$$

where sg is specific gravity, h is a consistent measure in feet or meters, and the subscripts o and w represent oil and seawater, respectively. For oil with a sg = .86, and seawater at 1.025, the column height ratio, h_o/h_w = 1.025/0.86 = 1.192. In other words, the column of oil with a density that produces equal pressure will be 19 percent higher than the balancing height of sea water. This is a significant difference, and it suggests that tankers might adopt loading schemes or install systems that take advantage of these differences in specific gravity, to minimize the potential loss of cargo from groundings.[4]

The way most tankers are loaded at present, the cargo level establishes a hydrostatic pressure at the tank (and ship's) bottom that is greater than external sea pressure. Thus, if the tank is breached, then cargo flows out. (In fact, the external pressure equals the tank pressure at a point several feet to 15 feet or more below the ship's bottom, depending on cargo density.) But if a tanker carried somewhat less cargo, to establish a hydrostatic balance at or several meters *above* the tank bottom, water would tend to enter the ship through the hole in the hull (as long as the highest point of damage was below the hydrostatic balance level). The implementation of this concept, hydrostatically balanced loading (or hydrostatic control), is discussed in Chapter 5. Cargoes with specific gravities close to that of water can be expected to form an emulsified interface; this is likely to increase the pollution potential of hydrostatically balanced loading schemes for such cargoes.

If cargo and seawater are in hydrostatic equilibrium at the ship's bottom, then for any distance above that point on the ship's side, the cargo and seawater pressures will be out of balance. Thus, the amount of the height of the oil column that must be lost to reachieve balance, say x, following rupture of the tank's side (as might occur in a collision a distance, y, above the bottom) will be given by $x = 0.192y$. For specific examples of the effects of hydrostatic pressure on oil cargo, see Figure 3-7.

This discussion of hydrostatic balance assumes equal atmospheric pressures at the surfaces of both seawater and oil cargo. As the latter is enclosed in a tank, it can be subject to other pressures. If there are overpressures, as required currently for IGS, oil outflow would be increased in a tank breach. Conversely, reduced ullage space pressures would reduce oil outflow. This suggests another means of controlling oil outflow in groundings; implementation of this concept is discussed in Chapter 5.

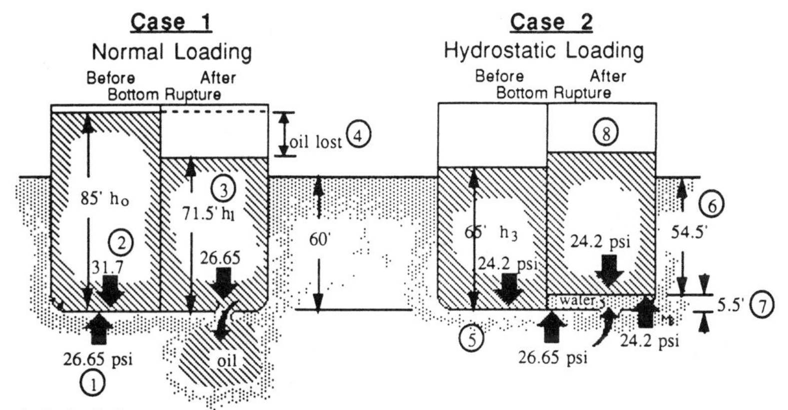

Case 1 Calculations

① Sea water pressure at 60' depth is: $60 \times \frac{1.025 \times 62.4}{144} = 26.65$ psi

② Pressure at bottom of 85' column of oil (.86 sg) is: $85 \times \frac{.86 \times 62.4}{144} = 31.7$ psi

③ Oil height to balance 26.65 psi sea pressure is: $26.65 \times \frac{144}{.86 \times 62.4} = 71.5'$

④ Case 1 oil cargo loss for hydro balance is: $\frac{85 - 71.5}{85} = 15.9\ \%$

Case 2 Calculations

⑤ Pressure at bottom of 65' column of oil (.86 sg) is: $65 \times \frac{.86 \times 62.4}{144} = 24.2$ psi

⑥ Sea water depth to balance 65' column of oil is: $24.2 \times \frac{144}{1.025 \times 62.4} = 54.5'$

⑦ Height of water entering tank to balance pressure is: $60 - 54.5 = 5.5'$

or $\frac{(26.65 - 24.2)144}{1.025 \times 62.4} = 5.5'$

⑧ Height of oil in case 2 after breaching bottom is: $65 + (60 - 54.5) = 70.5'$

NOTE: Oil s.g. = 0.86
Sea Water s.g. = 1.025
Fresh Water weighs 62.4 lbs/ft^3
Both cases assume ship does not change draft appreciably

FIGURE 3-7 How hydrostatically balanced loading works.

This figure depicts two tankers, each with a 60 foot draft. Case 1 shows the ship loaded with 85 feet of .86 specific gravity oil in the left tank. The hydrostatic oil pressure at the tank portion is greater than sea pressure. The right hand tank bottom is breached and oil runs out until the hydrostatic pressures are balanced. Note that the height of remaining oil is still above the waterline but that 15.9 percent of the original cargo was lost to the sea. (The example assumes the ship draft did not change significantly.)

In case 2, the ship is only loaded with 65 feet of oil. Breaching the right hand tank results in water entering to a height of 5.5 feet to establish an equilibrium with the oil. No oil escaped. This is equivalent of hydrostatic loading to a level of 5.5 feet inside the ship's bottom.

From case 1, it is shown that hydrostatic loading to the level of the bottom, to minimize cargo outflow, would mean carrying 15.9 percent less cargo than shown in the left hand tank.

There is one more mechanism that can increase oil outflow following a collision, *after* hydrostatic balance is established. Consider a situation in which some oil remains in a tank, after the balancing process, *below* the damage to the tank side. A disturbance, such as wave action or ship rolling, may throw seawater into the tank. The heavy water will sink, as it will not float on the lighter oil; the water therefore must displace the oil, which then rises and exits through the hole. This is obviously a much slower process than oil outflow driven by unbalanced hydrostatic pressure. Even without ship or water motion, some exchange of oil and water will occur as the result of differences in specific gravity.

Finally, it should be noted that a few liquid cargoes have specific gravities greater than that of seawater. In such cases, the cargo's lack of buoyancy will override hydrostatic phenomena and result in all cargo above a submerged hole "dropping out" of the opening. The cargo will be replaced by seawater to the level of the waterline. Consequently, the advantages associated with hydrostatic control scheme are applicable only to vessels carrying cargoes of specific gravity significantly less than water. This excludes vessels carrying, for example, feed stocks, caustic soda, or asphalt.

DIFFUSION AND DYNAMICS OF FLUID/VESSEL MOTION

When no membrane, such as hull plating, separates two fluids, such as oil and seawater, there will be some mixing at the surfaces where the fluids are in contact. The degree of mixing depends on the particular situation.

Consider the typical series of events when a cargo tank is breached at its bottom. If the oil has a hydrostatic head exceeding that of the seawater at the point of damage, then the oil escapes relatively quickly, especially if the tank opening is large. In fact, for the larger holes, oil outflow may be limited by the rate at which the ship's cargo venting system allows air to enter the expanding void space above the surface of the oil. For a large hole, it may take as little as two or three minutes to lose 20 percent of the cargo. Because most of this will be lost almost immediately, there is obviously insufficient time to take steps to repair the hull, even temporarily, to prevent this escape.

Once hydrostatic equilibrium is established at the opening, smaller amounts of oil can continue to escape as the tide falls (if the ship is aground) or as the ship rolls and pitches because of wave action. In this context, tide-related actions are measured in hours, offering time for use of on-board cargo transfer systems (discussed in Chapter 5). While such losses are driven in part by hydrostatic variations, over a longer time there is likely to be some mixing of oil and water inside the cargo tank due to hydrodynamic effects, such as turbulence in the oil and/or water. In high-energy collisions or groundings, violent mixing of oil and water can occur at the outset, creating a fluid with a specific gravity that lies between those of the oil and seawater. This may result in some oil outflow beyond that associated with simple hydrostatic effects.

When ullage space is sufficiently large, cargo sloshing can be a problem. If either or both the ship's rolling and pitching frequencies approach the natural frequencies of the fluid in partially filled cargo tanks, then higher wave forces than exist with smaller ullage spaces could develop within the tanks. Such forces are capable of damaging internal bulkheads.

SUMMARY

From the preceding pages, it is clear that tank vessel designers and operators must be aware of a variety of complex and interrelated physical phenomena. The topics highlighted in this chapter are of particular relevance in this study, as they become important factors in assessment of design alternatives intended to reduce the risk of pollution resulting from casualties.

Several findings can be drawn from the brief discussion of the physical basis of the damage process occurring in collisions and groundings, fire and explosion hazards, corrosion, hydrostatic pressure, and dynamics of fluid/vessel motion. These findings are as follows.

In considering the damage protection afforded by double bottoms, double sides, and double hulls, protection appears to be greatest when outer hull is as thick as possible, even at the possible expense of the inner skin. Beyond such guidelines for conventional hull structures, there appears to be great potential for diminishing the extent of hull damage, and even preventing the initiation of plate rupture in higher energy groundings and collisions, through proper selection of external hull materials and/or innovative structural design.

Hydrostatic balance can be achieved at some point along the hull of a cargo tank—at the bottom, for example—and this can reduce or eliminate oil outflow. Reduced pressure in ullage spaces can achieve hydrostatic balance with greater oil cargo depths. But long-term buoyancy naturally drives all oil from below a side hull opening beneath the waterline.

Corrosion, metal fatigue, and the threat of fire and explosion require fastidious and continuous attention, because of both their potentially disastrous consequences and the difficulty of prevention.

With this theoretical discussion as a prelude, the following chapter will discuss engineering considerations related to tank vessel casualties.

NOTES

1. A recent analysis (Lloyd's Register of Shipping, 1990) indicates that outflow occurs in less than 6 percent of all incidents.

2. The committee's pollution-control analysis of alternative tank vessel designs, conducted by Det norske Veritas and detailed in Chapter 5, was based on this methodology.
3. Also covers existing product tankers having high-capacity washing machines for cleaning tanks.
4. In actual practice, these calculations would be adjusted to account for inert gas pressure, if present.

REFERENCES

Gray, W.O. 1979. Requirements for Inert Gas Systems. Paper presented at IMCO Tokyo Seminar on Tanker Safety and Pollution Prevention, February 19-23, 1979.

Jones, N. 1990. Some Comments on the Collision Protection of Ships with Double Hulls. Paper presented at meeting of the Committee on Tank Vessel Design, Washington, D.C., June 5-7, 1990.

Lloyd's Register of Shipping. 1990. Statistical Study of Outflow from Oil and Chemical Tanker Casualties. Report conducted for the American Petroleum Institute, Washington, D.C. Technical Report STD R2-0590.

Minorsky, V.U. 1959. An Analysis of Ship Collisions with Reference to Protection of Nuclear Power Plants. Journal of Ship Research 3:1-4.

Vaughan, H. 1977. Damage to Ships Due to Collision and Grounding. Paper published by Det norske Veritas, Oslo, Norway, August 1977. DnV 77-345.

Wierzbicki, T., E. Rady, and J.G. Shin. 1990. Damage Estimates in High Energy Grounding of Ships. Paper presented at meeting of the Committee on Tank Vessel Design, Washington, D.C., June 6-7, 1990.

4

Engineering Considerations

Even with improvements in crew training, tank vessel operation, and navigation, accidents will occur. The effects sometimes can be mitigated by design. This chapter addresses engineering considerations related to tanker design as they affect oil outflow in collisions and groundings. Some of the same considerations apply to ocean-going barges. Arguments frequently offered for or against particular design features are addressed here from an engineering viewpoint.

In the context of the full report, this chapter reflects practical issues resulting from the physical phenomena explained in Chapter 3. Further, these discussions are intended to provide an understanding of issues that are basic to any evaluation of tank vessel designs. The overall assessment of specific designs mentioned in the following pages can be found in Chapter 5.

The topics to be discussed in this chapter include: hull strength; tank proportions, arrangements, and stability; salvage concerns; and safety of life.

HULL STRENGTH

When a tanker is damaged by collision or grounding, with resultant outflow of cargo, the hull obviously has failed. Its strength was insufficient to survive the trauma. Hull failure also can result from improper loading, design or construction flaws, corrosion, explosions, or fires. To understand all of these situations, a brief review of structural design principles will be useful.

Design Loads, Stress Analysis, and Scantling Selection

The loads considered in tanker structural design include ship and cargo weight, internal and external hydrostatic pressure at all angles of heel and trim,[1] wave loads, and dynamic components resulting from induced motions. (A more detailed discussion can be found in Appendix B.) These loads are variable and many act simultaneously. They can be broken down into the following components:

- internal/external liquid pressure differentials;
- shear forces generated by the non-uniform distribution of ship weight, cargo, and buoyancy along the length of the vessel, both in still water and in waves; shear is important in webs and girders (support framing), as well as side structure and longitudinal bulkheads;
- hull bending moment[2] caused by unbalanced longitudinal distribution of buoyancy and weights;
- wracking (twisting) loads, tending to distort the hull's shape, caused by oblique seas, asymmetrical loading, etc.;
- impact loads such as the slamming of the ship's forward body during heavy pitching;
- the weight of water on deck caused by heavy seas;
- sloshing loads on bulkheads caused by impact loads of cargo motions in partially filled tanks; and
- thermal loads caused by unequal temperatures such as from heated cargo.

Many of these loads can be present at any given time, even if a ship is in still water without wave loading and dynamic effects. Loads commonly cause a combination of tensile and compressive forces in both plate and girder structures; shear forces are imposed simultaneously on some components. With the presence of both compressive and shear forces, the possibility of buckling must be considered. This is increasingly true as ships increase in size, because plating and beam structures generally do not increase proportionally in thickness.

Ship designers have been aware of these principles for decades, but historically, in practice, they were limited to simplified definitions of loads. Designers compensated by applying time-tested margins, which allowed for imperfect theoretical understanding of loads. Margins also made up for design shortfalls and allowed for corrosion, material flaws, fabrication error, and operational mishaps.

In the 1960s, the availability of computers, coupled with improved understanding of oceanographic data, enabled naval architects to define more precisely the dynamic loads experienced by a ship, and to evaluate the response to those loads. In short, designers had a more rational basis for analysis and were able to produce more efficient designs. Traditional margins—or "safety factors"—were pared back, and ship structure became relatively lighter, or less robust.

In the meantime, the size of tankers increased dramatically, as described in Chapter 2. To compound the stress on ship structure, competitive pressures of the 1980s placed an additional premium on lighter (cheaper) designs, and classification societies further relaxed traditional margin requirements where a rational basis for design could be demonstrated. Because use of high-tensile steel could reduce hull weight further, this material proved increasingly popular. This introduced new concerns, e.g., corrosion margin, fatigue life, and tighter construction requirements. The net result was that the proportion of non-revenue-producing hull structure in tankers dropped dramatically over a 20-year period (as was shown in Table 2-1). As a result, modern hull structure is relatively less able to resist unexpected loads such as from grounding or collision.

The history of engineering provides a number of examples of design refinement and rapid growth in structure size leading to novel and unforeseen difficulties. A land-based example goes back to the early 1940s when natural gas pipelines were first being built to bring southwestern gas to the industrial Northeast. Several catastrophic failures occurred to these new lines (causing deep and long furrows along the right-of-way), which were the first to come north and by far the longest lines then in service. Knowledge of the principles of fracture mechanics grew with the awareness that the combination of long pipeline sections between connections and the cold weather environment made ideal conditions for fracture propagation.

With that image as a prelude, the engineering difficulties posed by advances in tanker design will be discussed in the following pages.

Design Margins

As noted previously, modern tankers have less design margin (Ferguson/Lloyd's Register of Shipping, 1990; Iarossi/American Bureau of Shipping, 1990) than they did in the past, whether it be the ability to accept flaws in design details, material quality, manufacturing errors, corrosion, or operations. Overall, they are less robust than their predecessors.

A major consequence is increased risk from corrosion—higher induced structural deterioration. All shipbuilding steels corrode at the same rate unless effective protective measures, such as coatings and/or anodes, are applied and maintained. Thus, if the thickness of plates and stiffening members is reduced, because of changes in design criteria and materials, there is less margin left to corrode before the structure is subject to possible failure. Large tankers built in the late 1950s and early 1960s commonly had deck and bottom plate thicknesses on the order of 30-35 mm. These

TABLE 4-1 A Comparison of Segregated Ballast Areas in a 260,000 DWT Tanker

Tanker Type	Total Area of Ballast Spaces* (flat plate area without stiffeners)
Pre-MARPOL	7,000 square meters
MARPOL	23,000
Double Bottom	34,000
Double Sides	39,000
Double Hull	64,000

* Figures are rounded off to the nearest thousand.

plates now are typically only about 20 mm thick. Wastage of unprotected steel proceeds at a constant rate; at 0.5 mm per year, for example, an unprotected 35 mm plate will be reduced to 30 mm after 10 years, a reduction of 14 percent. By contrast, the exposed portion of the 20 mm plate, also reduced by 5 mm in ten years, will be reduced by 25 percent. Thus, adequate corrosion protection is much more critical in newer ships than in older ships.

Compounding this concern is the fact that MARPOL tankers, due to the volume of segregated ballast tankage, have larger areas susceptible to the maximum corrosion rates than pre-MARPOL ships. Tankers with double bottoms, double sides, or double hulls have even more area in void/ballast tanks susceptible to corrosion, as shown in Table 4-1. (Crude oil cargo tanks, which are protected by oily residue, rarely experience as much wastage as ballast tanks.)

As ships grew larger and the design process became "more efficient" in establishing primary or global features, secondary and tertiary structural details, which had not been problems before, became increasingly critical (U.S. Coast Guard, 1990c). Furthermore, the types and numbers of such details multiplied; each detail adds to the potential for inadequate design analysis or fabrication error. Additionally, before computer-assisted finite element analysis (FEA) was possible, detailed analysis of these components was very difficult. In fact, even now, FEA of all structural details in a large tanker is a major undertaking, requiring a long time to complete and/or great expense (U.S. Coast Guard, 1990c).

Ironically, current rules allow a designer to lighten the outer shell scantlings (especially the bottom shell) of a double hull, because the inner sides and bottom carry some of the primary structural loads (Donald Liu, letter to the committee, May 10, 1990).

Other difficulties have been brought on by use of higher-tensile steels, as follows:

• The hull is more flexible and deflections are increased. This places additional strains on non-high tensile secondary components.

• Higher-tensile steels generally do not have improved fatigue performance over mild steel (Ferguson/Lloyd's Register, 1990; Iarossi/ABS, 1990). Initially small, innocuous cracks can grow to significant proportions, and unexpected problems can evolve, particularly in longitudinal connections to transverse web frames.

• High-tensile steel structures require more precise fabrication techniques and are less forgiving of fabrication errors. The higher the stress and the thinner the material, the greater the susceptibility to failure from either stress concentrations or misalignment.

In sum, higher design loads and extensive use of high-tensile steels per existing rules can lead to early fatigue or corrosion problems with some pollution potential. Without proper maintenance and inspection, these problems can escalate into major structure failure with massive pollution potential. This is not to say that high-tensile steel should not be used; its use, however, must be integrated carefully into the whole structural design.

The growth in tanker size, the reduction in design margins, and, to a degree, the greater use of high-tensile steel, have increased the need for careful study of structural design details. A related concern is that extensive employment of double-hull vessels (expected as a result of the Oil Pollution Act of 1990) would add a large number of additional structural details to be designed, fabricated, installed, and inspected.

Classification societies are examining their strength standards (Ferguson/Lloyd's Register, 1990), and some shipbuilders and owners are reconsidering their extensive use of high-tensile steel (Royal Institute of Naval Architects, 1990). It is imperative to follow through on such efforts. The problem of selecting the best structural materials, from the point of view of delaying and/or reducing damage from accidental loads, is discussed in Chapter 5.

Inspection and Maintenance

Inspection and maintenance are relatively more important in large, modern tankers, particularly those employing high-tensile steel. It is nearly impossible to describe, especially to someone who never has been on board a large tanker, just how extensive and complicated proper inspection and maintenance can be. A single tank in a VLCC has been compared to a Gothic cathedral; within one tanker there are literally tens of thousands of intersections of crossing structural members, each of which has several points of attachments, and all of these joints are susceptible to cracks. An indication of the area that must be inspected on a pre-MARPOL 250,000

TABLE 4-2 VLCC Inspection Factors in Cargo Tanks

PER 250,000 DWT VLCC (Pre-MARPOL)	
Vertical Height to Climb for Survey	10,700 m/35,000 ft
Tank Section Area to Inspect	300,000 m^2/74 acres
Total Length of Welding	1,200 km/750 miles
Total Hand Welding (included in above)	390 km/240 miles
Total Length of Longitudinal Stiffeners	58 km/36 miles
Flat of Bottom Area	10,700 m^2/2.6 acres
1.0 Percent Pitting	85,000 pits (each 0.40 mm diameter)

SOURCE: Exxon Corporation, 1982.

DWT VLCC is detailed in Table 4-2. (A detailed example of an inspection and maintenance plan is provided in Appendix C.)

Inspecting such a ship, if done all at once (such as on a 20-day ballast voyage) might require 1,000 to 2,000 person-hours. (Even more daunting is the job of inspecting the structure inside the void/ballast tanks of a double-hull tanker.) For comparative purposes, it is interesting to note that the U.S. Coast Guard spent 25,479 "experienced" person-hours conducting 3,359 tanker hull inspections of all types over the two-year period FY 1988 and FY 1989—an average of 7.6 person-hours per inspection (from Table 2-2). Clearly, the Coast Guard cannot do more than sample the hull condition of a large tanker; the owner, therefore, must be relied upon to conduct complete inspections. Quality of inspections could become even more of an issue as double hulls become more common.

Adequate inspections can be accomplished. They do not require new principles, but they will require additional resources and changes in current practices. In addition, the efficiency of inspection can be facilitated by careful consideration of tank dimensions and access, and attention to design details.

TANK PROPORTIONS, ARRANGEMENTS, AND STABILITY

Size and arrangement of cargo tanks are significant issues in design of double bottoms, double sides, and double hulls (the alternative designs proposed most frequently to reduce pollution risk). Each of these alternatives has proponents and detractors. Detractors often claim these designs create stability problems; this section examines these issues. A brief discussion of tanker stability and damage survivability will provide the framework for the analysis.

Stability Criteria

All tankers have loading limits that assure sufficient reserve buoyancy to accept flooding of any compartment, or a specified number of adjacent compartments, without sinking. These criteria may be the controlling factors in establishing the maximum allowable loaded draft. Strength considerations also influence loading limits.

In addition, every ship must have sufficient "damage stability" to withstand the effects of the flooding of any (or specified combination of) spaces, including the off-center loading involved, while retaining sufficient range of stability to heel or roll a specified number of degrees beyond the damaged equilibrium point. The extent of damage ships must be able to survive is listed in Appendix A. The following criteria especially have a major impact on design:

- side damage is assumed to extend transversely to one-fifth of the beam, or breadth, of the ship (B/5); and
- bottom damage is assumed to extend vertically to one-fifteenth of the beam (B/15).

These criteria are silent as to allowable outflow of oil. However, as described in Appendix A (page 199), MARPOL does limit the size of oil tanks, in effect limiting outflow in case of tank rupture.

Ballast Requirements

All tankers must carry sufficient ballast, when sailing without cargo, to ensure a minimum mean draft for seaworthiness. The amount decreases in a relative sense as ships get larger. The ballast must be distributed to ensure propeller immersion, with trim limited to 1.5 percent of length. The minimum ballast capacity is fixed by a tanker's dimensions.

MARPOL requires tankers to carry ballast in segregated tanks (SBT), so a large part of the ship must be dedicated to clean tanks that are unavailable for cargo. One effect is that these tankers tend to have deeper hulls, with cargo extending higher above the water line to compensate for the volume lost to ballast tanks.

Bulkhead and Ballast Tank Dimensional Considerations

Most conventional tankers in the 50,000 to 250,000 DWT range have two longitudinal bulkheads, dividing the cargo space into three tanks across.[3] The bulkheads typically are spaced at one-third to one-fifth the beam (B/3 to B/5), due to the damage-stability assumptions noted previously: The ship

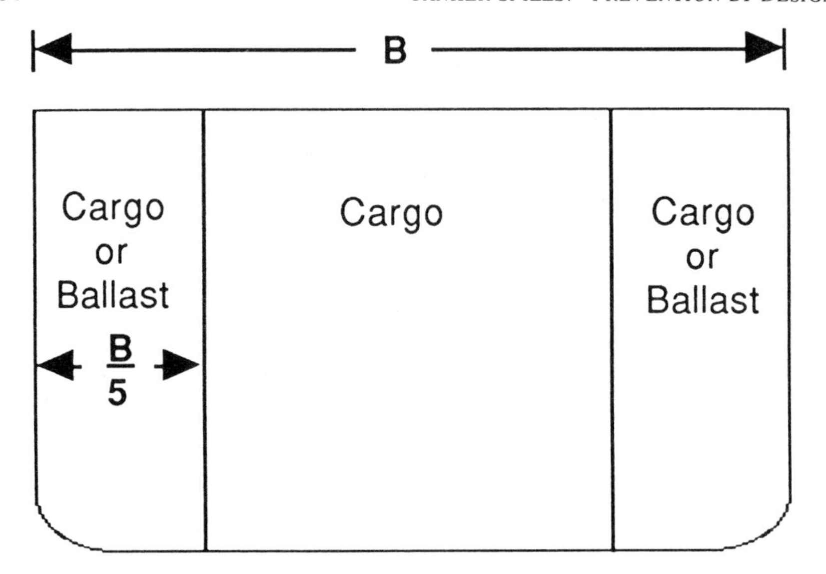

FIGURE 4-1 Transverse section of conventional MARPOL tanker.

must survive side penetration to a distance of B/5. Conventional MARPOL tankers usually carry the ballast in a series of wing tanks with a width of B/5. (See Figure 4-1.)

Neither minimum segregated ballast requirements nor damage-stability assumptions are altered for ships with double bottoms, double sides, or double hulls. The obvious design approach is to make use of the ship's required segregated ballast volume in arrangements that minimize additional impact on cargo capacity.

Double bottoms usually are set at a height of B/15 or more, except in the very large ships. The void space usually is used for ballast; there is no penalty in cargo capacity compared to other SBT designs. Actually, a B/15 inner bottom provides insufficient volume to satisfy segregated ballast requirements, so such designs require one or more pairs of side ballast tanks (fewer than a conventional MARPOL tanker).

For double sides, with a single-bottom cargo section, the B/5 width imposes a heavy penalty on cargo carrying capacity. The loss can be as much as 20 percent over and above the MARPOL segregated ballast requirement. Therefore, for double-side designs, tanks are sized to provide only the standard segregated ballast volume; the width of such side ballast tanks typically is B/9 to B/11. Of course, the designer must ensure that the ship has ade-quate stability to withstand collision damage extending into the inner tank, due to the B/5 penetration criteria. But this burden is far less onerous than losing more cargo capacity.

For double-hull designs, the height of the double bottom is usually B/15, driven by the rule governing extent of vertical damage. The balance of the required segregated ballast, when distributed along the entire side, requires double sides with a width of about B/15 also (by coincidence). This arrangement results in little or no penalty in cargo carrying capacity compared to the MARPOL segregated ballast requirement.

A study of how various combinations of double bottoms, double sides, and double hulls affect cargo capacity (and stability) can be found in Appendix D. The study focuses on a generic 35,000 DWT tanker. Because B/15 for this 89-foot-wide ship is less than 2 meters, the rules require a minimum 2-meter double side. For ships of this size, designers often provide 2-meter double bottoms—or slightly more than B/15. Nonetheless, the double-bottom, double-side, and double-hull designs have comparable cargo capacity, essentially unchanged from the MARPOL design. (See Appendix D.)

A minimum spacing of about 2 meters between hulls should allow adequate access for inspection. For larger ships, where a B/15 double bottom is so high that inspection and maintenance become difficult, a designer might choose to limit the inner bottom to a height such as 3 meters.[4] This approach has the further advantage of offsetting any reduction in cargo carrying capacity, and/or creating additional side protection (to fulfill ballast requirements) without further impact on cargo capacity. Other designers might favor higher double bottoms, to assure that all structure, such as framing, can be kept within the double bottom (this usually is facilitated, on larger ships, by heights greater than 3 meters).

For smaller tankers, 40,000 DWT or less, accident data indicate that a B/15 double bottom is very effective in preventing outflow in groundings (Card, 1975). There is no such casualty database for larger double-bottom tankers. An outflow analysis conducted by Det norske Veritas (DnV) (Appendix F, Table 3.3.1) suggests that exceeding the B/15 double-bottom height provides diminishing incremental protection against outflow during groundings of larger tankers. On the other hand, for VLCCs and larger tankers, outflow protection against collisions continues to improve with increasing distance between sides. A B/15 double side offers protection against low- and medium-energy collisions, but increasing the space between the double sides enhances this protection.

The key factor in determining outflow, of course, is the relative frequency and severity of groundings and collisions in U.S. waters. In any event, from a pollution-control standpoint, the DnV data suggest that, for tankers larger than about 120,000 DWT, some of the SBT volume dedicated to B/15 double bottoms might be more profitably invested in wider double sides. More study is needed to determine the best proportions of SBTs in larger double-hull tankers. In the meantime, in the interests of both ease of inspection and maintenance and improved overall pollution control, inner bottom heights of less that B/15 may be appropriate for larger tankers.

Segregated Ballast Tank Arrangements

Figure 4-2 shows schematically various configurations of tanks in double sides, double bottoms, and double hulls. The "C" and "B" signify cargo and ballast tanks, respectively. Figure 4-2a is a double-bottom arrangement. Most ships without double sides will have three tanks across because of the B/5 penetration rule. Double-side ships, Figure 4-2b, can be more flexible in tank arrangement as long as stability remains sufficient with a penetrated cargo tank side.

Figure 4-2c is a common double-hull ("L" tank) arrangement; it has the disadvantage of potentially large off-center weights if damaged. The "U" tank arrangement, Figure 4-2d, avoids that problem but doubles the total added weight in most damage cases. "U" tanks also can have a large free surface, involving some complications when loading and discharging cargo.

The arrangements in Figures 4-2e and 4-2f reduce the amount of off-center and total added weight for most damage scenarios, by separating the side and bottom ballast tanks. Figure 4-2e is more resistant to grounding damage while Figure 4-2f is more resistant to collision damage. The former provides improved structural continuity of the double bottom. The disadvantages of these two "DB + DS" arrangements over the "L" or "U" tank arrangements are added cost, and complications of access and ventilating the double bottom. However, the complications are no greater than for the double-bottom design, Figure 4-2a.

The number of longitudinal bulkheads shown in the double-side and double-hull examples is not intended to be literal. The number of internal bulkheads (up to three for the largest tankers) is dictated primarily by cargo outflow limits, the number of petroleum grades to be carried, structural configurations (on very large tankers), and the free surface effects on stability. For tankers up to about 150,000 DWT, current criteria generally can be met with no internal longitudinal cargo tank bulkheads, i.e., a single cargo tank between wing bulkheads.

Actual tank arrangements, of course, are selected by owners and designers to best meet these requirements. The various alternatives should be studied to determine the best proportions for a given trade and the risks encountered in that trade.

The foregoing description addresses the more conventional design approaches to double bottoms, double sides, and double hulls. Another tank arrangement scheme receiving considerable attention is the intermediate oil-tight deck (IOTD), where the cargo section is divided by a deck or a flat, horizontal bulkhead, with oil carried in both upper and lower chambers. This is essentially a double bottom of great height, loaded with oil. Figure 4-3 is a schematic of such a design. In addition to reducing the volume of oil exposed to a bottom rupture, the lower section has initial hydrostatic advantages over conventional MARPOL single-hull tankers. Coupled with

C = Cargo
B = Ballast

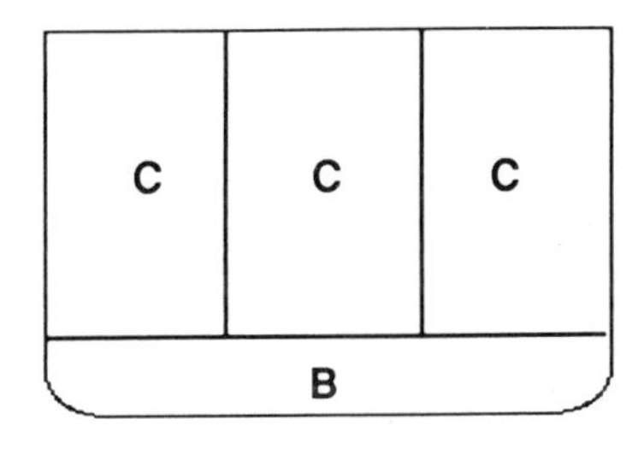

a. Double Bottom

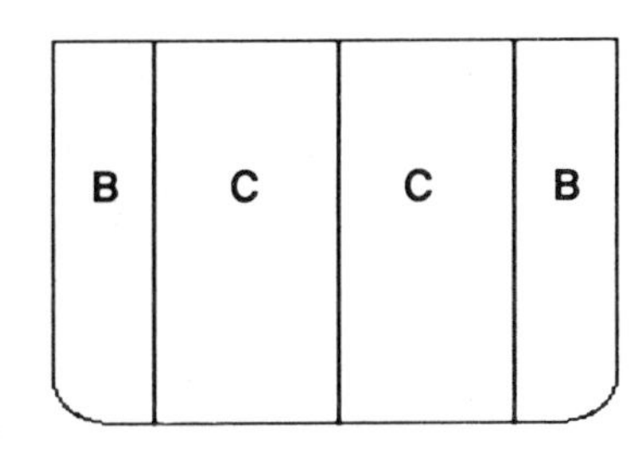

b. Double Sides

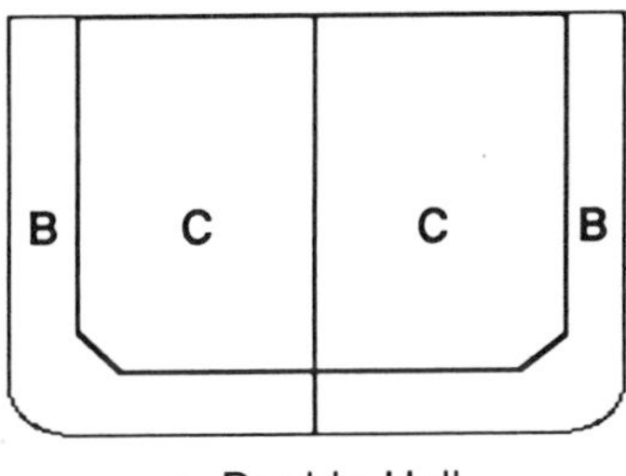

c. Double Hull
"L" Arrangement

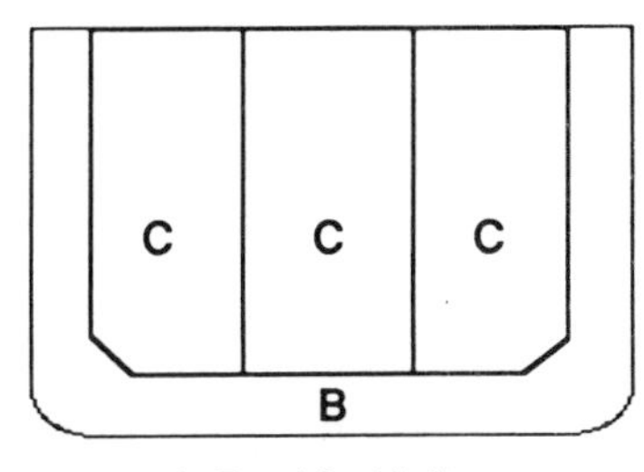

d. Double Hull
"U" Arrangement

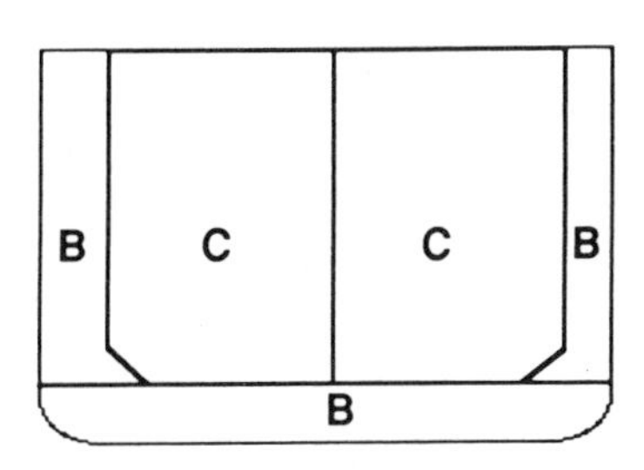

e. Double Bottom
Double Side

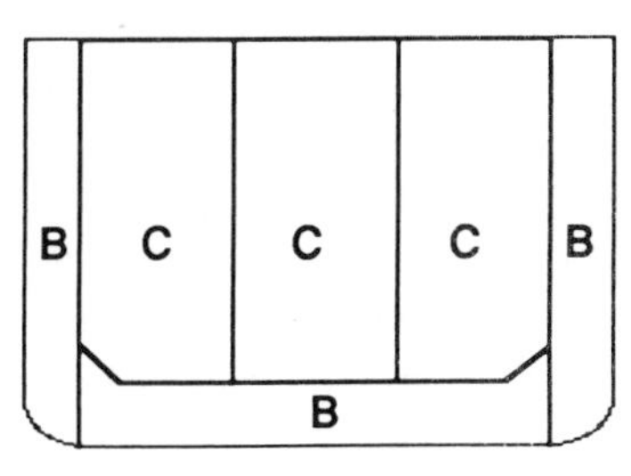

f. Double Side
Double Bottom

FIGURE 4-2 Tanker ballast tank arrangements.

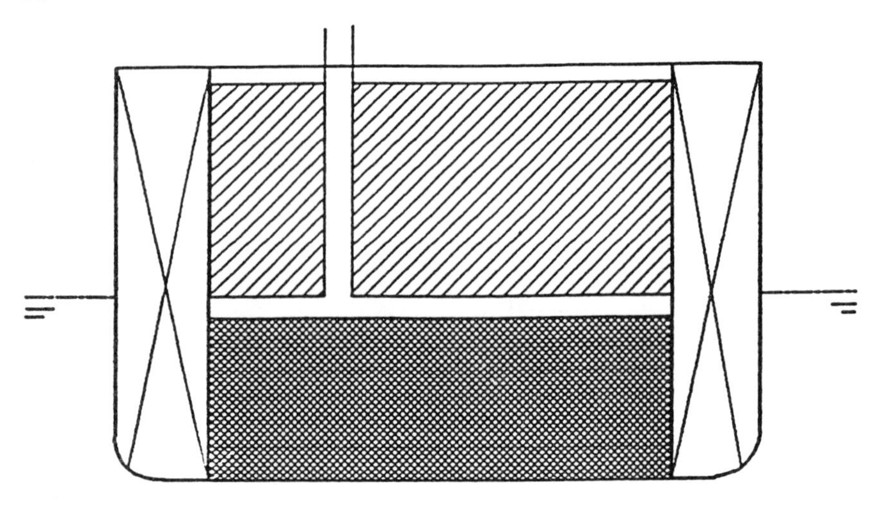

FIGURE 4-3 Schematic of intermediate oil-tight deck with double sides.

double sides, the IOTD also would provide protection against collisions. This concept will be evaluated in Chapter 5.

Stability of Double-Hull Tankers

A range of factors must be weighed in assessing double bottoms, double sides, and double hulls. These designs have operational and in some cases survivability advantages, but they also involve penalties in such areas as ease of inspection and maintenance, and possibly safety. A more comprehensive assessment of these designs may be found in Chapters 5 and 6. From an engineering perspective, however, all current stability criteria can be met, and exceeded, with properly subdivided B/15 double-hull designs.

All tankers must satisfy specified stability criteria to assure safe operation in heavy wind and sea conditions. They also must survive specified damage by meeting list,[5] flooded waterline, and range-of-stability criteria. As noted earlier, two of the damage criteria are the extent of vertical and horizontal penetration (B/15 and B/5, respectively), which influence the placement of double bottoms and double sides; other criteria involve the length of structural damage and the number of adjacent compartments that can be flooded. The details are not critical to this report, other than the fact that tankers exceeding 492 feet in length (about 15,000 to 20,000 DWT) must be able to survive flooding of two adjacent compartments.

Conventional SBT tankers are very stable; in fact, most such tankers far exceed the minimum IMO damage-stability requirements. This is largely because tankers typically have empty side ballast tanks separated by one or more loaded cargo tanks. When several cargo and ballast tanks are breached, the added off-center weight from flooded ballast tanks is in part compensated by the loss of weight due to oil flowing out of cargo tanks. By comparison, if a double-hull tanker incurs comparable hull penetration affecting only empty off-center SBTs, then there is no compensating loss of oil. The net increase in weight can be significantly more than for conventional tankers, and double-hull vessel damage stability, while meeting IMO standards, can be significantly less than for a conventional tanker. Thus, a double hull designed to meet current minimum damage stability criteria will not tolerate as much damage as a conventional MARPOL tanker—not because the former fails to meet damage stability requirements, but because the latter usually exceeds requirements.

Chevron Shipping Company has demonstrated these factors vividly. The company's report (abstracted in Appendix E) shows that the damage stability of an "industry standard" single-hull 130,000 DWT tanker greatly exceeds present criteria (Figure E-4 in the Appendix). A comparable double-hull tanker, with interhull spacing of 2 meters and fully loaded, also easily exceeds criteria (Figure E-5) but can withstand less damage than the single-hull version. When 96 percent loaded with heavy crude, the 2-meter double-hull ship will capsize with only three tanks flooded. (The ship has "J" ballast tanks and no internal cargo tank bulkheads, and it meets current criteria.) But the report goes on to demonstrate that, with imagination, a double-hull tanker can be designed to survive flooding of ballast tanks along an entire side of the ship.

The Seaworthy Systems study (see Table 4-3 and Appendix D) demonstrates that small tankers with double sides, double bottoms, or double hulls also can be designed to exceed by a wide margin the current damage-stability requirements. The most desirable characteristics are small equilibrium heel, large righting arm, and large range of stability. The table shows that all the designs considered (even under severe damage scenarios) have adequate stability because they have equilibrium heel that is less than the range of stability, and because they all have positive righting arm. However, some results are less desirable than others. Some of the double-bottom and double-side cases have a greater equilibrium heel, and lower range of stability, than is present with the single-skin case.

The foregoing discussion suggests that the damage-stability requirements for double sides, double bottoms, and double hulls should be increased, to approximate the stability achieved by comparable single-hull MARPOL tankers.

SALVAGE CONCERNS RELATED TO SBT TANKERS

This section addresses the frequent claim that double bottoms handicap salvage, as well as other salvage-related matters involving cargo systems,

TABLE 4-3 35,000 DWT Tanker 4 Compartment Damaged Stability Worst Case Scenario

Alternative	Damaged Scenario** (tanks damaged)	Equilibrium Heel (degrees)	Maximum Righting Arm (feet)	Range of Stability (degrees)
Single-Skin* (PL/SBT)	Nos. 6, 7, 8, and 9 Stbd Cargo Tanks	1	4.34	70
B/15 (2M) Double-Bottom	Nos. 5, 6, 7, and 8 Stbd Double Bottom	3	2.95	70
B/15 (2M) Double-Side	Nos. 4, 5, 6, and 7 Stbd Wing Tanks	6	3.26	70
B/15 (2M) Double-Hull	Nos. 3, 4, 5, and 6 Stbd Wing Tanks	7	9.50	70
B/5 Double-Bottom	Nos. 3, 4, 5, and 6 Stbd Double Bottom	13	0.40	17
B/5 Double-Side	Nos. 3, 4, 5, and 6 Stbd Wing Tanks	16	3.05	50
B/5 Double-Hull	Nos. 3, 4, 5, and 6 Stbd Wing Tanks	12	5.2	50

*No protectively located segregated ballast tanks were damaged to avoid running an inordinate number of damage conditions.

**The damage suffered is twice the current criteria (4 versus 2 adjacent tanks flooded).

In summarizing the discussion of tank proportions, arrangements, and stability, this table (derived from Table D-2) is instructive. It shows the results of calculations of the effect of severe damage on a single-skin tanker and a combination of double-bottom, double-side, and double-hull tankers that are typical of designs that could in practice be built. The following definitions apply:

- equilibrium heel is the angle of list (or tipping to the side) that the ship will have after damage has occurred.
- maximum righting arm is a measure of the relative reserve capacity of the ship to return to the equilibrium heel angle after damage has occurred.
- range of stability is the additional angle of heel the ship can accept and still return to the equilibrium position. When the range of stability is exceeded, the ship turns over. (For instance, the B/5 double-hull ship could roll to a total of 62 degrees and still return to its damaged equilibrium angle of 12 degrees).

emergency pumping capabilities, and on-board computers used to monitor loading.

Double Bottoms and Voids in Salvage

To examine the claim that double bottoms handicap salvage (and may increase oil outflow in a grounding), four types of grounding obstacles must be considered:

- Type 1: The pinnacle. In this scenario, the ship strikes a relatively small rock or pinnacle that cuts into the hull. Depending on ship momentum and the extent of interference by the obstacle, the vessel may pass over the obstruction with damage ranging from minimal to a long, narrow tear opening several hull compartments. If the ship passes over the obstruction, then it remains floating; if not, it is "stranded."
- Type 2: The rocky shoal. This obstruction, the big brother of the pinnacle, brings the ship to a halt. Damage will be extensive, although the longitudinal extent may not be as great as in the case of the pinnacle.
- Type 3: The submerged reef. This obstacle is characterized by a smoother contour than the rocky shoal; it may be mud, sand, coral, or loose rubble, with deep water on the far side. The ship may be brought to a standstill, or, if the interference is less than a foot or so (or the reef material is loose and moveable), then the ship may pass over the obstacle with little damage or flooding.
- Type 4: The sloping beach. The water depth shoals continuously over a broad area. Depending on the bottom material and ship speed, a vessel hitting the beach may sustain damage ranging from scratched paint to extensive rupture and tearing. If speed were more than 2 or 3 knots, the ship would remain stranded.

Tides play a major role in groundings. A ship that grounds lightly at low tide may float off with the rising tide. A ship stranded at high tide invariably will need help to be refloated.

The geographical location of a grounding also can be critical. For example, if a ship grounds on a pinnacle in the open ocean or on a coastline exposed to waves and ocean swells, then these forces can cause ship motions that can result in additional damage, or even loss of the ship in bad weather.

If the hull is breached, then flooding of voids such as inner-bottom or side-ballast tanks will make the ship heavier, leaving it harder aground. This is the basis for arguments that double bottoms complicate salvage. On the other hand, breach of a single-hull tanker usually will spill oil, making the ship lighter. Thus, in a Type 1 or Type 3 grounding, a single-bottom tanker would be lightened and would rise and pass over the obstacle. Even in the Type 4 grounding, the argument continues, the single-bottom ship theoretically would be easier to refloat, perhaps by backing off on a rising tide.

At issue is the frequency of Type 1 and Type 3 groundings where single-hull tankers passed over the obstruction or were refloated easily. The committee was unable to obtain data to define the relative risks of encountering the four bottom types described. Although there are reports of a few ships passing over obstructions while spilling only a modest amount of oil, the statistics are insufficient to counter the committee's view that most tankers, grounding at service speeds (14 to 16 knots), will remain firmly stranded—with or without the reduced-weight advantage of lost cargo.

An important point here is the definition of salvage. For purposes of the present study, salvage means voluntary assistance by outside professionals to save a ship and cargo in peril; thus, actions by the ship's own crew do not constitute salvage. These in-house actions, called damage control, generally will be useful to the salvor, especially if they have prevented further damage. In any event, regardless of the frequency of strandings requiring outside assistance, professional salvors routinely encounter such situations. And salvors do not consider double bottoms a handicap, for reasons to be described shortly; rather, salvors view double bottoms as helpful.

Another dimension of the salvage controversy is the claim that double bottoms can increase the amount of oil spilled. This argument is refuted by physics. A double-bottom tanker will not normally carry cargo at a higher level than will a comparable single-hull SBT tanker. And because of the bottom void, which will fill with cargo if the inner bottom is punctured, the hydrostatic head[6] of oil relative to the sea will be less than for a single-hull MARPOL tanker. Therefore, the outflow actually will be comparable, or even less, in a double-bottom tanker (the effect of hydrostatic pressure differentials was explained in Chapter 3).

The final claim to be considered here is that a stranded tanker with a flooded double bottom can be exposed to danger for an extended period of time, because of the increased off-loading required to refloat the casualty. During this added time, a sudden storm could interrupt the salvage operation and cause total loss of the ship and cargo. The single-hull tanker METULA, which was stranded in the Strait of Magellan in 1974, is offered as an example. Off-loading resources were limited, and, had the ship been fitted with a double bottom, an actual storm would have overwhelmed the ship before the required amount of cargo could have been off-loaded. However, the option of pumping some oil overboard (to prevent the sure spill of far more oil in the storm) would have been available. The possibility of losing the entire cargo from a stranded double-bottom tanker (where a single-bottom tanker would have survived) is remote.

In summary, there is insufficient evidence to refute the committee's opinion that the great majority of tanker groundings at service speeds will leave the tanker firmly stranded, and in need of significant outside salvage assistance whether or not the ship has a double bottom.

Salvage Procedures

To assess the impact of bottom and side voids on salvage, a basic understanding of salvage procedures is necessary.

Given a firmly grounded tanker, the task of the salvor is to determine as quickly as possible how, and when, to refloat the ship. The salvor considers not only tides, likely weather, and off-loading facilities, but also the extent of damage, hull strength, condition of propulsion and cargo systems, and the ship's particular features that can either assist or hinder salvage efforts. The following steps invariably apply in salvage of a stranded tanker.

Stabilizing the Ship. This is the first step in salvage; it prevents the ship from broaching or being driven farther aground. This generally entails "making the ship heavy," which also prevents the ship from incurring additional damage in often-violent hull motions due to usual sea conditions. Stabilizing the ship requires quick action and intact tanks (some intact spaces invariably must be flooded). Flooded double bottoms will help; intact inner-bottom and side ballast tanks are excellent repositories for additional weight. Hull stresses in the grounded condition must be addressed carefully, and this generally is beyond ability of the crew.

Removing Oil Cargo from Damaged Tanks. Loaded tankers, when stranded, invariably require off-loading of a significant quantity of cargo. Usually this involves removal of large amounts of oil floating on top of seawater that has entered through the damaged bottom, as well as removal of oil from intact tanks to minimize the threat of additional pollution (should the ship break up before or after being refloated). Removal of cargo is usually time-consuming; other tankers, or barges and tugs, must be marshaled, and weather may delay cargo transfer. Portable pumps and hoses always are required, unless the ship's cargo system is intact and operable. Generally, cargo must be removed from some intact tanks to allow for proper distribution of load and hull stresses after refloating. Environmental concerns may require removal of all remaining cargo before refloating, although this may mean leaving the ship stranded for a longer time. During cargo removal, compensating weight in the form of seawater ballast must be added to keep the tanker stabilized.

Lightening by Removal of Water. With the required amount of cargo removed, and the ship firmly under control of tugs and/or anchors and cable, ballast is pumped quickly out of intact tanks and sea water is blown out of flooded tanks with compressed air. (Water can be pumped much faster than oil, and it usually can be pumped overboard. If the tank tops are intact, "blowing" water out with compressed air is even faster.)

Bottom and Side Voids in Salvage

From the preceding, it follows that bottom voids and narrow wing tanks (or a full double hull) are valuable in salvage operations for four reasons: (1) If flooded, they immediately make the ship heavier and assist in stabilizing the hull on the strand; (2) The bottom voids likely will have protected at least some of the cargo tanks, not only limiting cargo loss but also preserving those cargo tanks for easy cargo removal and/or ballasting; (3) Flooded bottom voids, and usually narrow wing tanks, are strong enough to permit quick water removal by "blowing" with compressed air. (The deck over large cargo tanks is generally strong enough to withstand the pressure of blowing out only limited amounts of water.); and (4) Intact voids can be completely filled with ballast, and if necessary can be cleared relatively quickly, with submersible pumps, if the ballast system is inoperative.

The committee could not identify salvage-related concerns that should limit the use of properly designed double hulls.[7]

Cargo Systems

One factor influencing oil outflow in an accident, and subsequent salvage activity, is the tanker cargo system. There are two basic cargo system designs. The traditional system centers around pumps located in a pump room between the engine room and the last cargo tank. The complexity of the system is determined by the extent of cargo separation desired (to permit the carrying of different types of cargo). A variation on this system utilizes sluice valves, through which cargo flows by gravity toward the rear tanks, where the cargo pumps take suction. This system is not suited to tankers carrying multiple cargo types (combinations of piping and sluice valves have been used on crude carriers for this purpose). Figure 4-4 shows conventional cargo system arrangements.

Segregation requirements led to development of a second type of cargo system based on deep-well or submersible pumps located in the tank bottoms. In many deep-well pump tankers, each tank has its own pump and there is no horizontal piping. Sometimes one pump serves either a pair of tanks via a sluice valve or several tanks through appropriate piping. Such piping is placed above the main bottom structure, but the length of the main runs is significantly shorter than for conventional systems. Figure 4-5 is a schematic of a deep-well cargo piping arrangement; this system eliminates the need for a pump room.

Tanker cargo systems can minimize or—if damaged—encourage oil outflow in an accident. Because key components are located close to or at the bottom, fairly modest hull damage can disrupt the system, not only for the tanks involved, but also for intact tanks. Valves can be jammed, pumps can be crushed, piping systems can be broken. Other effects of the accident might force shutdown of the ship's engineering plant, eliminating power for operating the cargo system. Such events have a significant impact on short-term loss of oil after a grounding, as well as on salvage of a stranded tanker, as noted previously.

Emergency Cargo Pumping Capability

Cargo systems are not now designed with the possibility of an accident in mind. Lightening the casualty, therefore, usually requires removing oil

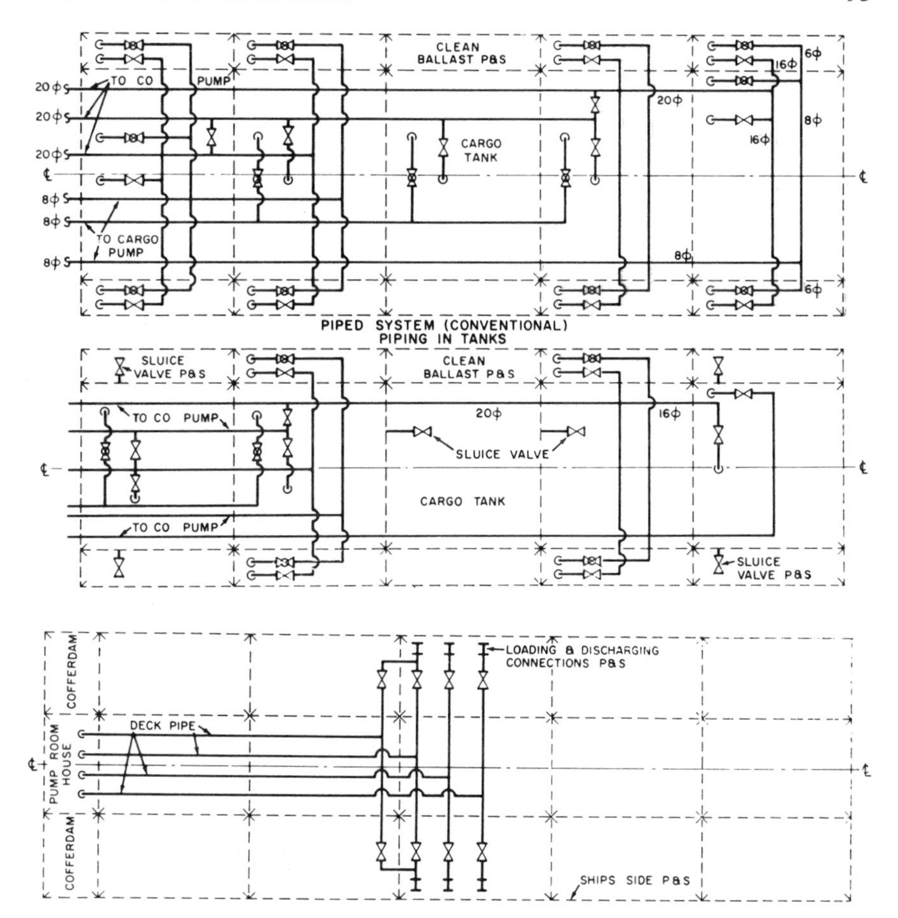

FIGURE 4-4 Conventional cargo systems. Source: Society of Naval Architects and Marine Engineers, 1980.

by portable pumps, always a slow and tedious process. Submersible pumps generally are used; their size and weight is limited by what can be airlifted and crew-handled, and can fit through a 12.5-inch "Butterworth" tank cleaning opening. (Some tanks fitted for crude oil washing may have no opening other than the main access, where ladders will interfere with pump placement.) Oil removal is particularly difficult in a punctured tank where the oil-water interface rises as oil is removed. The height of the pump may require frequent adjustment, especially as the tide rises and falls.

Cargo piping should not be located in a double bottom void. To do so risks cargo loss even when the inner bottom remains intact. Conversely, a properly located cargo piping system in a double-bottom tanker is less susceptible to damage than that in a single-bottom tanker.

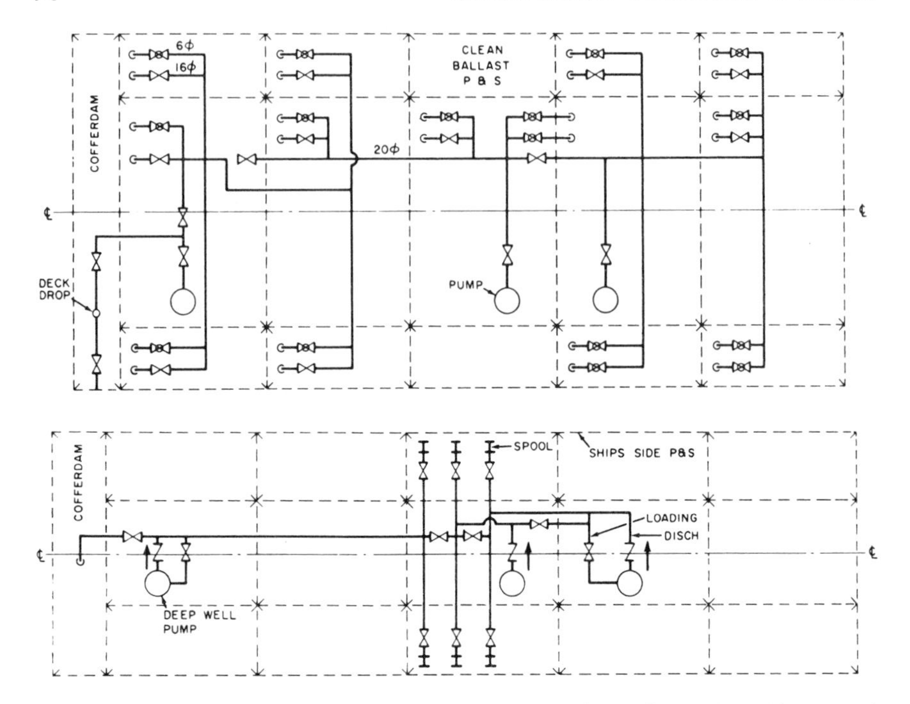

FIGURE 4-5 Deep-well cargo system. Source: Society of Naval Architects and Marine Engineers, 1980.

Another concern with conventional cargo systems is that once hydrostatic equilibrium is achieved, the cargo system, even if operable, will pump only water, once water reaches the suction level (assuming water flows in faster than oil can be pumped out). A separate system with alternate suction points, perhaps at quarter-tank height and at the level of the loaded water-line, would enable an otherwise intact cargo piping system to rapidly transfer or off-load cargo from breached tanks. In addition, improved (damage-resistant) details of cargo systems at suction points should ensure operability in the event of bottom distortion (short of rupture). Alternate pumping sources also would be useful if, for instance, the pump room were flooded or burned. Such capability would facilitate the transfer of cargo from a damaged tanker to another ship, or within a ship to intact empty tanks.

Salvage Calculations and Computer Programs

Computer use is universal in naval architecture. Besides enabling designers to produce "more efficient" structural designs, computers have facilitated studies on hydrostatics, stability, and damage stability. However, the predicted behavior of ships sustaining high-energy grounding damage sometimes has been significantly more pessimistic than that calculated by salvors.

This discrepancy arises because some computer programs calculate damaged stability assuming all cargo flows out of damaged tanks, to be replaced with seawater to the equilibrium water line. In fact, this methodology is required by IMO damage-stability criteria. While providing a conservative result for routine damage-stability studies for comparative design purposes, this method is inappropriate for tanker salvage scenarios involving major damage. It ignores the beneficial buoyancy of the oil remaining in the ship. In the case of the EXXON VALDEZ, for instance, the potential net buoyancy of the 42,500,000 gallons of oil remaining inside the hull was approximately 32,000 tons—or more than the 28,973-ton weight of the entire hull! Tankers seldom sink if the hull remains even reasonably intact. Witness the aircraft attacks on shipping in the Persian Gulf during the war between Iran and Iraq, where not one loaded tanker was sunk, and the more recent example of the MEGA BORG, which survived a major, week-long fire off Galveston, Texas, losing only about 10 percent of its cargo. Extreme care is required when using typical naval architecture computer programs in salvage operations and related analyses.

SAFETY OF LIFE

International resolutions related to ships traditionally have focused on safety of life, whether it be lifeboat requirements devised after the TITANIC catastrophe or load-line criteria resulting from losses of ships and entire crews due to overloading.

The committee's task was to evaluate alternate tanker designs intended to mitigate pollution from collisions and groundings. However, the accompanying risks of fire and explosion, and the personnel risks of operating and maintaining alternative tanker designs, cannot be ignored, any more than the pollution that could result from these hazards. This section addresses these aspects of alternative designs and their potential impact on the lives of the men and women who work on these ships.

Explosions and Fires

Segregated ballast tanks, mandated for new ships over 20,000 DWT, are by definition either empty or filled with seawater ballast. Particularly when empty, they are subject to rapid corrosion unless preventative measures are taken. Such methods include anodic protection, protective coatings, and possibly inerting (all described in Chapter 3). At least one wall of a segregated ballast tank is shared by a cargo tank. Any failure of this structure, due to corrosion or cracking, can allow oil to leak into the ballast tank, creating a risk of explosion. Potential sources of oil vapors are multiplied if cargo piping is routed through segregated ballast tanks (though this is seldom done and can be proscribed).

The alternative tanker designs proposed most frequently involve double bottoms and/or double sides. These designs increase significantly the void areas adjacent to cargo tanks. Assuming the risk of leaks is proportionate to the bulkhead area of ballast tanks shared with cargo tanks, double bottoms and sides increase significantly the risk of encountering explosive atmospheres. This is a common argument against such designs.

Existing data cannot predict precisely the increased risk of personnel hazard and explosion posed by greater amounts of segregated ballast tankage. On the positive side, there is a growing fleet of SBT, double-bottom, and some double-hull tankers, some of which have been sailing for 20 years or more. As far as the committee can tell, these tankers have not had a greater problem with either personnel casualties or fires/explosions than conventional pre-MARPOL single-hull tankers. However, this is not to say that the safety problem of additional void spaces can be ignored. Crew training will continue to be vital, along with careful design, inspection, and maintenance.

SUMMARY

This chapter has made a number of points that will figure in the forthcoming assessment of alternative tank vessel designs. Also worth noting are several findings that will contribute to the committee's overall recommendations for reducing pollution risk.

In general, modern tankers are proportionally less robust than their predecessors; this trend could be reversed through modification of classification society rules. To their credit, classification societies have commenced evaluation of these rules. More research should be done on the proper use of high-tensile steels in tanker construction. More research also is needed to improve resistance of tanker structure to collision and grounding damage, and to enhance residual strength following major damage.

In tanker design, the aims of pollution prevention, stability, strength, salvage, safety, inspection, and maintenance can be conflicting. However, from an engineering standpoint, all these interests can be satisfied in design of double bottoms, double sides, and double hulls, relying on present technological capabilities. All current stability requirements can be met, and exceeded, with properly subdivided B/15 double-hull designs. Furthermore, these designs generally enhance salvage. And, given the extra maintenance and inspection warranted by the increased void space, vessels with double bottoms/sides can be operated safely over a vessel's normal lifetime.

Consideration should be given to increasing the damage-stability requirements for double-hull tankers, to approximate the stability actually achieved by comparable single-hull MARPOL tankers. The proportions of double hulls also deserve further study, to optimally satisfy all interests including pollution control, stability, maintenance, safety, and cost. In addition, the outer shell of double-bottom, double-side, and double-hull ships should be no more vulnerable to penetration than conventional single-hull MARPOL tankers.

In ships with double bottoms and/or double sides, cargo piping systems should be kept out of the voids, to improve damage resistance and to reduce risk of oil leakage and explosion. Pollution control also could be enhanced through modifications to conventional cargo systems to facilitate the pumping of cargo from a damaged tank to other tanks or another ship. Alternate pumping sources for cargo systems also should be explored.

NOTES

1. Heel is the extent of a vessel's incline or tilt to one side; trim is the position of a vessel with reference to the horizontal, or the difference in draft forward and aft.
2. Hull bending moment is the summation of the weight, buoyancy, and dynamic forces that tend to bend the hull.
3. Tank width (and therefore bulkhead spacing) is affected by cargo outflow limits and by the "free surface" effect. The latter refers to the free movement of liquid in a partially filled tank; this has a significant negative impact on transverse stability. The effect is proportional to the cube of the tank width (w^3). Adding longitudinal bulkheads can improve stability significantly. For instance, the potential free surface impact of one 60-foot-wide tank is four times that of two 30-foot-wide tanks of the same length.
4. A recent Japanese study (Ministry of Transport, Japan, 1990) examines various large hull designs, several of which have inner bottom heights of less than B/15. Chevron Shipping Company (see Appendix E), an experienced operator of double-hull tankers, likewise promotes double sides/bottoms of B/15, but not less than 2 meters and not required to be much more than 3 meters.
5. A ship's tilt to one side in a state of equilibrium (as from unbalanced loading).
6. Pressure at the bottom of the loaded cargo tank.
7. Responses to an informal survey of salvors were unanimously in favor of double bottoms.

REFERENCES

Card, J. C. 1975. Effectiveness of Double Bottoms in Preventing Oil Outflow from Bottom Damage Incidents. Marine Technology 12(1):60-64.

Chevron Shipping Company. 1990. Double Hull Tanker Design. Paper prepared for Society of Naval Architects and Marine Engineers convention, sent to Committee on Tank Vessel Design, NRC, Washington, D.C., October 1990.

Exxon Corp. 1982. Large Tanker Structural Survey Experience. Paper published by Exxon, New York.

Ferguson, J.M. 1990. Structural Integrity and Life Expectancy. Paper published by Lloyd's Register of Shipping, London.

Iarossi, F. (president, American Bureau of Shipping). 1990. Getting Ready for a Shortening Life. Fairplay International September 20:37.

International Association of Independent Tanker Owners. 1990b. Quest for the Environmental Ship. Draft of paper to be published by the Association, Oslo, Norway.

Liu, D. 1990. Letter to the Committee on Tank Vessel Design, NRC, Washington, D.C., May 10, 1990.

Ministry of Transport, Japan. 1990. Prevention of Oil Pollution. Report prepared for IMO Marine Environment Protection Committee, received by Committee on Tank Vessel Design, NRC, Washington, D.C., November 1990. Toyko.

Royal Institute of Naval Architects. 1990. The Naval Architect. September:E357,E381,E385.

Society of Naval Architects and Marine Engineers. 1980. Ship Design and Construction. New York: SNAME.

U.S. Coast Guard. 1990c. Report on the Trans-Alaska Pipeline Service (TAPS): Tanker Structural Failure Study (draft). Washington, D.C.: U.S. Department of Transportation.

5

Design Alternatives

This chapter will introduce and explore tank vessel designs and operational alternatives that are intended to mitigate pollution in an accident. The committee's technical assessment is based on how the various characteristics of these designs influence safety, operation, and cargo outflow in an accident. The physical phenomena and engineering issues underlying this analysis were discussed in the previous two chapters.

The designs considered in this chapter were chosen from several dozen alternatives gathered or solicited from various sources and, in a few cases, suggested by committee members. Proposals ranged from the conceptual to the tested and operational. Some originated in the United States, others in one of several foreign countries; some were suggested by individuals, others by major international corporations or industry associations. All of the suggestions were reviewed, but not all were included in the formal evaluation. The exclusion criteria are explained in the following paragraphs.

Proposals were screened to eliminate the clearly impractical esoteric visions, as well as those judged beyond the bounds of industry compatibility. The surviving proposals then were rendered into 17 broad examples—designs generalized enough to encompass the key concepts and details.

For this chapter, the alternatives were grouped according to common physical principles. This resulted in the following three categories: secondary "barriers" to oil intermingling with water; the mitigation of pollution via "outflow management" techniques; and the reduction of pollution potential through "increased penetration resistance." Operational options for "accident response" were grouped separately; these options can be employed with most hull design alternatives. An itemized matrix listing all of these alternatives and operational options can be found in the next section of this chapter.

The committee evaluated the 16 alternatives based on the following criteria: (1) previous studies or documented experience with the design; (2) concerns about engineering, safety, and practicality derived from the experience of committee members; and (3) theoretical effectiveness of the design in preventing/mitigating pollution in collisions and groundings. As part of its analysis, the committee judged each design according to its developmental status; those not ready for immediate use were evaluated in terms of future promise.

The first half of the chapter covers general technical aspects of each design. Based on these considerations, some of the alternatives were eliminated from further committee consideration. The second half of the chapter mathematically assesses the pollution-mitigation potential of the remaining alternatives, and several possible combinations of these designs, in regard to groundings and collisions. For this section, the committee made use of a study developed by Det norske Veritas (DnV), contained in Appendix F. (This study will be referred to as the DnV analysis.) The methodology is explained in relevant sections of that report.

The designs and combinations assessed for pollution-mitigation potential also are subjected to cost-effectiveness analysis, described in Chapter 6.

THE MATRIX

The matrix presented on the following pages was prepared by the committee to combine, in one document, all of the technical considerations discussed in Chapters 3, 4 and 5. The matrix is intended to apply to both tankers and barges. The following description is intended to assist in understanding of the document.

Column 1 (Alternative Description) lists the design alternative considered in that row. Column 2 (Effectiveness) indicates, in a general sense, the type of accident in which the proposed alternative will be effective, in terms of controlling cargo outflow. If an alternative is effective in the accident type noted, then a dot (•) appears in the proper column. The terms "high" and "low" damage severity reflect the relative speed of the vessel prior to the accident. For collisions, speed refers to the ramming vessel; effectiveness applies to the vessel that is struck. The matrix does not grade the relative effectiveness of the various designs; that subject is taken up in the latter half of this chapter.

In Column 3 (Implementation), the committee has indicated technology status and technical constraints. Regulatory and financial constraints are not considered in the first two sections. The last three sections of Column 3 indicate whether the alternative, in the committee's view: (1) could be applied to new tankers at the present time (within the next two years); (2) should be considered at some time in the future; or (3) could be applied (retrofitted) to existing vessels. A dot (•) in the column indicates applicability, and a question mark (?) indicates technical applicability but economic difficulties. A blank space (no dot) indicates no applicability, from a technical standpoint. Retrofitting involves some unique complications, which will be discussed in the text as applicable.

In Column 4 (Concerns), the headings follow the format of Chapter 4 through the subject of explosions. A dot (•) in the appropriate section indicates that the design alternative, to some extent, creates that concern. A general explanation of these concerns can be found in Chapter 4.

The last four sections refer to the following concerns:

- Safety Downgrade—A dot in this section means that, to some extent, existing safety practices or requirements are diminished. (An example is an alternative that prevents the use of inert gas systems.) In all cases, safety downgrade probably could be overcome, but the committee felt it important to note the concern so that alternatives would not be viewed as panaceas.
- Operations Complexity—Because crew size has been reduced and crew fatigue has played a role in some pollution incidents, the committee felt that alternatives requiring extra vigilance in operation should be identified.
- Design Integration—A dot indicates alternatives involving either new technology or new design practices. This concern is highlighted to ensure that the advantages afforded by an alternative are not offset by some additional problem.
- Rules and Regulation—A dot indicates alternatives requiring interpretation or revision of existing rules.

The last section of Column 4, entitled "Comments," is intended to summarize the major technical or operational concerns to give the reader a sense of the priorities.

Column 5 (Pivotal Argument) is the committee's summary of all preceding columns; the comments represent the major arguments for and against each design alternative. These arguments are restricted to technical and operational matters.

Finally, Column 6 (Warrant Committee's Economic Evaluation) indicates which alternatives will be pursued in the benefit/cost assessment (Chapter 6). Those alternatives requiring significant time and/or research and development to implement, and those lacking sufficient information for a benefit/cost assessment, are indicated by dots in the "No" section.

TANK VESSEL

1. Alternative Description	2. Effectiveness				3. Implementation				
CONTROL METHOD: Barrier	Grounding		Collision		TECHNOLOGY STATUS (YEARS TO DEVELOP)	Constraints	Applicability		
	Damage Severity		Damage Severity		CONCEPT / RESEARCH / DEVELOPMENT / EXISTING		NEW CONSTRUCTION		RE-FITS
	HIGH	LOW	HIGH	LOW			NOW	FUTURE	
1. *Protectively Located Segregated Ballast* (Marpol Tanker) Ballast tanks isolated from cargo tanks. Located to restrict possible outflow. This is current regulation.		•		•	Existing	No constraints - present standard	•	•	•
2. *Double bottom* A non-cargo space between the cargo tank bottom plating and the ship's hull bottom plating.	•	•		•	Existing	Structural and weight-complications w/refit	•	•	?
3. *Double Sides* A non-cargo space between the cargo tank side plating and the plating of the ship's hull.		•	•	•	Existing	Structural and weight complications w/refit	•	•	?
4. *Double Hull* A non—cargo space between the cargo tank and the hull.	•	•	•	•	Existing	Structural and weight-complications w/refit	•	•	?
5. *Resilient Membrane* A tough,pliable, nonstructural barrier separating the cargo from the ship's structure and acting to maintain separation of cargo and water in the event of being breached.		•		•	Concept - minimum 10 years to develop	Technology not sufficiently developed to support even a 'proof of concept' case.			

DESIGN ALTERNATIVES

4. Concerns										5. Pivotal Argument		6. Warrants Committee Economic Assessment	
STRENGTH	MAINTENANCE	STABILITY	SALVAGEABILITY	EXPLOSION	SAFETY DOWNGRADE	OPERATIONS COMPLEX	DESIGN INTEGRATION	RULES & REGULATIONS	Comments	For	Against	YES	NO
		•	•	•					Damaged stability concerns if ballast tank is ruptured.	Has eliminated a major cause of world's oil pollution (ballasting of oil tanks). Inexpensive and in existence.	Does not effectively handle damage conditions. Will not reduce oil outflow unless only ballast tank is ruptured.	•	
	•	•	•	•					Explosion concerns due to gas in voids.	Will prevent pollution in groundings where inner hull is not breached. It can be effective in limiting or preventing pollution in case of low to moderate damage.	Does not assist in collision damage. Possible increased pollution potential due to collision because salt water ballast (SWB) in D.B. will reduce amount of protectively located segregated ballast. Explosion concerns due to gas in voids.	•	
	•	•	•	•					Explosion concerns due to gas in voids.	Will prevent pollution outflow in collisions where inner hull is not breached. It can be effective in limiting or preventing damage in low to moderate damage cases.	Does not assist in grounding damage. Loss of buoyancy and resultant heel may cause increased ground reaction. Increased structural maintenance. Explosion concerns due to gas in voids.	•	
	•	•	•	•					Design depth of double hull could provide structural and capacity restraints.	Will prevent pollution or mitigate extent of immediate pollution in the event of all but the most severe accidents.	Increased structural maintenance. Explosion concerns due to gas in voids.	•	
	•			•	•	•	•	•	No technical support for the concept Required materials, operations, design demands not investigated.	Simplicity of concept. If problem of integrating membrane into ship's structure can be overcome, this solution could be quite beneficial.	The practical limitations of material properties. Impact of operational functions and the behavior of resilient membranes in contact with complex shapes and arrangements are virtually unexplored.		•

TANK VESSEL

1. Alternative Description	2. Effectiveness				3. Implementation				
CONTROL METHOD: Outflow Management	Grounding		Collision		TECHNOLOGY STATUS (YEARS TO DEVELOP)	Constraints	Applicability		
	Damage Severity		Damage Severity		CONCEPT / RESEARCH / DEVELOPMENT / EXISTING		NEW CONSTRUCTION		RE-FITS
	HIGH	LOW	HIGH	LOW			NOW	FUTURE	
6. *Passive Control Hydrostatically Balanced Loading Concept* The establishment of the potential oil/water hydrostatic equilibrium at a height above the highest designed, or incurred for, or incurred point of damage. Would limit (via loading criteria) the cargo's head pressure at tank bottom to equal to or less than draft's water head at tank bottom such that bottom damage results in less oil outflow.	•	•			Existing.	No modifications required.	•	•	•
7. *Intermediate Oil Tight Deck* A structural deck running the full length of the cargo area about 1/4 to 1/2 the depth above the bottom.									
7a. *Independent Tanks* The top and bottom tanks would be independent of each other. Would require upper and lower cargo piping.	•	•		•	Development — minimum 2 years to incorporate.	Design of piping system and strength of int. deck. Outflow performance uncertain— testing needed.		•	?
7b. *Convertible Tanks* The top and bottom tank would be integrated through sluice valves.	•	•		•	Development — minimum 2 years to implement.	Need to develop and test system.		•	?
8a. *Mechanically Driven Vacuum* The ullage space is subject to a mechanically induced vacuum such that the combined vacuum plus cargo head favors water inflow in the event of bottom damage.		•			Development — minimum 4-5 years to implement in full-size ship tests.	Need to prove practicality of providing tight deck and fail safe valving.		•	•

DESIGN ALTERNATIVES

4. Concerns										5. Pivotal Argument		6. Warrants Committee Economic Assessment	
STRENGTH	MAINTENANCE	STABILITY	SALVAGEABILITY	EXPLOSION	SAFETY DOWNGRADE	OPERATIONS COMPLEX	DESIGN INTEGRATION	RULES & REGULATIONS	Comments	For	Against	YES	NO
		•	•	•	•			•	Operating with all slack tanks could create a free surface problem.	Mitigates pollution in the first few hours after the accident; instantaneously on-line. Low cost option to retrofit.	Reduces deadweight of vessel which increases cost to transport and requires more vessels. Given sufficient time after accident and no other response, the pollution effect would be the same as a vessel carrying the same amount of cargo. Vessel may be subject to high sloshing loads, which is an especially critical problem in retrofit application. May be difficult to administer operationally.	•	
	•	•		•		•	•		Magnitude of piping produces operational concerns.	Would mitigate pollution in high-energy accidents except for the most catastrophic ones.	Complex. Cracks in intermediate deck, or open cargo valves would void the function of the intermediate deck.	•	
	•			•		•	•		Magnitude of valving produces operational concerns.	Would mitigate pollution in all but the most catastrophic accidents.	Operationally burdensome and complex. Hydrostatic isolaton of upper and lower tanks could be in jeopardy.		•
•	•		•	•	•	•	•	•	Major operations hazard. Existing deck structure insufficient. Requires constantly maintained tight ship.	Mitigates pollution outflow from groundings.	Adversely affects vessel and personnel safety. Generates serious operational problems (i.e. vapor disposal, maintenance & reliability).		•

TANK VESSEL

1. Alternative Description	2. Effectiveness				3. Implementation				
CONTROL METHOD: Outflow Management (cont.)	Grounding		Collision		TECHNOLOGY STATUS (YEARS TO DEVELOP) CONCEPT / RESEARCH / DEVELOPMENT / EXISTING	Constraints	Applicability		
	Damage Severity		Damage Severity				NEW CONSTRUCTION		
	HIGH	LOW	HIGH	LOW			NOW	FUTURE	RE-FITS
8*b*. *Hydrostatically Driven Vacuum (Passive)* A vacuum which occurs due to the run out of oil cargo.		•			Development — minimum 2 years to implement.	Need to prove effectiveness and environmental safety of chemicals.		•	•
8*c*. *Imaginary Double Bottom* A passive vacuum system coupled with a water layer below the oil cargo.		•			Development — minimum 2 years to implement.	Need to prove practicality of providing tight deck and fail safe valving.		•	•
9. *Smaller Tanks* Increase compartmentalization to reduce oil outflow exposure.		•		•	Existing		•	•	
9a. *Service Tank Location* Position all oil service and oil waste tanks clear of the vessel's hull in a defensive location relative to hull damage.	•	•	•	•	Existing.		•	•	

DESIGN ALTERNATIVES

4. Concerns: Strength	Maintenance	Stability	Salvageability	Explosion	Safety Downgrade	Operations Complex	Design Integration	Rules & Regulations	Comments	5. Pivotal Argument: For	Against	6. Warrants Committee Economic Assessment: YES	NO
•	•		•	•	•	•	•	•	Major operations hazard. Existing deck structure insufficient. Requires constantly maintained tight ship.	Mitigate pollution outflow for groundings.	Adversely affects vessel and personnel safety. Generates serious operational problems (i.e. vapor disposal, maintenance & reliability).		•
•	•		•	•	•	•	•	•	Major operations hazard. Existing deck structure insufficient. Requires constantly maintained tight ship.	Mitigate pollution outflow for groundings.	Adversely affects vessel and personnel safety. Generates serious operational problems (i.e. vapor disposal, maintenance & reliability).		•
	•			•			•		Increased maintenance and piping.	Smaller tanks reduce pollution for a specified amount of hull damage. Effective for smaller vessels and barges.	Major capital and operating expense. Will allow slightly less pollution than existing PL/SBT in event of major hull penetration.	•	
	•			•					No negative impact on existing protocols.	Positive pollution preventing requirement easily achievable on new buildings.	Inefficient use of space. Only addresses minor pollution source of ship's outflow.		•

TANK VESSEL

1. Alternative Description	2. Effectiveness				3. Implementation				
CONTROL METHOD:	Grounding		Collision		TECHNOLOGY STATUS (YEARS TO DEVELOP)	Constraints	Applicability		
Increased Penetration Resistance	Damage Severity		Damage Severity		CONCEPT / RESEARCH / DEVELOPMENT / EXISTING		NEW CONSTRUCTION		RE-FITS
	HIGH	LOW	HIGH	LOW			NOW	FUTURE	
10. *Internal Deflecting Hull* An internal, forward, inner lower and bottom tank structure of exceptional strength similar to existing icebreaker hulls. Conventional outer hull would combine to form partial double hull. Upon grounding the vessel rides up and the inner hull deflects the vessel away from the obstacle. Existing SBT is retained.	•	•		•	Development–minimum 2 years to develop.	Need to develop structure and perform model tests.		•	
11. *Grinding Bow* Forebody bottom structure to be double bottom with internal transverse structure designed to act in a grounding as a "rock rasp". Thereby grinding down underwater obstacles and allowing safe override by remaining structure.	•	•		•	Concept–minimum 10 years to develop.	Need to develop & test both materials and structure.		•	
12. *Unidirectionally Stiffened Bottom Structure* Bottom structure designed to have "crushable" transverses in combination with longitudinal structure such that bottom structure moves (in the event of grounding) as a unified pliable member.	•	•		•	Research–minimum 3 years to develop.	Existing crushable structures would need to be reviewed and expanded. Need to model test structure.		•	
13. *Honeycomb Hull Structure* Utilize high energy absorbing deep honeycomb steel structure, sandwiched between steel plates.	•	•		•	Research–minimum 5 years to develop.	Need to develop structure and perform model tests		•	
14. *High Yield Steel Bottom Structure* Construct bottom plating and structure using high yield type steel in conjunction with design stress limitations associated with mild steel.		•		•	Existing.		•	•	
15. *Concrete* Construct hull using reinforced concrete internally molded to outer steel cargo tank structure.		•		•	Concept–minimum 10 years to develop.	Need to examine methods of utilizing concrete for this application.		•	
16. *Ceramics* Place ceramic coating on the exterior of the hull.		•		•	Concept–minimum 10 years to develop.			•	•

DESIGN ALTERNATIVES

4. Concerns: STRENGTH	MAINTENANCE	STABILITY	SALVAGEABILITY	EXPLOSION	SAFETY DOWNGRADE	OPERATIONS COMPLEX	DESIGN INTEGRATION	RULES & REGULATIONS	Comments	5. Pivotal Argument: For	Against	6. Warrants Committee Economic Assessment: YES	NO
				•			•			Provides both a forward secondary barrier to grounding induced damage as well as acting to divert vessel from continued damage. Provides superior protective support in the area most prone to grounding.	Does not address side damage Will cause considerable damage if the vessel rams another vessel.		•
				•			•	•	Success largely dependent upon strength of rock.	Provides secondary bottom structure which must be breached before cargo is exposed to the sea. In groundings, could provide better protection of cargo areas.	Practical materials and applications seem unlikely.		•
•				•			•		Could compromise bottom strength in a seaway.	Provides structural support for grounding without losing cargo capacity.	Does not address collision. Research needed to insure structure would protect against plate tearing.		•
	•			•	•	•	•	•	Increases safety concerns without improving upon conventional double hull.	Provides secondary structural barrier via high impact absorbing structure. Could be less weight and cost than conventional double hull.	All of the negatives of double hull structure plus a quantum increase in risks associated with uninspected voids adjacent to cargo tanks.		•
							•			Raises the severity of bottom impact that can occur without hull rupture. Should be considered in combination with other options.	Single barrier defense which, once breached, offers no improvement over conventional single hull design. Repair at yards unfamiliar with the material is a concern.		•
•	•						•	•		Raises the severity of bottom impact that can occur without incurring oil outflow.	Unique torsion and tension properties of concrete do not match those required to manage ship structural dynamics.		•
	•			•			•	•	Effects of adhering ceramics to the hull.	Raises the severity of bottom impact that can occur without incurring oil outflow. Could conceivably provide a smoother cleaner hull surface.	Torsional and impact properties of ceramics are not understood for this application. Could be cost prohibitive.		•

TANK VESSEL

1. Alternative Description	2. Effectiveness				3. Implementation				
CONTROL METHOD: Accident Response	Grounding		Collision		TECHNOLOGY STATUS (YEARS TO DEVELOP) CONCEPT / RESEARCH / DEVELOPMENT / EXISTING	Constraints	Applicability		
	Damage Severity		Damage Severity				NEW CONSTRUCTION		
	HIGH	LOW	HIGH	LOW			NOW	FUTURE	RE-FITS
17. *Enhanced Information Processing* Incorporation of pipe and structure sensing devices connected to a central processing facility that would advise crew immediately of any off-limit structure and cargo system integrity, and would recommend pollution mitigating action consistent with overall vessel limitations.	•	•		•	Concept – minimum 5 years to develop.	Considerable wiring, sensing and control system design and testing.		•	
18. *Towing Fittings* Installation on board vessel of easily access fittings to facilitate emergency hookup to service and tow/tug boats.		•		•	Existing.		•	•	•
19. *Distressed-Ship Cargo Transfer System* The cargo tanks are fitted with a high suction dedicated auxilliary pipe/ valve system that could transfer oil out of the upper part of any damaged tank to undamaged ballast tanks. Transfer would be used to equalize oil/water hydrostatic head above the damaged area.		•			Development – minimum 2 years to implement.	Need to develop sluice control system.	•	•	•

DESIGN ALTERNATIVES

4. Concerns										5. Pivotal Argument		6. Warrants Committee Economic Assessment	
STRENGTH	MAINTENANCE	STABILITY	SALVAGEABILITY	EXPLOSION	SAFETY DOWNGRADE	OPERATIONS COMPLEX	DESIGN INTEGRATION	RULES & REGULATIONS	Comments	For	Against	YES	NO
	•		•	•	•	•	•			Improved response time and salvage decision process. Conceptually attractive.	Malfunction/erroneous signal could worsen situation. For data display to be positive factor would require integrity and reliability improvements. Not currently available for use in marine/oil environment.		•
									Hardware addition.	Improves stabilizing and movement of damaged vessel. Easy to implement.	Fittings are not a major issue. Vessel currently fitted at both ends and sides with mooring fittings.		•
	•	•	•	•	•	•	•	•	Safety issues raised by introducing hydrocar-bons into noninerted spaces.	Potentially mitigating pollution system offering simplicity, ease of installation, modest capital cost and no deadweight impact with respect to existing tonnage.	Potential in leakage of explosive gases into voids; impacts I.G.S. and tank venting. High reliability required for rare usage.		•

BARRIERS

All of the design concepts in this category *act initially to prevent the loss of cargo containment integrity* during the sequence of an accident that ruptures the vessel's primary, or outer, hull. Barriers are, in effect, secondary physical obstacles to the loss of cargo containment integrity. They are inevitably passive (requiring no action by personnel or machinery to initiate) and, once brought into play, their performance is both irreversible and uncontrolled other than by accidental events.

A barrier can take the form of an inner hull or a resilient non-structural membrane. The concept, application, and operational characteristics of various types of inner-hull secondary structures are well known. Those protecting against groundings generally are referred to as double bottoms, those guarding against collisions as double sides, and those protecting against both as double hulls (or double skins). For tankers, the voids between the outer hull and inner structure almost always are utilized as ballast tanks, in order to immerse the hull to meet regulatory requirements tied to propulsion and maneuvering. Ballast is less likely to be carried in the voids of barges, as their propulsion source is independent of the vessel.

A double structure may not extend throughout the entire vessel. For example, while a double bottom generally is combined with some double-side tankage, a double-sided vessel seldom features a double bottom because sufficient ballast can be carried in the side tanks to meet regulatory requirements.

Double bottoms, double sides, and double hulls are in use today, albeit to a moderate degree. (Approximately 600[1] tankers over 10,000 DWT—or 20 percent of the world fleet of approximately 3,000[2]—have some form of double structure.) Their protective advantages must be weighed against increased risk of explosion, higher susceptibility to risks associated with the vagaries of workmanship, and increased construction and operating costs. Engineering issues related to double structures, including strength, stability, and safety, were discussed in Chapter 4.

An alternative to structural barriers is a tough, resilient membrane (or tank liner), a concept akin to the inner tube within a bicycle tire. This idea, in numerous forms, has the advantage of simplicity. Whether protecting against bottom damage or side damage, the principle is the same: The pliable membrane activates and adjusts to fit the structural dislocation caused by the accident. In so doing, it becomes a physical barrier between the oil cargo and the water. The membrane is not now in use, due to a host of practical obstacles that have confounded its inventors and proponents over the years.

1. Protectively Located Segregated Ballast (PL/SBT or MARPOL) Tanker

This design (see Figure 5-1) is considered the baseline alternative, as it is the current international standard and represents about 15 percent of the current world fleet (10,000 DWT and larger); about 35 percent of the fleet has segregated ballast, either with or without protective location.[3] MARPOL design provisions and additional requirements for tankers trading in U.S. waters were described in Chapter 2.

2. Double Bottom

Double bottoms (see Figure 5-2) are intended to provide the maximum protection for all but the highest energy groundings. Even for grievous damage, double bottoms could protect tanks on the periphery of the damage area.[4] A double bottom of the MARPOL-required depth (2 meters or B/15, whichever is less) will not hold sufficient ballast to meet regulatory requirements for the ballast voyage, so some side wing tanks are designated for ballast. The arrangement reduces the side hull area protected by ballast tanks, compared to the MARPOL tanker.

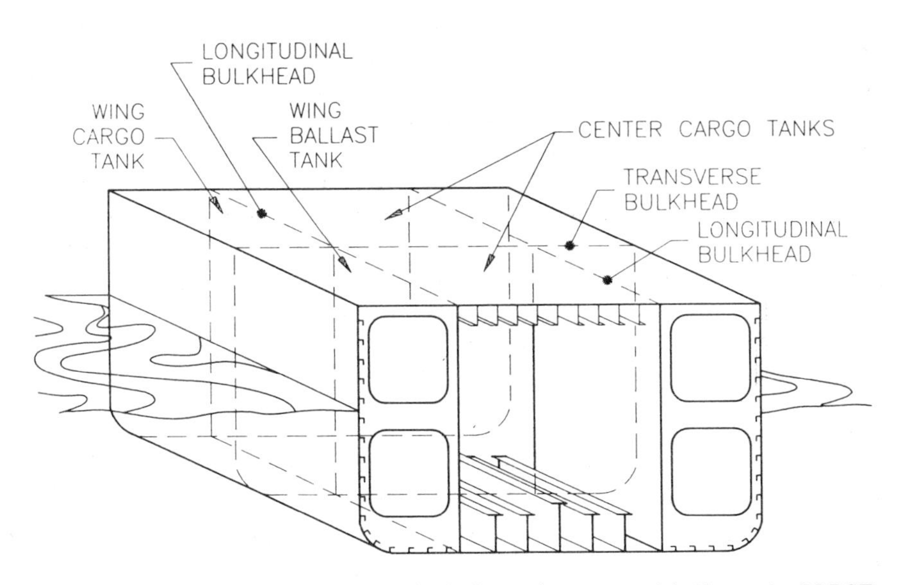

FIGURE 5-1 Alternative 1: Protectively located segregated ballast—the PLBST tanker is separated by longitudinal bulkheads into wing and center tanks. Transverse bulkheads separate one set of tanks from another. (MARPOL Tanker.)

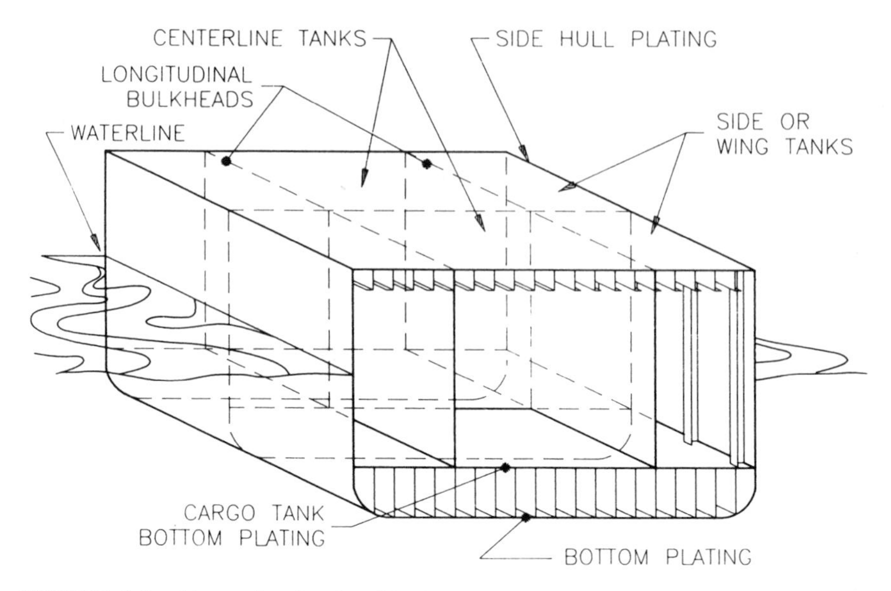

FIGURE 5-2 Alternative 2: Double bottom—the void space between the cargo tank bottom plating and the bottom hull plating.

Among its attributes, a double bottom provides a very smooth inner cargo-tank surface compared to the baffled and egg-crate-type tank bottom in a conventional tanker. Cargo flow (or any liquid drainage) to the discharge suction is optimized, facilitating tank cleaning. The cargo transfer system could be placed in the double-bottom void to enhance pumpout efficiency; however, such a system, besides being a source of potentially explosive hydrocarbon gas in those empty tanks, most likely would be damaged in a grounding. Therefore, from a damage and outflow standpoint, cargo transfer piping should be placed within the cargo tank. This arrangement also is better suited to the transfer of cargo from a damaged tank.

Drawbacks to a double bottom include increased risks associated with poor workmanship, corrosion, and obstacles to personnel access. Proper construction, inspection, and maintenance are required. Conceptually, any extra void space increases the risk of fire or explosion, although, as noted in Chapter 4, there is no hard evidence of greater incidence. The risk is reduced by the periodic clearing of tanker void spaces with ballast (this may not be the case on barges). This process, however, increases the risk of ballast tank corrosion.

Double bottoms do not increase oil outflow in grounding, although some critics have made this claim. The committee's reasoning was explained in Chapter 4, in the discussion of perceptions about salvage of double-hull tankers.

Should a double-bottom vessel suffer damage to the outer hull only, seawater would fill the void and the vessel would increase draft, trim, and heel accordingly. However, as noted in Chapter 4, tankers are exceptionally stable compared to other large vessels and, loaded or not, are tolerant of most hull damage.

A double bottom offers one definite advantage in groundings where the obstacle reaches high enough to pierce both the outer and inner hulls: The extra structure may diminish the longitudinal extent of damage by absorbing energy and perhaps will reduce the number of tanks affected.

In the committee's view, the double bottom can prevent outflow in some accidents and deserves further consideration.

Retrofitting Tankers and Ocean-Going Barges

Retrofitting an existing single-skin vessel with a new double bottom would cost an estimated 25 percent of full replacement value.[5] Thus, retrofitting a 30-year-old U.S.-built tanker of 40,000 DWT would cost about $15 million, or about 25 percent of its replacement cost. For tankers, additional regulatory costs would be involved as a double bottom would be considered a "major" conversion; the vessel would have to be upgraded in safety and other areas. These costs, plus renewal of wasted hull steel and machinery upgrades and rebuilds, could bring the total retrofit cost to 40 to 50 percent of replacement cost. Such an investment would have to be analyzed carefully in light of the longevity of the tanker. As tankers normally are depreciated over 20 to 25 years, the owner of a 15-year-old tanker faces the dilemma of the retrofit cost plus the non-depreciated value of his tanker, balanced against the remaining life of the tanker.

Due to their flat profile, barges could be retrofitted with double bottoms at a reasonable cost (15 to 20 percent of replacement cost).

3. Double Sides

Double sides (see Figure 5-3) can be found in some crude oil and product tankers. As with double bottoms, the minimum MARPOL-required width is 2 meters or B/15, whichever is less; this requirement may not provide sufficient ballast capacity. Therefore, because no other void space exists, the double-side tanker employs a greater tank width (B/7 to B/9) so as to meet ballast requirements.

Because the cargo is separated from the outer side hull by a void, the design offers good protection from collisions. Double sides also offer some of the advantages of double bottoms, because the wide side tanks protect as much as 15 to 20 percent of the outboard region of the bottom. This design could be improved, in terms of limiting outflow following a grounding that

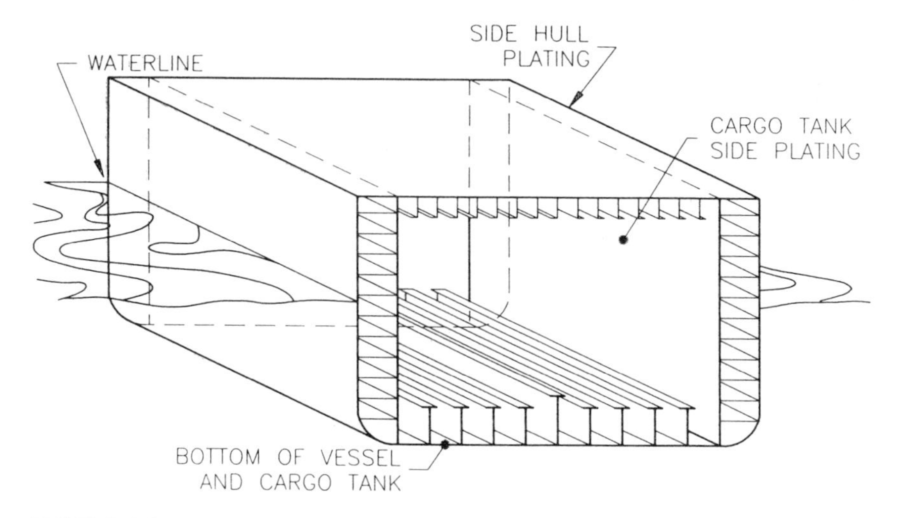

FIGURE 5-3 Alternative 3: Double sides—the void space formed between the cargo tank side plating and the side hull plating. Note: In Figures 5-3 and 5-4, longitudinal cargo bulkheads have been omitted for clarity.

breached the hull, by the use of hydrostatically balanced loading. This operational alternative is discussed later in this chapter (alternative #6).

The double-side configuration improves cargo operations somewhat, due to the smooth inner sides of the cargo spaces (the complex "structure" is on the ballast side of the bulkheads). The smooth surface reduces cargo clingage, enhances tank coating life, and simplifies tank cleaning and inspection. However, if the cargo bottom is damaged, the cargo transfer system could be rendered inoperative (as its suction opening, located close to the bottom hull, would be under water). In this respect, the double-side tanker is no better than the single-skin vessel.

After incurring side damage, a double-side vessel is highly susceptible to asymmetric flooding. However, as with double bottoms, the extent of settling and heeling depends on several factors, principally ballast tank arrangement. The question of heel is somewhat different in a double-side vessel, in comparison to a double-bottom design, due to the outboard placement and capacity of the ballast tanks. However, as discussed in Chapter 4, double-side tankers can be designed to be stable.

Retrofitting Tankers and Ocean-Going Barges

Retrofitting double sides to an existing vessel, combined with other additional U.S. regulatory requirements, could cost more than 50 percent of the replacement value of the vessel. With the relatively low incidence of colli-

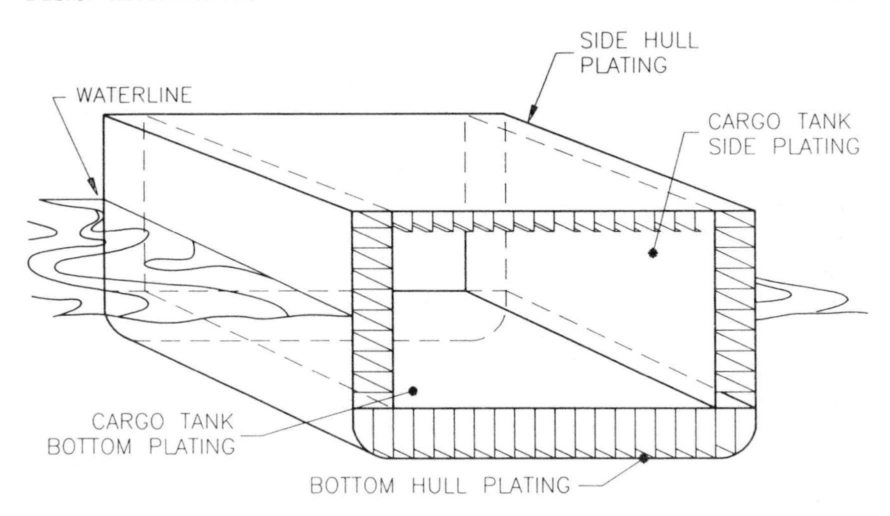

FIGURE 5-4 Alternative 4: Double hull—the double hull is a void formed between the vessel's outer side and bottom hull plating and the cargo tank plating both at the sides and bottom.

sions in U.S. waters, double sides would appear to be an unattractive retrofit option from both economic and pollution-abatement standpoints.

Retrofit for barges may be more desirable if a detailed assessment of accidents reveals that pollution was caused as a result of side damage. Retrofitting a barge generally is simpler and less costly than for tankers, due to less complex hull and structural formation and absence of requirements for concurrent upgrading of crew and accommodation-related standards.

4. Double Hull

For preventing oil outflow in low-energy collisions or groundings, the double hull is, logically, the most effective design (see Figure 5-4). Some product tankers, combination carriers, and a few crude carriers already have a double hull, which essentially is a combination of a double bottom and double sides. The width of the side tanks may be less than in the double-side design however, because the required ballast can be divided among the side and bottom spaces. Consequently, the double-hull vessel may have a lower threshold of sustainable side damage than a double-side vessel in a major collision.

From the cargo operations point of view, the double hull is the best of the barrier designs. With inner tank sides smooth, the flow of cargo to the suction inlets is unhindered, side clingage is minimized, tank cleaning is optimized, and cargo tank coatings are easier to maintain. With the cargo transfer system in the cargo tanks, the double-hull design retains cargo transfer capability after an accident, as long as the inner hull is not damaged and the cargo transfer system still works.

At the same time, however, double hulls magnify all the risks noted for double bottoms and double sides related to construction, maintenance, inspection, and safety. The large structural area subject to cracks, for example, puts high demand on access for inspection. All of this implies high standards on the part of designers and builders, and added vigilance on the part of operators, builders and classification societies.

The pollution-control performance of a double hull in some high-energy accident scenarios could be improved by the use of hydrostatically balanced loading. But this compound design alternative would involve additional concerns, as described later in this chapter (in the discussion of alternative #6) and in Chapter 6.

In the committee's view, the double hull would prevent outflow in many accidents, and the related concerns are manageable if the following conditions are met:

- inspection access;
- minimum inter-hull spacing requirements;
- minimum outer-hull plate thickness;
- exclusion of cargo piping from ballast spaces;
- constraints on the longitudinal extent of each double hull void;
- maintained corrosion-prevention systems; and
- damage-stability requirements that assure residual (post-accident) damage stability consistent with that of single-hull vessels subjected to the same accident scenario.

Retrofitting Tankers and Ocean-Going Barges

Retrofitting double hulls on existing vessels would be possible but very expensive; the previous discussions of double-bottom and double-side retrofit give some indication of the cost and other concerns. In addition, potential difficulties could result from combining new and old structural materials. Replacement of the entire ship forward of the machinery space would entail fewer technical difficulties and would require less time in a shipyard; however, it might be more expensive than retrofitting a full double hull.

5. Resilient Membrane

In this concept, not yet in use, cargo would be separated from the outer hull by a resilient membrane sheeting (see Figure 5-5). In theory, a breach of the outer hull would activate the membrane; during normal passage, the

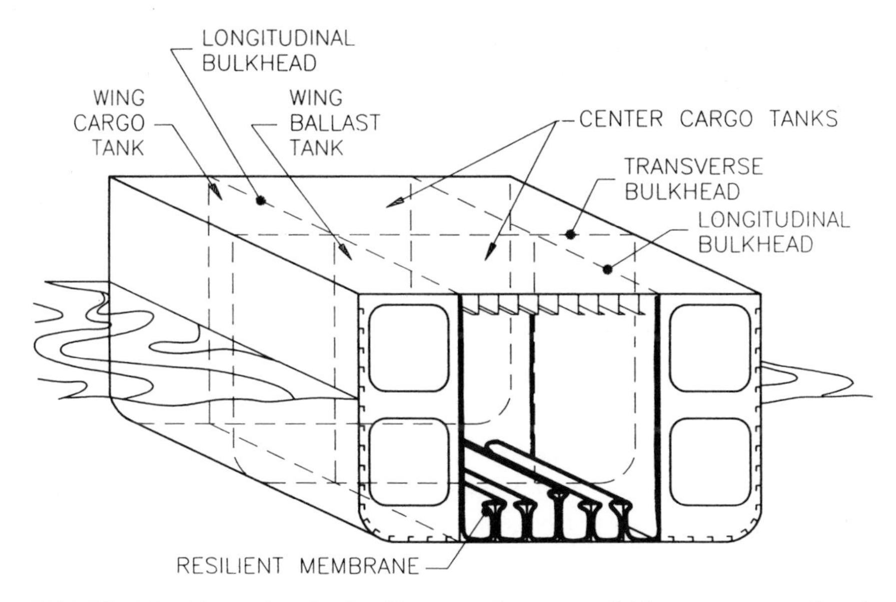

FIGURE 5-5 Alternative 5: Resilient membrane—a pliable, non-structural tank liner.

membrane would be inactive, integrated into the ship's design to allow unhindered operations.

The membrane, in some form or other, has become a familiar concept over the years. Large, inflatable, compliant bags employing various materials have been tested and used on a limited basis for storage and marine transport of petroleum products. But the committee could find no evidence that this concept has been utilized successfully in a cargo tank. Its total absence in the tank vessel industry likely is due to practical obstacles, which have been insurmountable so far.

Internal tank structure and equipment are not physically conducive to the fitting of liners. Piping, pumps, heating coils, washing machinery, ladders, operating rods, and gauging equipment all thwart the freedom of motion required by a liner. Furthermore, operating practices are not compatible with liner characteristics. Cargo pumping requires a free cargo flow into the suction openings; cleaning requirements introduce concentrated hydraulic forces that would tear the membrane; maintenance and repair requires unhindered access to structure and systems; and normal inspections must cover the very areas hidden by an inactive membrane.

Other drawbacks noted in previous evaluations are equally difficult to overcome. Liner strength factors, folding characteristics, fatigue, and abrasion remain major concerns. Furthermore, all these factors suggest that liners are unlikely to perform in the event of fire, explosion, or significant bottom damage. The membrane might be torn easily by the sharp edges of a damaged hull.

Although membranes are not in use today in cargo tanks, they might be suitable in smaller tanks having fewer structural, systems, or operating constraints. Fuel oil or waste oil tanks may be environments that, with modest research and development, would benefit from a pollution-prevention strategy incorporating membrane liners. However, in the committee's view, this concept would not become viable for cargo tanks in the foreseeable future and will not be considered further in this study.

OUTFLOW MANAGEMENT

While barriers would prevent the onset of cargo outflow, *"outflow management" restricts the amount of cargo subject to outflow.* Outflow management can be passive (smaller tanks) but generally is active, requiring either personnel or machinery to activate. Barriers and outflow management techniques are not mutually exclusive; in fact, an environmentally optimized tanker could draw upon a mixture of features, as noted previously.

Outflow management amounts to either a manipulation of the hydrostatic equilibrium acting on the oil/water interface (explained in Chapter 3), or downsizing or special placement of the cargo volume subject to outflow. In all cases, some initial oil outflow is expected, however slight.

Among the outflow-management concepts promoted by several sources are vacuum systems to retain cargo in a breached tank. These systems reduce atmospheric pressures on cargo surfaces, manipulating oil/water interface hydrostatics. There are two basic schemes. The first is active, employing mechanical systems to maintain low pressure in the ullage space so that, theoretically, little or no oil can escape. The second is semi-active, relying on reduced pressures induced by the falling oil level in a damaged tank. This scheme limits rather than prevents outflow.

6. Hydrostatically Balanced Loading Concept (passive)

This concept involves less-than-full loading of cargo tanks. An accident then would result in the damaged tank's oil/water interface hydrostatics favoring water inflow, or, at least, limiting cargo oil outflow. The physical basis of this concept was explained in Chapter 3. (See Figure 3-7, Case 2.) This condition is achieved by the carriage of ballast simultaneously with cargo and/or the light loading of all cargo tanks.

This concept is unique among the alternatives considered because it could be achieved almost immediately on most ships. Hydrostatically balanced loading (or hydrostatic control) does not require new structural arrangements or major physical changes to most existing ships.

Hydrostatic control involves six major drawbacks. First, it is an operational approach requiring continued compliance in order to be effective. Thus, it is not equivalent to a structural solution in terms of actual pollution reduction. Monitoring would be difficult, although a precedent for monitoring compliance through regulation was set with the LOT system (described in Chapter 2).

Second, cargo would be lost after a hull breach; the amount would depend on tidal variation, sea state, and damage location. Over time, there also is likely to be some intensive mixing of oil and water inside the cargo tank due to hydrodynamic effects, including turbulence in the oil and/or water, surface tension dynamics, and associated stratified flow near the hull rupture opening. Loss of cargo may be aggravated when a moving vessel is penetrated. Little is known about these phenomena under large-scale conditions, but the effect could be to encourage oil outflow. However, there is some evidence from accidents that the amount of oil actually lost closely approximates the loss indicated by calculations based on hydrostatic balance, without accounting for violent ship motion.

Third, there are technical problems related to cargo sloshing and the loss of stability due to the "free surface" effect (discussed in Chapter 4). Baffles and special design of transverse structures can ameliorate sloshing to avoid pitch resonance with ship motions; ballast and cargo distribution for specific classes of ships also may obviate much of the sloshing potential. Similarly, dividing cargo spaces into smaller tanks would reduce free surface effects. The forces resulting from sloshing during light loading may be a significant concern for some existing vessels, particularly if they are weakened by corrosion. Vessels would have to be checked individually; those unable to operate with reduced cargo loads either would have to be restricted from U.S. waters, or would require strengthened or additional bulkheads. Preliminary data (Lloyd's Register of Shipping, 1989) indicate that about 15 percent of the current tanker population would be so affected. New tank vessels and combination carriers, especially ore/bulk/oil (OBO) carriers, also would need to be evaluated, as they may have been built to standards preventing partial loading.

The fourth major drawback is economic: Cargo is traded for ballast. A ship would carry 15 to 20 percent less cargo, depending on its design and on the level of cargo required for hydrostatic control. As a result, up to 15 percent more ships (and/or larger ships) would be required to carry the equivalent cargo, and, as a result, the frequency of collisions and groundings likely would increase.

Fifth, if a cargo tank is breached *above* the hydrostatic balance level, oil will escape relatively quickly, especially if there were a large hole. In fact, for large holes, the limitation may be the rate at which the ship's cargo venting system will allow air to enter the expanding ullage space. With a hole equivalent to 1 percent of tank area, a tank may lose 20 percent of its cargo in only 3 to 5 minutes. Obviously, that is insufficient time to take physical measures to curb outflow. Setting a potential oil/water interface equilibrium at a height above a probable point of damage can be accomplished by passive or active means, as described in the following design options.

The sixth concern relates to practicalities of tanker operations. Ships frequently have multiple port discharges, and sometimes loadings, particularly where ships are being lightered. If a ship were to ground while partly discharged, the hydrostatic balance might have been disturbed, and, therefore, the beneficial effect of the concept would be lost. The protection against such a circumstance would be to lower the cargo level in all tanks while the ship is being discharged; from the standpoint of cargo handling and logistics, this may be impossible. Also, dropping all tanks to a hydrostatically balanced level in light-loaded ships could induce instability or aggravate sloshing, especially in OBOs.

In the unusual situation where cargo oil density is greater than that of water, hydrostatic control would not be applicable, as explained in Chapter 3.

Hydrostatic control, in the committee's view, offers a means to quickly reduce pollution potential of many existing vessels, and, despite operational limitations, it merits further consideration.

7a. Intermediate Oil-Tight Deck (independent tanks)

Hydrostatics favoring water inflow can be achieved by dividing cargo tanks into separate upper and lower chambers (see Figure 5-6). This greatly reduces the cargo head acting on the lower chamber and the hull bottom plating (the lower chamber's cargo head equates to its height, rather than the full depth of the vessel). In the event of bottom damage, hydrostatics would strongly favor water inflow, at least initially; in addition, the reduced volume of the lower chamber (compared to a full-depth tank) would limit the amount of oil subject to outflow.[6] Coupled with a double-side arrangement of ballast tanks, the design also could provide good protection against outflow resulting from collisions.

An intermediate oil-tight deck (IOTD) would reduce slightly the available cargo cubic capacity of a standard MARPOL tanker. Besides the steel weight increase, roughly equivalent to that for double hulls, the need to vent lower tanks and to arrange for personnel access would reduce the volume of cargo spaces. The amount depends on design details. However, the intermediate oil-tight deck results in less loss of cargo capacity than simple hydrostatic control, with potentially much greater protection against oil outflow.

The IOTD, which is not a new concept, has a number of attractive at-

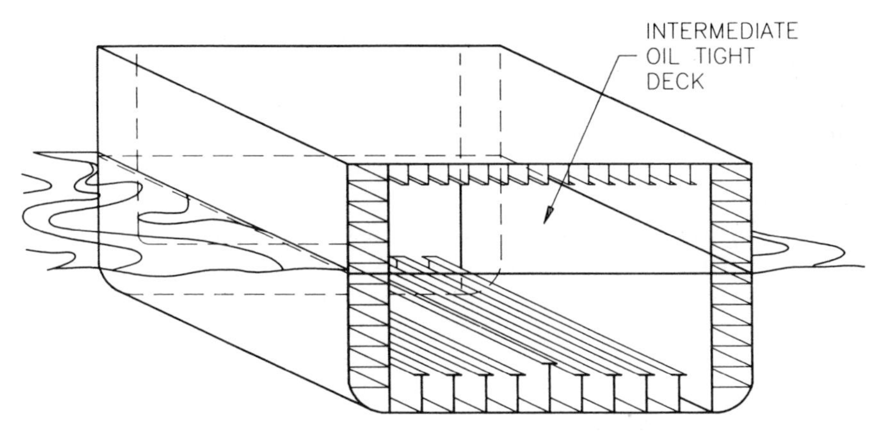

FIGURE 5-6 Alternatives 7a and 7b: Intermediate oil tight deck.

tributes, particularly when coupled with double sides (IOTD w/DS). (See Figures 4-3 and 5-6.) The following discussion compares the theoretical performance of the IOTD w/DS with that of other design options, particularly the double hull.

To comply with MARPOL space requirements for segregated ballast tanks, the double side-voids in an IOTD w/DS vessel would be about twice the width of those in a double-hull vessel; therefore, the IOTD w/DS would offer greater protection to cargo tanks in some collision scenarios.

The IOTD w/DS provides significant protection against oil outflow in the event of bottom rupture; the hydrostatic pressure favoring water inflow would be greater than in simple hydrostatic control. Furthermore, the wide side tanks would protect against grounding damage extending to the side of the ship. Overall, in theory, the IOTD w/DS likely would provide less protection than the double hull in low-energy groundings,[7] but greater protection in high-energy groundings (when damage penetration would pierce the inner hull).

When the bottom of an IOTD w/DS is breached and seawater enters the lower chamber, the oil, providing it does not have access to the double-side voids, is lifted into the ullage space (as little as 2 percent) and forced up into the vent/access spaces. Hydrostatic balance is quickly established with the seawater only slightly above the tank bottom. In one Japanese 280,000 DWT design (Ministry of Transport-Japan, 1990), the cargo level could be as little as 0.5 meters above the bottom of the ship, making the design subject to possible outflow from vessel list or trim and the dynamics of oil/water mixing. If the ship remained afloat, vessel rolling could cause some leakage. Small-scale tests, however, indicate that the outside pressure of the sea is more than enough to hold all oil in the vessel even with significant heeling or rolling. A smaller IOTD w/DS design would be subject to oil outflow from a receding tide following a grounding. In all cases, leakage would be significantly less than for simple hydrostatic loading. Nonetheless, an IOTD w/DS design is likely to lose some cargo following a low-energy grounding whereas a double-hull design would not (provided the inner hull remained intact).

An IOTD is more complex operationally than a double-hull tanker or any other design involving permanent structural pollution controls. Operator error when loading could negate the hydrostatic advantages of the IOTD design; cargo transfer systems must be properly controlled and separation of cargo in upper and lower tanks must be ensured.

Structurally, the IOTD design has both advantages and disadvantages. Ballast tanks in an IOTD w/DS have less surface area than in a double-hull, implying less risk of corrosion or other structural problems that could lead to oil leakage into the void/ballast spaces (posing an explosion hazard). The wider side tanks also mean that an IOTD w/DS design should be easier to construct, inspect, and maintain than a double hull. The IOTD w/DS has more cargo tank structure to inspect and maintain than either a single-hull MARPOL tanker or a double-hull tanker, but risks associated with structural deterioration or failure stemming from corrosion may be diminished. This follows from location of the intermediate deck at or about the neutral axis. Hence, the deck structure and its components will be subject to less primary hull bending stress, and will be less prone to failure than vital structure and components of some other designs.

The effect of an undetected leak or structural failure in the intermediate cargo deck would be less hazardous than a leak from a cargo tank to a ballast tank. (The cargo areas separated by the intermediate deck each would be subject to inert gas control and monitored for pressure changes.)

Overall, comparisons of the IOTD w/DS and the double hull based on structural and operational concerns are imprecise.

The principles of the IOTD w/DS design are simple and straightforward, and they do not entail any extreme departure from current shipbuilding practice. However, the concept has many permutations, with more undoubtedly to follow. Much work remains to be done to determine the most desirable proportions and arrangements.

Retrofitting. Conceptually, retrofit of an intermediate deck to an existing MARPOL vessel would require less steel than retrofit of a double bottom. However, a major modification for the cargo handling, ventilation, and cleaning systems would be required. The deck retrofit also could be carried out afloat, whereas adding a double bottom would mean extended drydock time. The deck installation in the wing tanks would require considerable fitting around existing structure, and/or slotting of that structure; this would increase the cost, as would the extra piping, vent and/or truck systems. Thus, the deck retrofit would cost roughly the same (25 percent of replacement value) as would the double-bottom retrofit. Likewise, depending on the age and condition of the base vessel, and interpretations of regulations, the total retrofit cost could approach 40 to 50 percent of the replacement cost of the tanker. An exception would be the retrofit of an intermediate deck within the existing double-side vessel (the deck would only be fitted in the adjacent side, center cargo tanks). This could cost as little as 15 percent of total tanker replacement cost.

7b. Convertible Tanks

Convertible tanks are a variation on the intermediate oil-tight deck. The cargo tanks have an intermediate deck, which contains remote-activated, securable openings connecting the upper and lower chambers. During conventional cargo handling operations, the valves would be open. Once the tanks were loaded, the valves would be closed, thus dividing the cargo.

A bottom accident would expose only the lower chamber to the sea, and, for the same reasons as noted for the intermediate oil-tight deck, hydrostatics would dictate initial water inflow rather than cargo outflow.

This concept poses the same difficulties as the intermediate oil-tight deck and has the additional drawback of extreme dependance on reliability of valve operation, which demands diligent maintenance and repair. For these reasons, in the committee's view, this concept is not worthy of further consideration in the present study.

Summary—Hydrostatic Control Alternatives

Hydrostatic control, while not yet in use, appears to be an effective concept for reducing pollution. Furthermore, hydrostatic control could be implemented immediately on many tankers; barges are more problematic because the system requires attention from the crew, and barges generally are unmanned.

The intermediate oil-tight deck, especially in combination with double sides, is attractive. Some members of the committee consider the IOTD w/DS ready for implementation on new vessels and view it as at least equivalent to the double hull in overall performance. Other committee members, however, consider such findings premature; they recommend deferring a decision pending a thorough evaluation by the IMO. The results of such a study could be used to determine whether the IOTD w/DS design should be accepted as a substitute for the double hull in reference to the mandate of OPA 90.

Retrofitting of the IOTD may be possible for existing double-side tankers. While hydrostatic control is not practical for barges, the IOTD concept could be considered as a means of limiting outflow resulting from groundings; it would be, in essence, a high double bottom.

8a. Mechanically Driven Vacuum (active)

Vacuum systems, as shown in Figure 5-7, work by reducing atmospheric pressures on cargo surfaces, thus manipulating oil/water interface hydrostatics. The mechanically induced vacuum has the benefit of not requiring major structural changes to existing tank vessel designs, provided the deck and bulkhead structure is adequate to handle the stress imposed by the pressure differentials. The system may be triggered either automatically or by the crew. The vacuum is created by securing all openings to the cargo tank(s) while simultaneously withdrawing air from the ullage space. Depending on the cargo type, the vacuum system design characteristics, and the contribution of hydrostatic control (if any), the force restraining oil outflow can be made to vary in accordance with the perceived risks.

The committee is aware of only one instance in which a prototype of this type of equipment was fitted on a tank vessel and tested under controlled conditions. The idea is believed to be fraught with practical concerns, even though the concept and principles are straightforward. It is not a question of whether the system can be made to work; it can, with an appropriate

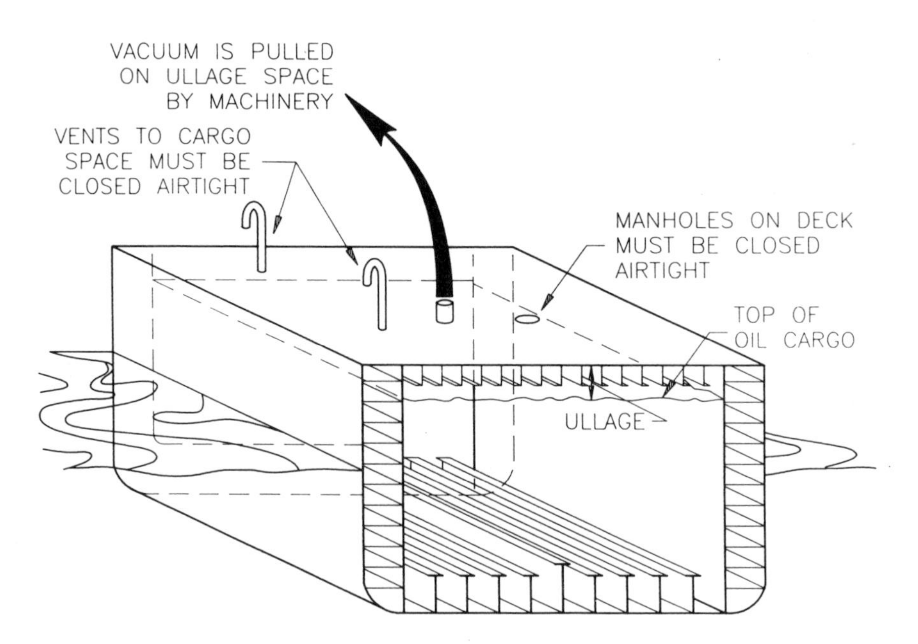

FIGURE 5-7 Alternative 8a: Mechanically driven vacuum (active).

expenditure of funds. The real issue is whether it can be made to function on the one occasion it may be needed, after years in service with an unpredictable quality of maintenance. This system also would require modification of inert gas systems (IGS is discussed in Chapter 3) through a change in international conventions, because this safety system depends on overpressure of cargo tanks. Cargo, vent, and inerting systems would need to be automated, and fitted with elaborate sensing and monitoring instruments. Backup safeguards also would be required to prevent inadvertent activation of the vacuum system (i.e., during cargo discharge), which could rupture tanks.

In one version of this concept, tankers would operate with the vacuum permanently active, so that no crew intervention or automatic activation would be required after an accident. In this design, theoretically, no outflow would result from a grounding. But such a system appears to involve two major problems. First, a system would have to be developed to collect hydrocarbon boil-off vapors, which would be created rapidly in a vacuum. Vapor recovery systems for tankers have been under discussion for approximately 20 years, but they have not yet been implemented successfully to any major degree except in specialized trades (liquified petroleum gas, for example). Second, once a system has been developed, there is the question of vapor disposal. Liquefaction is very expensive, and burning of the vapors requires special modifications of propulsion engines, or boilers.

The committee is not aware that the full range of potential operational problems related to a vacuum has been addressed. Moreover, even with all necessary modifications, the machinery required to effect the vacuum (unless utilizing a stand-alone power source) would require that the ship's main machinery space be functional. That is not assured in the case of significant bottom damage.

In the committee's view, the mechanically driven vacuum system poses significant problems, although research aimed at enhancing the concept should not be discouraged. This alternative will not be considered further in this study.

8b. Hydrostatically Driven Vacuum (passive)

This system, as shown in Figure 5-8, capitalizes on the physical effects of initial cargo outflow following a grounding. By securing all openings to cargo tank(s) with applicable vacuum relief safeguards, the tight ullage space will incur a progressive drop in pressure consistent with outflow of cargo. Theoretically, outflow would cease when the combined ullage vacuum and cargo head no longer exceeded the hydrostatic head of water.

This concept has several inherent disadvantages. It poses the same difficulties as the mechanical vacuum in terms of interface with the ship's struc-

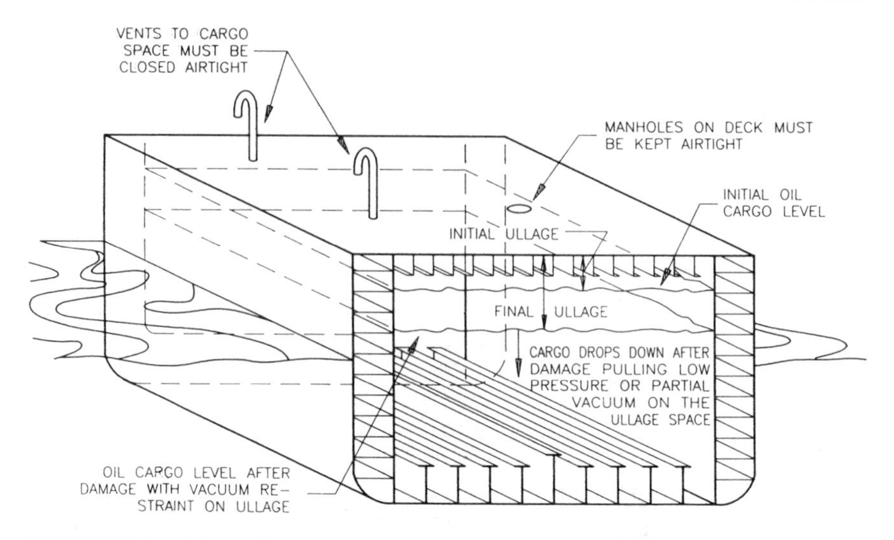

FIGURE 5-8 Alternative 8b: Hydrostatically driven vacuum (passive).

ture and existing safety systems. Secondly, without instantaneous identification of the damaged tank(s), and closure of all vent lines and piping to the tank(s), the system would not react quickly enough to attain anywhere near the vacuum required.

If a passive vacuum system depends on automatic closure of all tank vents from a remote location, or on pressure and/or liquid-level sensors in tanks, malfunction of this system (shut-off of all vents during normal cargo discharge) could cause a catastrophe. With a ship discharging at full rate, the vents absolutely must remain open, or the IGS must be operating. Otherwise, a major collapse of the deck and/or other structure would ensue, with a strong likelihood of explosion and fire—all occurring in a port. Finally, the allowance for initial cargo outflow negates the limited advantages of this concept over the mechanically driven vacuum system. The concept will not be considered further here, but it deserves further research and development.

8c. "Imaginary" Double Bottom

A variation on the hydrostatically driven vacuum principle or hydrostatic control alternatives is the "imaginary" double bottom, where chemically treated water is placed in the bottom of the cargo tank to a depth of 1 to 2.5 meters (see Figure 5-9). Upon grounding, only this layer of treated water would escape from the hull as the vacuum is established; thus, conceptually, no oil would flow out.

The imaginary double bottom concept has limited application. This con-

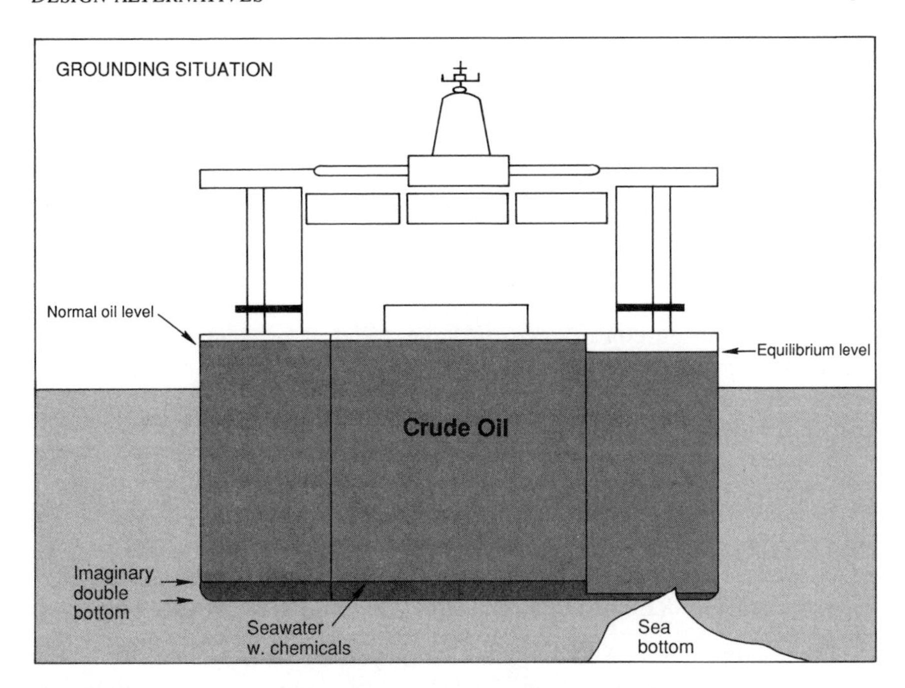

FIGURE 5-9 Alternative 8c: Imaginary double bottom concept.

cept could not be used with any products that would be contaminated by water. Crude that must be heated could not be carried with a water layer underneath; otherwise, it would be impossible to heat the cargo.

The committee will not consider this concept further, but it deserves further consideration for possible research.

Summary—Vacuum System Alternatives

A vacuum has natural advantages, and development of a practical, fail-safe, vacuum system would be desirable. The committee encourages research along these lines. Proposals for manually closing a ship's vents when the ship is in shallow waters, or for connecting IGS branches to cargo tanks, may have merit. However, there is still a concern with additional cargo vapor pressure build-up.

In the committee's view, the various vacuum concepts lack sufficient cost data to be considered further in the present study.

9. Smaller Tanks

This concept represents a return to past practice. Dividing the cargo section into smaller tanks, with additional bulkheads, would limit the volume of cargo exposed to a given point of damage. This concept would significantly reduce pollution in collisions, and it also would be effective in groundings.

The committee was able to obtain cost data for this option for new construction, but retrofit costs are highly dependent on base vessel configuration. Only the new construction alternative will be pursued further in later chapters.

9a. Service Tank Location

While pollution-control efforts have focused on restricting outflow from cargo tanks, most tankers carry 2 to 5 percent of their cargo deadweight in heavy fuel oil. In fact, all ships carry fuel oil in tanks subject to collision or grounding damage.

In new tanker designs, fuel tanks could be placed in defensive locations so as to minimize fuel oil pollution. In the committee's view, no further study is required.

INCREASED PENETRATION RESISTANCE

These proposals aim to limit pollution risk by making the ship's hull more resistant to damage. Where "barriers" would prevent any pollution, and "outflow management" would limit short-term pollution, *"increased penetration resistance" would try to control the worst case.* The proposals do not have a common principle, and they address diverse problems.

The pollution risk associated with tank vessel accidents is related to both material characteristics and the reaction characteristics of the design details. For example, tripling the scantlings of all lower hull structure would raise the threshold of grounding impact that could be tolerated before hull rupture. Alternatively, the same result could be achieved by selecting either a more damage-tolerant steel or a more grounding-tolerant bottom structural configuration.

Conventional tank vessel designs reflect an economic balance of widely available material and human resources. In all cases, designs tend toward the practical and are influenced strongly by historical factors, as well as by recognition that tank vessels may need structural repairs at remote locations. Unique or revolutionary designs, exotic or unusual materials, and unconventional work practices traditionally have been resisted, pending incentives that would support their research, development, and integration into the shipbuilding and ship repair industry.

10. Internal Deflecting Hull

In this design, shown in Figure 5-10, the forward 10 to 20 percent of the vessel would incorporate an inner double hull made of massive slab-type

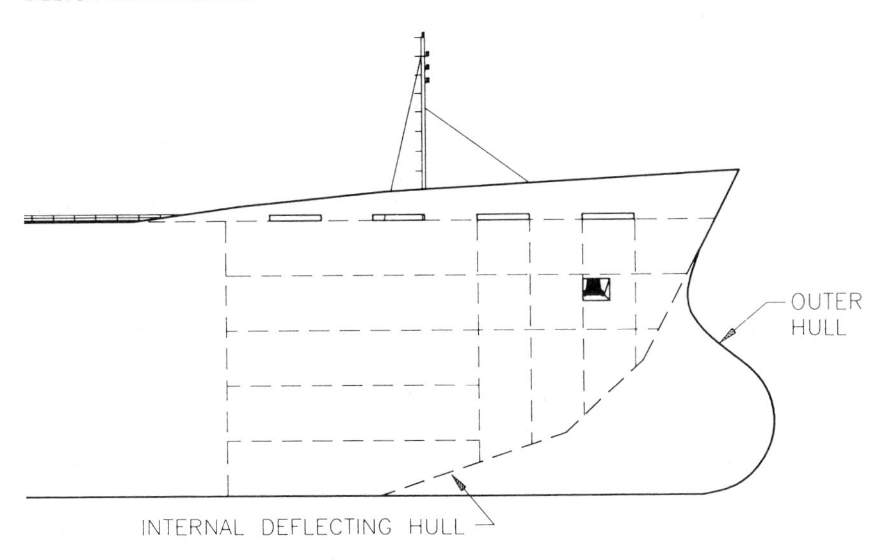

FIGURE 5-10 Alternative 10: Internal deflecting hull.

structure of exceptional strength, configured much like the lower bow plating (forefoot) of an ice breaker. The outer hull, made of normal-grade steel plating, would be shaped (as now) to suit the most efficient hydrodynamic form. In addition to shielding cargo with the resultant voids, the vessel, upon grounding, would initially ride up upon (and/or have its course deflected by) the inner slab-shaped structure. This would deflect damage from the remaining 70 to 80 percent of the hull.

This design provides local hull protection for that portion of the hull (the forebody) most likely to be damaged in groundings. On the downside, the risks associated with double-bottom voids, including fire/explosion hazard from gas leaks and flooding in event of hull breach, also apply here. Another concern is increased potential damage to vessels struck by a ship with a deflecting hull.

While a deflecting hull might be suited to small tankers and barges, the committee believes that in large tankers, considering their mass and speed, the amount of energy the hull would have to withstand to deflect the vessel would be enormous. The questionable effectiveness of this diversion might compromise protection of aft structure from damage.

This concept does not exceed the bounds of existing technology nor analytical evaluation techniques, although it would require development. But it presupposes a specific accident scenario[8] and in doing so, limits itself. Extensive cost and effectiveness evaluation, and small-scale testing, would be needed to provide even a preliminary assessment of various versions of this concept; the committee will not consider the concept further.

11. Grinding Bow

In this design, shown in Figure 5-11, the forward 20 to 30 percent of the vessel would incorporate a double bottom with an internal transverse structure specifically engineered and constructed to suit rock-grinding criteria. In a grounding, the most vulnerable cargo tanks would be protected by a double bottom. In moving over the obstacle, the rasp-like transverse structure would grind down the material, thus creating a clear and unobstructed passage for the remaining 70 to 80 percent of the vessel (or a smooth bed for the ship to come to rest on).

The partial double bottom would involve the related increased risk of explosion. A greater practical concern is that the grinding-bow concept never has been studied, as far as the committee can ascertain. It is an entirely unproven application of technology, and would be applicable only to the pinnacle or submerged-reef types of grounding described in Chapter 4.

In accidents, this design conceivably could minimize the extent of hull penetration and the resulting, largely uncontrolled destructive forces that mangle ship structure. But the concept is not sufficiently developed to be considered further in the present study.

12. Unidirectionally Stiffened Hull Structure

The initiation of plate tearing requires greater stress than the continuation of plate tearing. Stress buildup leading to plate tearing occurs as a local function of the plate's resistance to the deflecting force or obstacle. In recent tank vessel designs, highly rigid transverse bottom stiffening attached to more flexible bottom plating creates points of peak stress as a moving vessel grounds (see Figure 5-12). If the bottom transverse structure were designed to fail early in the stress build-up, or if conventional transverse structure were eliminated altogether, the bottom would flex uniformly and thus could avert onset of plate tearing. Although the bottom structure would deform significantly, it would not necessarily fracture.

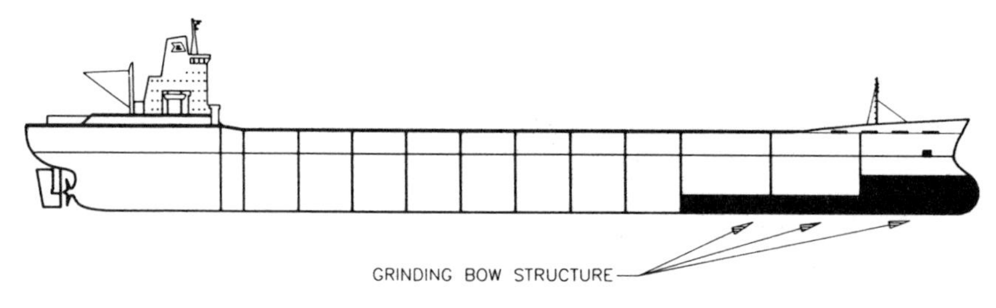

FIGURE 5-11 Alternative 11: Grinding bow—lower bow structured to sustain impact and grinding criteria.

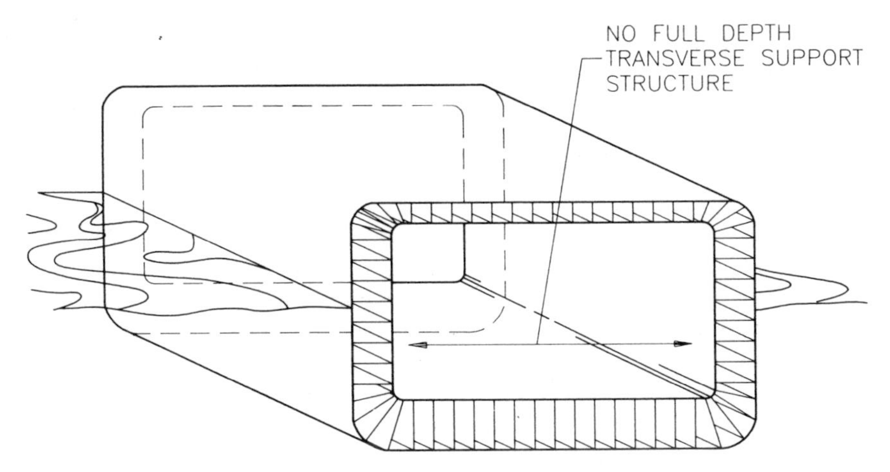

FIGURE 5-12 Alternative 12: Unidirectionally stiffened structure.

This concept has been applied to some ships. Its integration into tank vessels has required a new design approach; a Japanese shipyard has such tankers under construction for the product trade. Elimination of much transverse material would lower structural redundancy significantly and could increase susceptibility to catastrophic collapse in heavy weather or minor groundings.

Nevertheless, the proposal warrants strong research and development support, as its characteristics would enhance any existing or proposed design in terms of pollution prevention. The committee will not consider it further in the present study due to insufficient design and effectiveness data.

13. Honeycomb Hull Structure

This concept entails a "sandwich" double-plate construction of high-density material bonded to the inner and outer hull plating of a double-hull design. While this concept is intended to provide both a high-yield and an energy-absorbing hull structure, the committee found no evidence that the energy dissipation involved in the displacement and crushing of the honeycomb structure would provide significantly increased resistance to failure in comparison to conventional structures, such as the double hull.

This concept would require extensive research and testing. It involves novel material requiring unique production technology, and it introduces design and operational requirements never previously addressed by the industry. Maintenance and repair requirements would be unique, inspection plans would require special development, and the risk of cargo leaks into voids and the related explosion hazard would be considerable.

Evaluation of this proposal requires considerably more support documentation than is available. The committee will not consider it further.

14. High-Yield Steel Bottom Structure

The behavioral characteristics of steel can be controlled by its composition and manufacturing process. Conventional shipbuilding steels generally are chosen to suit minimum strength criteria, balanced against the shipbuilder's best economic interest regarding production, labor skills, and steel cost. Some steels provide significantly better impact resistance than those generally used in tanker construction, but they require more costly fabrication. These steels (such as the HY80 steel used in submarine hull construction), if used as hull plating on otherwise conventional single-hull tankers, would increase resistance to grounding and collision damage.

However, the use of exotic materials would require a commensurate investment in production facilities and personnel training. At present, worldwide repair resources, in terms of both materials and skill, are not readily available.

This concept will not be considered further in the present study, but research and development should be encouraged. Research and development needs concerning materials for tank vessel hulls and structures are discussed in Chapters 4 and 7. The use of high-yield steel is one aspect of the overall research requirement for:

- more robust and uniform-strength scantlings;
- greater use of steels having high strain energy absorption and high strain-to-rupture characteristics; and
- other materials or structural configurations that will enhance the resistance of the hull girder to rupture.

15. Concrete Hull Structure

Concrete has been used before in the marine industry. During World Wars I and II, a small number of moderate-size prototype concrete ships were constructed. Since then, concrete has been utilized periodically in offshore storage facilities, barges, and in some yachts. The benefits include good compressive qualities, relatively high tolerance to salt water, potentially seamless construction, and adaptability to fluid forms. In sufficient mass, concrete hulls may offer greater resistance to the onset of hull failure from groundings than would single-hull steel designs.

Concrete structure would seem to offer advantages in inland barge structure, where hull bending loads are limited. Such barges could be built economically with less-skilled labor than required for steel construction. The added mass provided by concrete would allow vessels to sustain minor damage without rupture. As some barges are restricted in dimension and draft, the added mass would be detrimental to cargo deadweight capacity.

Considerable research is required to sort out the engineering issues. The greater concern, however, would be the willingness of the industry to adopt the necessary facilities and support infrastructure worldwide. Given these conditions, the committee will not consider this concept further.

16. Ceramic-Clad Outer Hull

A thick ceramic cladding might increase significantly hull surface hardness, so as to withstand greater impact without fracture. In addition, a ceramic surface could reduce resistance to vessel movement over minor obstacles, thereby increasing hull tolerance for minor groundings.

This technology is novel and unresearched. Other than noting that maintenance likely would require development of specialized facilities, the committee cannot evaluate this proposal other than by educated judgment. There are no data that encourage development of the concept at present. The committee will not consider it further.

ACCIDENT RESPONSE

The previous three categories dealt with structural hull designs that could prevent or mitigate pollution at the time of an accident. This category includes concepts that could mitigate pollution by aiding response to an accident. These options involve less drastic revisions than the structural designs discussed previously. They are, in effect, "add-ons" that could be combined with other designs to enhance overall effectiveness. Although the committee will not consider these proposals in outflow-reduction or economic assessments, some nonetheless warrant consideration for use.

17. Enhanced Information Processing

In this concept, the ship's hull, tank, and pipe structure would incorporate extensive sensing devices connected to a central processing facility. This monitoring system would alert the crew to any unacceptable condition: It could confirm structure and cargo system integrity, analyze response alternatives, and recommend action to mitigate pollution. In theory, quick and accurate knowledge of the status of the ship and its cargo following an accident could promote effective response. The following data would be useful, for example: Fluid levels in all cargo, ballast, and fuel tanks together with level trends; the draft of the tanker and trends in draft change; extent of flooding of cargo or ballast piping; the status of engine room function; and tidal and other environmental data.

Some of the technology for this concept exists. Most modern tankers already have some type of computer(s). Most classification societies require stress computers to calculate loading, strength, trim, and stability. Many ships also have, for example, computerized cargo load and discharge sequence programs. In addition, programs exist that measure "damaged" tanker stress, trim and loading, and most modern tankers have installed liquid-level gauges that can be read in a central control room. However, these computers are stand-alone units and rely on prior calculations stored within. No sensors are used to update or provide input to the computer programs.

The committee believes that all of the hardware and software necessary to implement this concept exists. One classification society has even developed a prototype "black box" recorder, similar to those used on aircraft to record accident data. However, there are a number of problems inherent in an extensive computerized system. After a major hull rupture, cargo outflow can be very rapid, and the computer system would have to respond nearly as quickly. Furthermore, the most important unknown is often the extent of damage to the ship. This may not be detectable by computer. Even if all the necessary information is immediately available, the ship must retain its capability to transfer cargo from damaged tanks and must have intact, empty tank space available to receive it. Finally, performance of the system would depend heavily on maintenance and repair; malperformance would risk ship safety. Exposure to the marine environment and possible corrosive effects of cargo and ballast present maintenance problems for sensors and wiring.

The marine industry is becoming accustomed to the use of computers, but much of the industry still relies on experience, judgment, and past practice. With the increasingly broad application of computer assistance in work places, and the availability of most elements needed to enhance shipboard information processing, the committee considers this option worthy of further development. However, the committee was unable to evaluate the concept further within the scope of the present study.

18. Towing Fittings

Standardized towing fittings already are required by a number of tanker owners, although they are not required by any regulatory body. Towing fittings and their arrangements may enhance the ability of tugs to save disabled tankers before grounding or foundering. These fittings also assure that damaged tankers can be attached to positioning equipment, thus reducing the risk associated with uncontrolled movement of a stranded tanker.

Towing fittings can tolerate significantly stronger forces than standard mooring fittings; such forces are encountered when attempting to free and float a grounded and lodged vessel.

All tankers regularly trading in Alaska carry emergency towing wires, as a result of a near-grounding in that region a decade ago. The concept has been evaluated and recommended by the IMO.

In the committee's view, towing fittings should be mandated for tank vessels, although insufficient data exist to allow further consideration in this study.

19. Distressed-Ship Cargo Transfer System

Tankers that run aground often remain stranded, requiring a major off-load prior to refloating. The cargo system often is rendered useless, and even if not, pumping cargo from a breached tank is impossible when the suction is at the bottom of the tank. At present, it is usually necessary to rely upon portable pumps with relatively low capacity (compared to operational pumps), which must be brought to the damaged ship and lowered into the tanks for off-loading. This can be time-consuming; for a large tanker, the process might take a week or more, leaving the ship exposed to greater damage or possible loss.

This risk might be reduced by two basic improvements. First, a means allowing for rapid transfer of cargo from a breached tank to intact tanks, such as segregated ballast tanks, could reduce oil outflow from a stranded ship as the tide falls, or due to wave action or ship motion. To be effective, such a system would have to be available to use within an hour or two after stranding. Second, if a ship were equipped with a system for transferring cargo from breached tanks at a level well above the bottom, where oil and water were mixing, this might facilitate much more rapid off-loading or lightering of cargo. This would enable more rapid refloating of a stranded ship than is common with the small, portable salvage pumps now used.

The committee is aware of three transfer-system concepts, as follows:

- Each cargo tank could be fitted with permanent suction at about mid-tank height, capable of discharging oil from above the oil/water interface in a damaged tank. One proposal would have the mid-height suction line extended aft into the ship's pump room, where it could be installed so as to line up with the suction side of the segregated ballast system. A spool piece, not in place during normal operations, could be fitted following an accident, allowing the ship's ballast system to handle nearly immediate evacuation of oil from damaged cargo tanks into intact segregated ballast tanks.
- Depending on a ship's arrangement of cargo and ballast tanks, it might be possible to fit sluice or transfer valves between cargo tanks and adjacent or nearby segregated ballast tanks. In the event of damage to a cargo tank, these sluice or transfer valves would be opened, permitting the cargo to

flow by gravity (or by hydrostatic pressure in the IOTD designs) into the empty ballast tanks. Of course, before activating such a system, the crew would need to be sure that the receiving tank(s) were intact.

- The third option is a portable emergency cargo transfer system employing high-capacity submersible pumps, which could be lowered into damaged cargo tanks while maintaining a vacuum in these tanks (to hold in the cargo). Such a system, which would be stored aboard on a moveable dolly, could be used to transfer cargo either to intact segregated ballast tanks or to a lightering vessel.

The committee found these concepts promising, as they employ, for the most part, existing hardware and sound operational principles. However, insufficient details were available for the committee to judge how rapidly these concepts could be activated following an accident. Because the initial outflow from a large hole takes place in a matter of minutes, the committee does not believe any of the systems described are likely to make significant reductions in immediate outflow. However, these systems could be useful in mitigating subsequent outflow, and in hastening the refloating of a ship.

Only rough figures were provided regarding costs, suggesting that such systems might add 1 to 2 percent or more to the cost of a new ship. Even without more precise application and cost data, the committee feels these concepts deserve further consideration by industry, the Coast Guard, and the IMO.

SUMMARY OF INITIAL TECHNICAL EVALUATION

Following are the significant findings drawn from the foregoing technical assessment:

- Regardless of vessel design, the potential advantages associated with towing fittings argue for their serious consideration for all new and existing tank vessels.
- Distressed-ship cargo transfer systems deserve further consideration by industry, the Coast Guard, and IMO.
- The committee has eliminated from further consideration, regarding pollution control and cost, six design alternatives that, though promising in concept, lack the basic technical supporting data and involve major design and operational uncertainties. These include the resilient membrane, convertible tanks, deflecting hull, honeycomb hull, imaginary double bottoms, and ceramic-clad hull.
- A number of other proposals should be considered for further research and/or development. These are the mechanically and hydrostatically driven vacuum systems, grinding bow, unidirectionally stiffened bottom structure, high-yield steel bottom, and enhanced information processing. A design combination worthy of study is the longitudinally reinforced bottom struc-

ture, plus high-yield steel bottom plating and a grinding bow. In addition, the concrete hull structure for tank vessels requires research to resolve numerous application problems and to understand the lack of industry interest, reflected by the many decades of unsuccessful efforts to develop maritime markets.

• Based on results of its technical assessment, the committee determined that the following design alternatives should be assessed on the basis of their pollution-control effectiveness:

MARPOL tanker (as the reference vessel)
double bottom
double sides
double hull
hydrostatic control
intermediate oil-tight deck
smaller tanks

• In conducting a technical assessment of the above designs, it was apparent that the following three compound design alternatives may offer improved outflow reduction potential for a range of accident scenarios. Moreover, these compound designs employ proven or well-understood technology:

double sides with hydrostatic control
double hull with hydrostatic control
intermediate oil-tight deck with double sides

ASSESSMENT OF DESIGN ALTERNATIVES APPLICABLE TO BARGES

The committee's charge encompasses all ocean-going vessels 10,000 DWT and greater. Hence, ocean-going barges over approximately 5,000 gross tons are considered in this report.

The DnV study was prepared for tankers based on a tanker data base and could not be extended to cover barges in the time allowed for preparation of this report. However, as noted earlier, the technical assessment of design alternatives applies in some cases to barges as well as tankers.

The following discussion delineates how the various alternatives apply to barges and how application may be limited due to the unique nature of barges.

Barriers

The offshore barge can carry up to 50,000 tons deadweight and is configured much like a tanker of equivalent size (see Chapter 1, Figure 1-7).

Thus, all of the structural barriers would be applicable to offshore barges, with DnV pollution-effectiveness rankings similar to those of 40,000 DWT tankers.

Outflow Management

Barges are unmanned and are either towed or pushed by a power source (tugboat). Consequently, any outflow management alternative requiring manned intervention, or some automatic device electrically coupled to the tug, would be impossible to maintain in extreme weather conditions (when barge accidents are more likely to occur). Thus, the vacuum systems, and the oil transfer system using deck-operated valves, would not be applicable. The intermediate oil-tight deck and smaller tanks might be applicable.

Hydrostatic control appears to have limited, if any, application to towed barges. Barges carry a wide range of petroleum products of varying densities and properties. Thus, the cargo-loading levels required to achieve hydrostatic control would vary. Such precise control is at variance with barge crew manning levels and practices. Hydrostatic control may be somewhat safer and more feasible for some articulated barge/tug (pusher-tug and notched-barge) operations, subject to stringent cargo-loading controls.

Increased Penetration Resistance

The internal deflecting bow, grinding bow, high-yield bottom structure, honeycomb, and concrete structures would not be prime barge designs, because groundings usually take place either at low speeds (under power), or at very low speeds and unpredictable angles (when the barge is separated from the tug in heavy weather).

The longitudinally reinforced bottom structure, or any other design that manages bottom failure in high-energy accidents, would be poor choices, because barges are more prone to low-energy groundings.

Accident Response

Enhanced information processing would be difficult, as the system would have to be connected from the tug to the barge. Installation of towing fittings is routine on barges, and use of a distressed-barge cargo transfer concept poses no significant technical problem.

Smaller Offshore Barges and Inland Barges

While offshore barges of less than 10,000 DWT and inland barges are outside the scope of the present study, this brief discussion illustrates the dimensional constraints of small tank vessels.

Due to these dimensional considerations, the Coast Guard, in information circular NVIC 2-90 (1990), recommends the following:

- For vessels (tankers or barges) under 20,000 DWT, the width of a double hull can be reduced linearly from 2 meters for a 20,000 DWT vessel, to 1 meter for a 10,000 ton vessel.
- For vessels under 10,000 DWT certified for inland routes (primarily barges), the void space in a double hull must maintain a clearance of 2 feet. This width is considered the minimum for construction, inspection, and maintenance.
- Bow and stern compartments do not carry oil.
- The protective spaces formed by the double hull should not contain oil. (The committee notes that, for survivability purposes, it would be preferable not to vent these spaces, as they are not ballasted.)

POLLUTION CONTROL ANALYSIS OF SELECTED DESIGNS AND DESIGN COMBINATIONS

The next assessment the committee applied to the 10 selected design alternatives and compound designs addressed the potential for reducing oil outflow in collisions and groundings. To aid in the analysis of the seven basic design alternatives, the committee sponsored an addition to a study conducted by Det norske Veritas (DnV) for the Royal Norwegian Council for Scientific and Industrial Research (Det norske Veritas, 1990). DnV is a Norwegian classification society that has developed a methodology for calculating oil outflow. The committee conducted its own evaluation of the three compound design approaches based on its own estimates and calculations derived from the results of the DnV analyses.

The calculations made in this analysis are based on a number of assumptions, and the committee does not necessarily accept all of these as the most reasonable of possible assumptions. Therefore, the committee regards the specific percentages for reduction of oil outflow derived from this analysis as indicative examples. These results represent only one attempt to quantify reduction in outflow; application of a different set of assumptions might produce different results.

Using a probabilistic approach to past accidents derived from Lloyd's Register, the U.S. Coast Guard, and DnV databases, the methodology is based on the damage severity experienced in collisions and groundings. Using this damage for predictive purposes, outflow is estimated for each alternative design, first for collisions and groundings separately, and then combined using a 40/60 weighting (based on the collision and grounding frequencies for tankers worldwide, per DnV statistics). The committee recognizes that the 40/60 (collisions/groundings) frequency weighting is approximately correct for the number of tanker spill incidents over 10,000 gallons in U.S. waters, while the ratio of oil volume lost in collisions versus groundings approaches 20/80 (in volume). Nevertheless, the committee and DnV agree that the relative outflow rankings of each design would be the same regardless of which of these two sets of ratios are used.

For purposes of the committee's study, the model was applied to 21 design arrangements. Outflow was calculated or estimated for two tanker sizes: VLCC (13 arrangements) and 40,000 DWT (eight arrangements). In addition, DnV assessed the implications of the design arrangements for outflow performance of an 80,000 DWT tanker; this assessment was based on extrapolations of the results for VLCCs and 40,000 DWT vessels, where appropriate. These three sizes represent a cross-section of the sizes of tankers generally employed near the U.S. coastline. The VLCC is involved in offshore oil trade and is lightered by tankers in the 80,000 DWT range, as described in Chapter 2. The 40,000 DWT tanker is typical of coastal carriers distributing refined products among U.S. ports.

The DnV report (except for its appendices) is presented in Appendix F. In addition, the committee interpretations of the assumptions used and some of the DnV conclusions can be found in the Appendix (where the committee has little or no comment, the committee accepts the DnV statements in the context of their report). Following is a summary of the designs analyzed and the rankings of these designs in terms of oil outflow.

General Features of the Tankers

All the designs analyzed have the following main particulars[9]:

	VLCC	40,000 DWT
Length	315.0 m	190.1 m
Beam	57.2 m	27.4 m
Depth	30.4 m	17.8 m
Draft	20.8 m	12.9 m

Probabilistic Ranking of VLCC Tankers

Statements drawn from the DnV report are in italics (some are edited for brevity and style). Committee comments are in standard type.

An analysis of serious casualties carried out by DnV shows that, for tankers above 20,000 DWT worldwide, the probability of collisions is approximately 40 percent and for groundings approximately 60 percent. These percentages differ in other reports of worldwide accounts (some show nearly equal frequency) but approximate reports of major accidents in U.S. waters on a frequency basis; however, as noted previously, the relative rankings of various alternatives are not affected.

Using the 40/60 ratio, the average oil outflow for different designs in collisions and groundings, respectively, have been combined in Figure 5-13 for 5-knot grounding speed, and in Figure 5-14 for 10-knot grounding speed. In these figures, the total index has been given as well as the portions derived from collisions and from groundings.

The different VLCC hull design arrangements and variations, and how they relate to the committee's design alternative categories, are listed in Table 5-1; sketches of specific arrangements and more detailed data regarding ship parameters are provided in Appendix F.

Results—VLCC Analysis

Statements drawn from the DnV report are in italics (some are edited for brevity and style). Committee comments are in standard type.

The double side/double bottom (i.e., double hull) VLCC designs (arrangements 1A,1B) have the smallest potential oil outflow in collision and grounding, given the assumptions regarding damage location and extent. Compared with a modern conventional MARPOL VLCC, the double side/double bottom tanker is likely to spill, on the average, only about 33 percent of the standard amount of oil.

Comparing the two double side/double bottom designs, the alternative with two longitudinal bulkheads, and double sides (arrangement 1A) is likely

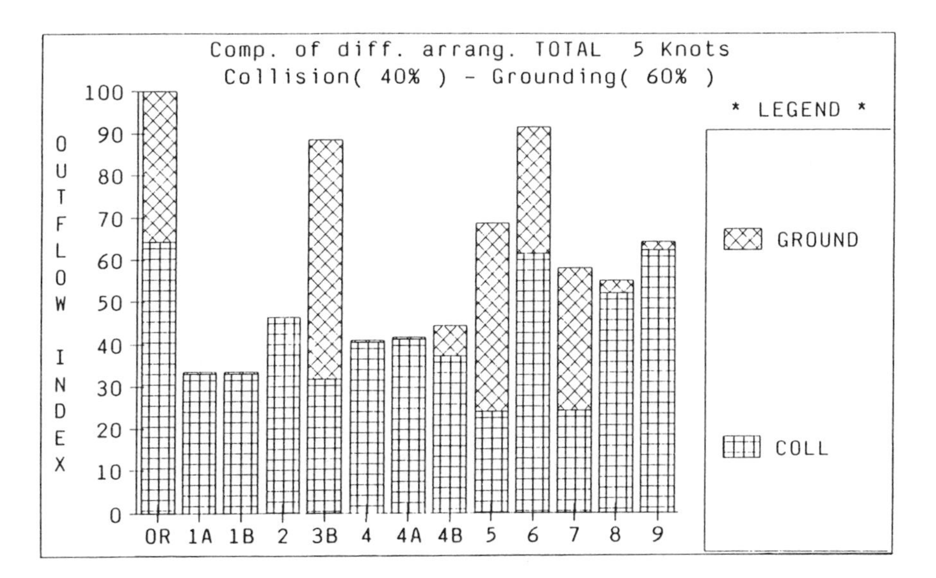

FIGURE 5-13 Combined ranking, 5 knots (low energy). Reference DnV Report, Figure 3.25, Appendix F.

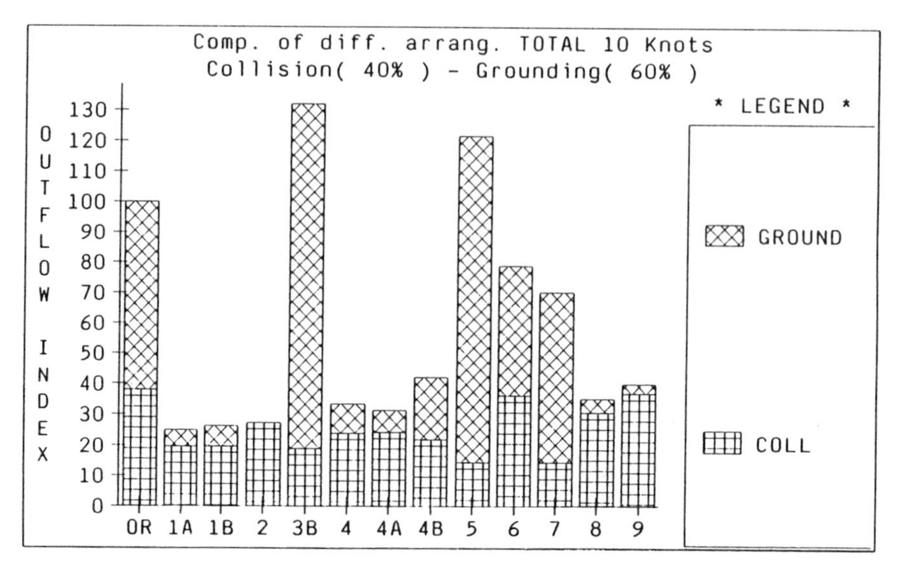

FIGURE 5-14 Combined ranking, 10 knots (high energy). Reference DnV Report, Figure 3.26, Appendix F.

to leak less oil in about 75 percent of all collisions than the design with only one longitudinal bulkhead at the centerline (arrangement 1B), although both designs have the same overall combined performance.

The amount of oil likely to escape from a VLCC with a double bottom for ballast and segregated ballast side tanks (arrangements 4, 4A and 4B), is about 40 percent of the amount that would be spilled from a modern conventional MARPOL VLCC. The positive influence of locating the side ballast tanks forward, with regard to potential oil outflow in collisions, is evident comparing arrangements 4B and 4A.

Single-bottom designs with wide wing tanks for ballast (arrangement 3B) perform quite poorly overall (combined ranking) due to the large volume of oil likely to escape in grounding. In fact, this design performs only slightly better than modern conventional VLCCs.

Reducing the tank size to half of MARPOL requirements (arrangement 7) reduces the potential average oil outflow (combined ranking) to about 60 percent of the outflow from a conventional VLCC.

The partial double-bottom VLCC (arrangement 5) does not perform too well in groundings. The basic reason is that the double-bottom height is too low to be effective, given the study's assumptions. Increasing the height, and reducing the side tank width, would make this design similar to arrangement 1A and 1B.

TABLE 5-1 VLCC Tanker—List of Alternatives/DnV Arrangements Analyzed

Committee Alternative	DnV Arrangement (Arrangement Code)
MARPOL Tanker (reference standard)	• Modern Conventional VLCC (OR)
Double Bottom	• Double bottom in whole cargo area (2) • Double bottom with ballast in side tanks (options 4, 4A, and 4B) • Double bottom in side tanks, single bottom in center tank (6)
Double Sides	• Double side, single bottom (3B) • Double side in whole cargo area and partial double bottom (5)
Double Hull	• Double side and double bottom, with two longitudinal bulkheads (1A), with bulkhead at centerline (1B)
Intermediate Oil Tight Deck	• Intermediate Oil Tight Deck (8)
Hydrostatic Control	• Hydrostatically balanced loading (9)
Small Tanks	• Tank size half of MARPOL requirements (7)

The double-bottom designs analyzed would spill no oil at all in about 85 percent of all groundings, whereas some oil always escapes from single-bottom designs, irrespective of tank size and the presence of a vacuum tank system.

A vacuum system reduces significantly the amount of oil escaping in ground-ings for the single-bottom designs analyzed. However, the total amount of oil lost from the modern conventional VLCC with a vacuum system is still about twice the amount escaping from the double side/double bottom design.

Comparing VLCCs with double sides to VLCCs with single sides, the former provides an effective barrier against oil outflow in 20 percent of the all collisions for arrangements 1A and 1B, and in 42 percent of all collisions for arrangement 3B. The protection is particularly effective for low-energy collisions with limited damage penetration. The influence of increasing the double-bottom height as compared with increasing the double-side width is shown in Table 5-2. For example, increasing the VLCC double-bottom height from 2.0 to 3.9 m (B/15) reduces the probability of oil outflow in groundings from 58 to 14 percent, while increasing the width of side tanks from 2.0 to 3.0m only reduces the probability of outflow in collision from 88 to 80 percent.

The specific conclusions drawn by DnV can be found in Appendix F (page 275); the committee's commentary can be found on page 300.

TABLE 5-2 Probability for *NO* Oil Leakage as Function of Double Bottom Height and Distance Between Double Sides*

GROUNDING (VLCC)

DISTANCE BETWEEN INNER/OUTER BOTTOM	PROBABILITY FOR NO LEAKAGE	
METERS	B/—**	PERCENT
2.0	28.6	41.7
2.4	23.8	52.6
3.0	19.0	69.0
3.9	14.7	86.0
6.6	8.6	99.8

COLLISION (VLCC)

DISTANCE BETWEEN INNER/OUTER SKIN	PROBABILITY FOR NO LEAKAGE	
METERS	B/—**	PERCENT
2.0	28.6	12.1
3.0	19.0	20.4
5.8	9.86	39.4
6.3	9.08	42.0

*Reference Table 3.3.1 in DnV Report, Appendix F.
**Ratio, beam to between hull width.

Probabilistic Ranking of 40,000 DWT Tankers

The 40,000 DWT designs chosen for the analysis are similar to the VLCCs. The objective in analyzing both very large (VLCC) and small (40,000 DWT) tankers was to detect trends, so the results could be applied, in a general sense, to other tanker sizes. The committee also selected the 40,000 DWT tanker for analysis by DnV because the size is typical in U.S. coastal activity.

Particular attention has been directed to investigating the influence of double-side width on oil outflow, as several 40,000 DWT product tankers have been built with narrow sides.

The arrangements listed in Table 5-3 were analyzed by DnV; Appendix F (page 278) provides a summary of the assumptions made regarding ballast capacity, side width, bottom height, and longitudinal bulkheads.

Sketches of the various arrangements are found in Appendix F (pages 281-285). Table 5-3 links the DnV arrangements to the general design alternatives considered by the committee.

The combined ratings for collision and grounding (using the 40/60 weighting) gives the combined ranking shown in Figure 5-15 for 5-knot grounding speed, and in Figure 5-16 for 10-knot ground speed.

TABLE 5-3 Small Tanker (40,000 DWT Tanker)—List of Alternatives/ DnV Arrangements Analyzed

Committee Alternative	DnV Arrangement (Arrangement Number)
MARPOL Tanker (reference standard)	• Modern 40,000 DWT SBT (1)
Double Bottom	• Double bottom and single sides (3)*
Double Sides	• Double sides and single bottom (2)
Double Hull	• Narrow (0.76 m apart) double sides and double bottom (4)*
	• Double sides (1.2 m apart) and double bottom (5)*
	• Wide (2.0 m apart) double sides and double bottom (6) *
	• Wide (2.0 m apart) double sides and double bottom, short tanks (7), no centerline bulkhead
Intermediate Oil Tight Deck	• Intermediate Oil Tight Deck (8)

*With centerline bulkhead.

At 5 knots, design arrangement 6 with wide double side and low double bottom achieves a rating of 57, followed by arrangement 8 (the intermediate oil-tight deck) with an rating of –65. Only design arrangement 2, the double side without a double bottom, has an index above the reference vessel (MARPOL). The ranking does not change for the 10-knots scenario; design arrangement 6 (wide double sides with double bottom) remains the best with a rating of 38, followed by design arrangements 8 (intermediate oil-tight deck), 4 (narrow double sides and double bottom) and 5 (mid-width double sides and double bottom).

Results—40,000 DWT Tanker Analysis

This analysis demonstrates the value of double bottoms in preventing oil outflow.

Table 5-4 shows the effect of increasing double-bottom height and double-side width on the probability of spilling oil. The probability falls rapidly as double bottom height is increased. By contrast, increasing the width of double sides does not have the same dramatic effect.

Estimated Oil Outflow from a 80,000 DWT Tanker

Based on the studies of VLCCs and 40,000 DWT tankers, DnV appraised the potential oil outflow from a 80,000 DWT tanker designed with several hull configurations. No supporting calculations were carried out. The conclusions drawn by DnV can be found in Appendix F (page 296); the committee's commentary can be found on page 302.

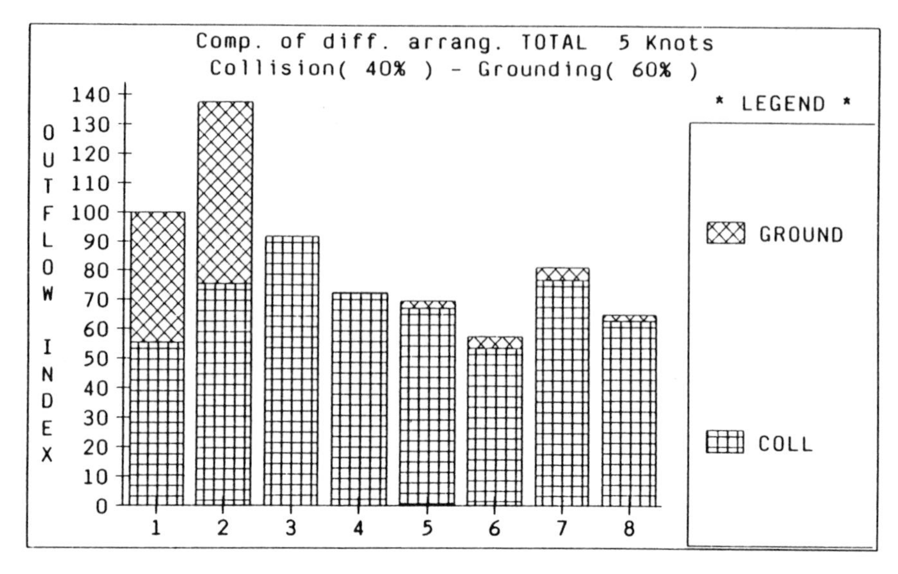

FIGURE 5-15 Combined ranking, 5 knots (low energy), 40,000 DWT. Reference DnV Report, Figure 4.18, Appendix F.

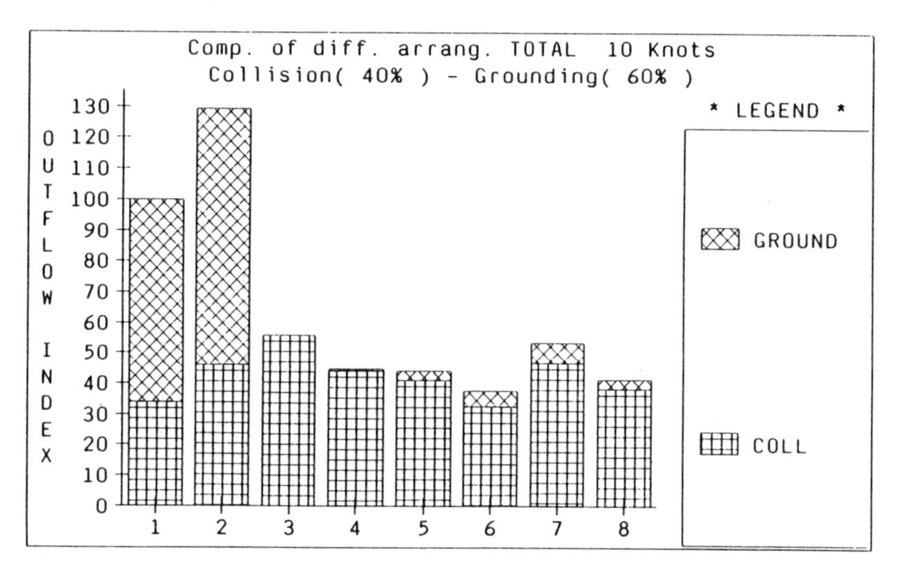

FIGURE 5-16 Combined ranking, 10 knots (high energy), 40,000 DWT. Reference DnV Report, Figure 4.19, Appendix F.

TABLE 5-4 Probability for *NO* Oil Outflow in Collision and Grounding*

GROUNDING (40,000 DWT)

DISTANCE BETWEEN INNER/OUTER BOTTOM		PROBABILITY FOR NO LEAKAGE
METERS	B/—**	PERCENT
1.83	15.0	85.1
2.0	13.7	90.0
2.6	10.5	98.4
3.9	7.0	99.9

COLLISION (40,000 DWT)

DISTANCE BETWEEN INNER/OUTER SKIN		PROBABILITY FOR NO LEAKAGE
METERS	B/—**	PERCENT
0.76	36	8.6
1.2	22.8	16.3
2.0	13.7	29.2
3.0	9.1	41.9

*Reference Table 4.3.1 in DnV Report, Appendix F.
**Ratio, beam to between hull width.

The Committee's Overall Conclusions from DnV Analysis

As noted earlier, DnV's conclusions, along with committee comments, are provided in Appendix F. The committee drew the following overall conclusions from the analysis of all tanker sizes:

- Double hulls provide significant overall protection against oil outflow in low-energy collisions and groundings.
- Wide tanks are likely to spill more oil than long, narrow tanks.
- Double sides protect against oil outflow in collisions, particularly low-energy collisions.
- Double bottoms protect against oil outflow in groundings, particularly low-energy groundings.
- The intermediate oil-tight deck, when combined with double sides, should provide protection against both groundings and collisions.
- Hydrostatic loading and/or smaller tanks should reduce oil outflow when used in conjunction with any design concept.

SUMMARY AND SIGNIFICANCE OF OUTFLOW ESTIMATES

The results of the DnV probabilistic study and the committee's own estimates for *oil outflow* relative to a standard of reference (100%), the

TABLE 5-5 Performance of Alternative Designs for Large Tankers (VLCC)—Oil Outflow

Design Alternative for VLCC (240,000 DWT) Tanker	Oil Outflow Relative to MARPOL* (100%) for Composite** of Collisions (40%) and Groundings (60%)	
	Low Energy (5 kn)	High Energy (10 kn)
Double Bottom (B/15)	42	37
Double Sides	88	130
Double Hull	33	26
Hydrostatic Control (passive)	62	40
Smaller Tanks (1/2 MARPOL)	58	70
Intermediate Oil-Tight Deck with Double Sides	32***	23***
Double Sides with Hydrostatic Control (passive)	32***	21***
Double Hull with Hydrostatic Control (passive)	30***	22***

*MARPOL standard tankers have protectively located segregated ballast tanks.
**Composite based on frequency of collision and groundings.
***Committee estimate (see Appendix K).

single-hull tanker with protectively located segregated ballast tanks (PL/SBT—MARPOL), are listed in Tables 5-5 and 5-6.

As noted earlier, the committee recognized the possibility of combining certain design alternatives to improve pollution control. The committee used DnV data to derive outflow estimates[10] for three combinations: double sides with hydrostatic control; double hull with hydrostatic control; and intermediate oil-tight deck with double sides. The outflow ratings for these combinations were not evaluated directly by DnV. Therefore, these oil outflow percentages (as shown by ***) are to be considered indicative examples of performance but less rigorous than the other percentages.

Outflow Performance Relationships and the Underlying Reasons

To understand the significance of these performance values, it is important to identify the major differences between designs, to understand the reasons for these differences, and to understand to what extent this knowledge can be used for decision making.

1. The performance benefits from each design alternative are significantly better for the big ships (VLCCs) than for the little ships (40,000 DWT). This is true at both high and low speeds.

The reason is that, in big ships, the size and numbers of cargo tanks are not greatly changed among design alternatives. This, in turn, is the result of both IMO hypothetical oil outflow requirements and structural demands in

TABLE 5-6 Performance of Alternative Designs for Small Tankers (40,000 DWT)—Oil Outflow

Design Alternative for 40,000 DWT Tanker	Oil Outflow Relative to MARPOL* (100%) for Composite** of Collisions (40%) and Groundings (60%)	
	Low Energy (5 kn)	High Energy (10 kn)
Double Bottom (B/15)	82	50
Double Sides	136	130
Double Hull	68	43
Hydrostatic Control (passive)	52	34***
Smaller Tanks	68	76***
Intermediate Oil-Tight Deck with Double Sides	57***	36***
Double Sides with Hydrostatic Control (passive)	70***	44***
Double Hull with Hydrostatic Control (passive)	61***	39***

*MARPOL standard tankers have protectively located segregated ballast tanks.
**Composite based on frequency of collision and groundings.
***Committee estimate (see Appendix K).

big tankers. In smaller tankers, the presence of a protective skin (i.e., double bottom or double side) allowed the designer of the DnV series to select a smaller number of much larger tanks. Neither hypothetical oil outflow nor structural demands impose significant limits on cargo tank size in these small ships. The result is that in any accident piercing the inner skin of a smaller tanker, a significantly larger amount of cargo is exposed to release than in the base ship.

2. The performance improvement for all alternatives appears better at high speed (10 knots) than slow speed (5 knots) for both big and little tankers.

The reason is that, at higher speeds, the longitudinal extent of damage is greater in groundings (although collision, as assumed here, is independent of speed). The greater the longitudinal damage in the bottom, the more cargo potentially exposed to spillage, and accordingly protective alternatives appear to be, and are likely to be more useful at higher speeds. In this regard, it is important to recognize that the base ship performance value of 100 percent represents a very different amount of oil spillage in the two ships. For the base VLCC at 5 knots, 100 percent represents grounding outflow of about 11,000 tons while at 10 knots, 100 represents grounding outflow of about 21,000 tons—or an increase of almost twofold.

3. Four cases (double hull, double hull with hydrostatic control, double sides with the intermediate oil-tight deck, double sides with hydrostatic control) appear to provide nearly the same pollution prevention benefit.

This is due to a significant assumption used in making these estimates, that is, that hydrostatic control works nearly perfectly with no loss due to tide, swell, or ship motion. This assumption may not be valid, and these estimates thereby may underestimate oil outflow.

These important apparent performance differences also can be explained, however, by recognizing that:

- The double hull is definitely very effective in low-energy groundings, but it cannot prevent some collision outflow due to narrow side tank voids.
- Because the double hull alone, in this analysis, prevents most of the outflow in groundings, adding hydrostatic control improves performance only marginally (outflow already is minimal); furthermore, this combination only improves performance in collision as a result of cargo tanks being less than full.
- In this analysis, double sides with the intermediate oil-tight deck is better than a double hull in collision as a result of wider side ballast tanks, and is presumed to prevent nearly all outflow in low-energy groundings and some outflow in high-energy groundings (because the analysis takes no account of factors such as tides, current, or ship motion).
- In this analysis, double sides with hydrostatic balance appears equivalent to double sides with the intermediate oil-tight deck for the above reasons, but, in fact, will not perform as well because tides, swells, and ship motion can be expected to ensure grounding outflow following the accidents.

Applicability of Results

From the preceding, the following observations can be made concerning the applicability.

These particular numbers, while providing a relative performance indicator for *one particular set of designs and assumptions*, should be regarded only as a sample of the type of analysis that can be made for these and other design possibilities (particularly with regard to number and size of cargo tanks in small ships).

Other assumptions about grounding and collision speeds, etc., also should be investigated.

Accidents other than grounding and collision (i.e., fire and explosion, structural failures) should be subjected to similar analysis.

Significance of Results

Because the committee made substantial use of data from the DnV study, as well as the committee estimates, it is important to understand the significance, limitations, and possible implications of these numbers before using them as the basis for the cost-benefit analysis discussed in Chapter 6.

These estimates are significant in that they show a substantial potential reduction in oil outflow in collisions and groundings for several alternatives for one plausible set of designs, and grounding and collision circumstances. However, they show that a wide range of results are possible even for the same hull design concept, when these designs are applied to ships of different sizes with different cargo tank configurations and sizes.

Due to a variety of limitations, the estimates clearly should be regarded as only a sample of the type of work that could be done with more comprehensive analysis. The limitations include use of one particular statistical casualty profile, a number of simplifying assumptions regarding oil outflow immediately following collisions, and the disregard for tidal and wave effects in the grounding outflow analysis.

The DnV study points out the relationship between tank arrangements and oil outflow. Although the DnV study did not directly address damage stability, oil outflow in an accident and damage stability are linked directly to particular tank arrangements. The results of the DnV analysis have two important implications, which are consistent with the committee's findings regarding damage stability discussed in Chapter 4.

First, for MARPOL SBT tankers (having excess freeboard and a large amount of empty ballast tank volume during loaded passage), providing improved oil outflow reduction under all plausible accident scenarios would require careful consideration of *both* ballast and cargo tank size and arrangements. While there is no question that the use of ballast tanks as protective spaces can reduce oil outflow, it also can have adverse consequences on stability. Similarly, protecting cargo tanks with ballast spaces outboard, or beneath them, does not mean that internal cargo tanks can be made larger than is common in single-skin ships without possible increase in oil outflow. This suggests that, for smaller ships, both damage stability assumptions and hypothetical oil outflow should be reconsidered by IMO and the Coast Guard.

Second, because all the analyses submitted to the committee, by DnV and others, were based on investigations of conventional tanker designs, it is difficult to predict how combination carriers (OBOs and ore/oil, or O/Os) would be affected under similar accident scenarios and outflow limitations. While it is clear that combination carriers, when carrying oil, must comply with all tanker regulations, the committee recognizes the following difficulties that may be encountered, particularly in OBOs:

- The very wide tanks in OBOs may lead to difficulty in hydrostatic loading, as OBOs cannot be sailed safely with slack tanks;
- The lack of longitudinal subdivision of OBOs also may lead to complications if hypothetical oil outflow criteria for ships smaller than VLCCs are adjusted downward;
- The narrow side tanks typical of OBOs also may lead to complications in complying with criteria applied to conventional tankers.

• The committee has not conducted studies of damage stability in OBOs but believes these ships must be able to comply with whatever criteria are deemed necessary for tankers;

• The committee questions whether hatch covers on OBOs and O/Os would be able to withstand forces imposed by a vacuum system, were that option to be applied.

While it might be argued that combination carriers need not comply with each specific requirement applied to tankers, the committee feels it would be wrong to waive regulations for combination carriers when this would result in unequal protection against oil outflow from accidents of any type. Such a loophole could encourage increased construction of combination carriers, to circumvent requirements designed to protect the environment.

NOTES

1. Tanker Advisory Center, Guide for the Selection of Tankers, 1990, New York.
2. Clarkson Research Studies Ltd., 1990.
3. Clarkson Research Studies Ltd., 1990.
4. The Coast Guard estimates that an 11.5' (B/14.4) double bottom on the EXXON VALDEZ would have reduced oil outflow by 60 percent at most, and by a minimum of 25 percent—still a significant figure (U.S. Coast Guard, Marine Safety Center internal memorandum, May 25, 1989).
5. The committee estimated retrofit costs based on information from shipyard sources. These costs are highly dependent on the base vessel.
6. Assuming that cargo tank lengths and the ship's beam are the same as in the equivalent deadweight MARPOL standard.
7. The committee was shown videotapes of small-scale tests using a plexiglass model of one IOTD w/DS proposal, in a simulated low-energy grounding. While the results were impressive some committee members remain skeptical about actual performance under a wide range of operating conditions. Of special concern is the dynamic condition of a vessel running aground at service speeds (14 to 16 knots).
8. On the average, it would be applicable to about 40 percent of accidents and 60 percent of outflow volume in U.S. waters—those events where major grounding damage and spillage (more than 30 tons) have been incurred (see Figure 1-11).
9. Additional parameters are detailed in Appendix F (pages 252 and 278).
10. Committee estimates for design combinations were derived using DnV data in the combined ranking values. The value in the estimates for collision performance for one design is added to the grounding performance for another alternative, producing a composite estimate for a case not shown directly by DnV. For example, to generate an estimate for double sides with hydrostatic control at 5 knots, the committee combined the performance in collision of the double-side case 3B with the grounding performance of the hydrostatic control case 9, to generate a combined ranking of about 32. A more complete explanation can be found in Appendix K.

REFERENCES

Clarkson Research Studies, Ltd. FAX to D. Perkins, National Research Council, Washington, D.C., August 31, 1990.

Det norske Veritas. 1990. Potential Oil Spill from Tankers in Case of Collision and/or Grounding: A comparative study of different VLCC designs. Report conducted for the Royal Norwegian Council for Scientific and Industrial Research, Oslo. DnV 90-0074.

Lloyd's Register of Shipping. 1989. Maritime Overseas Corporation 64,000 DWT Product Oil Carrier Sloshing Investigation. Report prepared for MOC, New York, October 1989. CSD 89/33.

Ministry of Transport-Japan. 1990. Prevention of Oil Pollution. Report prepared for IMO Marine Environment Protection Committee, received by Committee on Tank Vessel Design, NRC, Washington, D.C., November, 1990. Toyko.

Tanker Advisory Center. 1990. Guide for the Selection of Tankers. New York: TAC.

U.S. Coast Guard. 1989. Marine Safety Center internal memorandum, May 25, 1989.

U.S. Coast Guard. 1990. Navigation and Inspection Circular 2-90. Published by the Coast Guard, Washington, D.C., September 21, 1990.

6
Benefits and Costs of Design Changes

The preceding chapter narrowed the field of conceivable tank vessel designs to those that appear, from an engineering perspective, to warrant further consideration. This chapter identifies ways to address qualitative and quantitative information on the likely economic and environmental effects of these design alternatives. The data that are presented have many limitations, and much more work is needed on both benefits and costs of designs. Cost estimates are sensitive to assumptions about future traffic patterns as well as construction costs. Benefit estimates are even more elusive, depending on judgments about design effectiveness and spill consequences. Hence, the numerical findings should be taken as indicative at best.

THE ANALYTICAL APPROACH

If it were possible to place credible dollar values on all economic and environmental effects, a **monetized benefit-cost analysis** could be developed to show (1) how the present values of the alternatives' net benefits compare, and (2) how these comparisons would be affected by plausible changes in underlying assumptions (that is, by a sensitivity analysis). For two basic reasons, such precision is not possible. First, many of the major effects can be characterized only qualitatively, at best. Second, even where information is sufficient to allow quantitative estimates of the magnitude of a particular effect, the value of this effect in dollars is open to debate. For example, even if the number of miles of shoreline that would be spared from pollution by a particular design could be estimated, the dollar value related to this benefit is debatable.

Dollar values can be assigned, however, to most costs related to the various designs, and the resulting benefits can be expressed in non-monetary fashion, namely, as tons of oil *not* spilled. This permits **cost-effectiveness analysis**, which highlights the cost that would be incurred per unit of benefit achieved, that is, per ton of oil spillage likely averted. The fundamental challenge here is to quantify the reduction in spillage achievable with each design. One method for attempting to quantify the reduction is presented in the latter half of Chapter 5; the numbers will be summarized again in the appropriate section of this chapter. Again, the numbers derived from this exercise are not intended to be conclusive, but rather to provide indicative examples. Therefore, the results of the forthcoming cost-benefit analysis are by nature tentative, as they rely upon the use of numbers that could vary significantly given the application of different assumptions.

This economic analysis can be viewed from different angles, to gain additional perspective on the various design alternatives. For example, as benefit estimates are more conjectural than the cost estimates, the latter can be presented in a form of **"societal insurance" analysis**: that is, how much a design alternative will add to the delivered price of oil in exchange for some added (but unmeasurable) degree of protection against spills, much like an insurance premium. The ranking of designs that emerges, shown later in this chapter, still requires judgment about their relative effectiveness. But a societal insurance cost ranking can help (1) identify cost ranges, and (2) indicate designs with such similar costs that the choice should turn on other factors.

To put these costs in context, the committee gathered information on how society has valued past oil spills. Available data, while by no means comprehensive, will be presented to help in judging how "reasonable" the cost of the various designs may be. Also considered are the possible worldwide effects of design changes, in the form of dislocations in the oil transportation network.

Identifying the Base Case

Any analysis of benefits and costs of design changes must start with the identification of a base case, the situation likely to prevail if none of the contemplated changes are made. Each design alternative is intended to lessen the number and severity of oil spills; if effective, fewer of the casualties that occur will lead to oil spillage, and spills that do occur will be smaller.

The base case needed to evaluate design benefits, then, is some characterization of the spills that can be expected in the absence of mandated design changes.

Each design alternative also entails compliance costs, and, correspondingly, the base case needed for discussion of design costs is some characterization of existing tanker construction and operating costs. The base case for costs rests on an assumption that the oil tanker trade will be dominated for the foreseeable future by single-skin MARPOL vessels. Of course, the Oil Pollution Act of 1990 (OPA 90) requirement for double hulls, once implemented, will mean substantial changes in ship construction. Because such implementation has not yet occurred, for purposes of this analysis double hulls will be viewed as simply one of several design alternatives.

The base case for design benefits is drawn from data on accidents over the past decade (reviewed in Chapter 1). Of course, the future accident record cannot be predicted with certainty, but the historical record provides at least some indication. On one hand, historical oil spill data may overstate future risks, since some important risk-reduction measures have not yet been implemented fully. On the other hand, continued growth in tanker shipping of petroleum may increase the risk.

The volume of oil spilled annually from accidents larger than 10,000 gallons (or approximately 30 metric tons, for world spills) during the ten years 1980 through 1989 is shown in Table 6-1. To place these spills in perspective, they can be compared to total volumes shipped annually of roughly 600 millions tons in U.S. waters and 1,500 million tons worldwide.[1] These data represent the dimensions of the oil spill problem, or the base case for evaluation of design benefits.

Environmental Damages

The nature and extent of environmental damages arising from oil spills vary significantly, depending on the type of oil involved, the location and size of the spill, weather conditions during the spill, clean-up activities, and the efficacy of clean-up (National Research Council, 1985).

The assessment of environmental damage in economic terms remains a developing science. Historically, cost estimates have been based on clean-up and legal liability expenses. Less obvious costs related to restoration of natural resources, replacement of species, ecological damage, socioeconomic and other effects remain difficult to measure. While such costs historically have been excluded from estimates, there is a growing trend to include them.

The EXXON VALDEZ is likely to set a record as the most costly oil spill to date. Clean-up costs alone have exceeded $2 billion. The government *effort* to assess natural resource damages cost $35 million in the first year of an anticipated five-year effort.[2]

TABLE 6-1 Tons of Oil Spilled Annually,* 1980-1989—Spills over 10,000 Gallons or Approximately 30 Tons

	Best year	Average year	Worst year
In U.S. waters	300	9,000	40,000
Worldwide	5,000	100,000	360,000
Causes of spills in U.S. waters			
Groundings	0	5,750	37,000
Collisions	0	1,500	5,150
Explosions	0	650	6,600
Structural/other	0	800	3,250

*Data are rounded to the nearest 50 tons.

The previous record for the most costly spill was held by the AMOCO CADIZ, but this litigation continues and final damages are uncertain. In 1988, a decade after that spill, a U.S. court awarded damages of 720 million French francs (about $143 million), including interest (Gundlach, 1989). The AMOCO CADIZ litigation is further noteworthy for its length; 12 years after the incident occurred, and the appeals process has just begun.

Natural Resource Damage and Recovery

The major difficulty in assessing natural resource costs is that many of the more subtle effects of oil spills are neither well understood nor well documented. Short-term effects can be obvious; it is widely recognized that a major spill in a biologically rich area will result in the immediate deaths of marine mammals, seabirds, plankton, and fish. Other, chronic impacts presumably will persist for many years, affecting the reproductive success of some species as well as food sources.

Petroleum impact often has been judged based on either circumstantial or insufficient evidence, making unequivocal assessment of the effects of spills nearly impossible (National Research Council, 1985). However, the most important natural resource damages generally are placed in two categories: habitat loss or alteration, and loss of species. The ecological effects of oil on marine organisms are caused by either physical contamination (smothering or fouling) or chemical toxicity. For example, aromatic hydrocarbons are known carcinogens; they can cause cancerous lesions in the liver, lungs, and mammary glands of animals. Exposure to oil also causes reproductive failure in mammals and fish. Organisms throughout the marine food web, starting with the phytoplankton at the very bottom of the food chain, are subject to the effects.

Measuring the exposure of organisms and subsequent effects requires the expenditure of significant resources and time, and, with respect to certain aspects of the ecosystem, still may fail to reach conclusions of any certainty. At minimum, several years are needed to assess the full extent of environmental damage; the process may take decades. For example, 20 years after the spill of 16,000 tons of oil by the S.T. ARROW off Nova Scotia, much of the now-weathered oil still is believed to permeate shoreline strata. Whether this residual oil is biologically innocuous remains in contention. The local fishing industry reportedly rebounded within three years; however, populations of soft-shelled clams from oiled sites were still stressed six and seven years after the spill (National Research Council, 1985).

Recovery depends on population dynamics and ecological interactions of replacement species. Recovery in the water column appears to be very rapid, but oil and its residue may remain in sediment and shore substrate for many decades. Toxicity and malformation of fish eggs and larvae also have been observed over long periods of time. Repopulation by original species may be slow; near-shore benthic communities may not experience full recovery from a major crude oil spill for several years or more. Damages to wetlands may persist for decades if extensive erosion has taken place, or if there have been significant physical or biological effects.

In addition to the impact of oil itself on the environment, various spill clean-up methods may have adverse ecological effects.

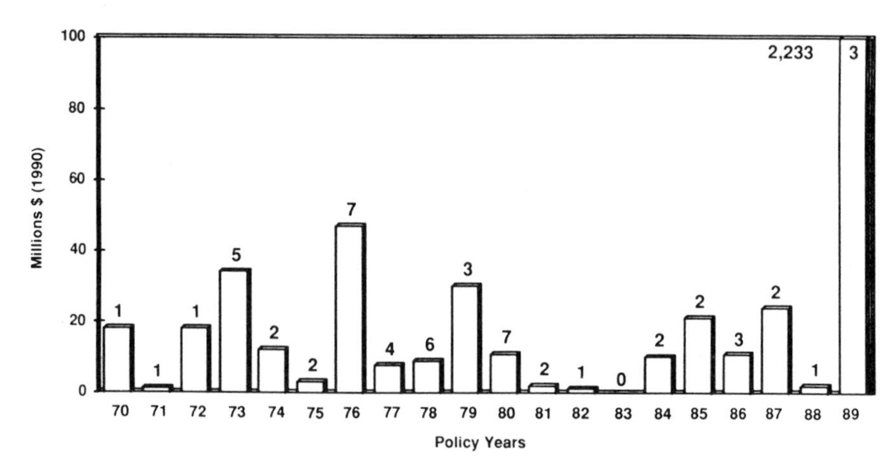

FIGURE 6-1 Major persistent oil spills by tankers—annual claims amounts. Source: Temple, Barker & Sloane, Inc. Note: The numbers of claims are shown above the columns; specific claim information is proprietary.

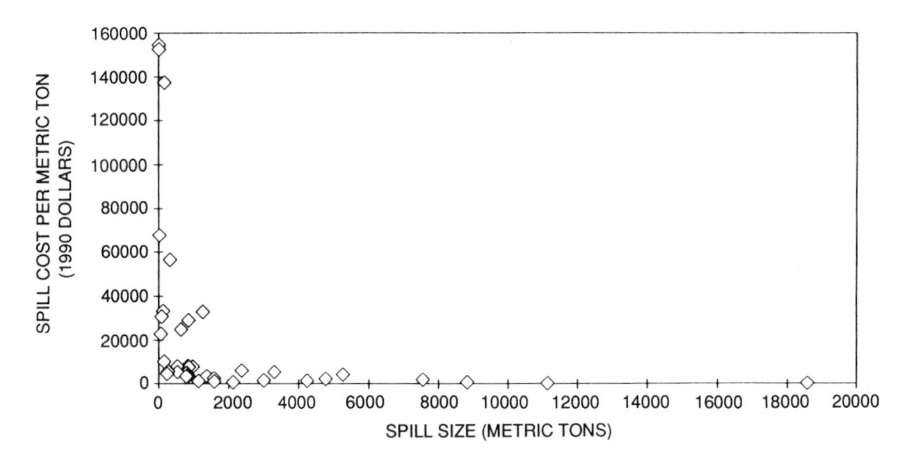

FIGURE 6-2 Major U.S. spills: claim estimates per metric ton versus spill size. Source: Temple, Barker & Sloane, Inc. [Analysis of a sample of 38 major persistent oil tanker spills (of a total of 64) in U.S. waters during the 1970-1980 period, resulting in clean-up plus damage and restoration costs in excess of $1 million each.]

Placing Dollar Values on the Cost of Spills

As noted earlier, an accurate translation of effects or damages from spills into monetary terms is a very complex task, if not impossible. Estimating the total cost is limited by uncertainties with respect to (1) effects on the marine environment, (2) validity of valuation techniques, and (3) the difficulty in valuing intrinsic benefits.

While not a complete representation of costs, legal claims payments are one example of how costs in practice are assigned to spills. Figures 6-1 and 6-2 show the claims pattern for major spills in U.S. waters over the past two decades. An analysis of 38 large spills (Temple, Barker & Sloane, Inc., unpublished data, 1990) found that, in 1990 dollars, claims have clustered at around $28,000 per ton of oil spilled; the startling exception, of course, is the EXXON VALDEZ case, in which costs could reach $90,000 per ton. It should be noted that the claims amounts paid are determined under international and domestic laws and, as such, reflect policy decisions and generally do not represent the full cost of a spill.

An example of covered costs is provided by the International Oil Pollution Compensation Fund (IOPC Fund), established in 1978 pursuant to a 1971 international convention. (The United States is not a party to this convention.) The fund generally covers claims for clean-up costs, consequential losses for owners or users of contaminated or damaged property, and claims for economic losses suffered by persons such as fishermen who depend directly on earnings from coastal or sea-related activities. The fund has not, however, accepted claims for damage to fishing grounds due to the after-effects of oil on fisheries, because the claimants have not been able to prove with sufficient data actual economic damage. Similarly, the fund does not cover claims for damage to the marine environment, unless the damage can be translated into direct market value.

The claims situation under federal law in the United States is governed by the OPA 90, which vastly increases spiller liability and penalties.[3] Claims under state law are governed by widely varying statutes of the several states, as well as common law. OPA 90 expressly does not preempt state laws in this regard. Prior to this legislation, claims for oil spill removal costs were made under Section 311 of the Federal Water Pollution Control Act (FWPCA), and claims for damages were made under general maritime law and common law. Regulations for determining damages to natural resources were promulgated by the Department of the Interior (DOI).

The DOI regulations favored estimating losses in "use values,"[4] as opposed to valuing injured or dead creatures or plants. The DOI excluded the cost of intrinsic values[5] (such as non-commercial marine mammals), unless there were no use values to measure. This approach, in effect, overlooked what may be a significant portion of total damage; a recent study by the American Petroleum Institute (API) found that "ignoring non-use (including intrinsic) values would lead to measures of damages that understate the true social losses." One empirical study cited by API suggests that intrinsic values comprise about half of the value of a resource (API, 1989).

In general, the DOI regulations took a more restrictive view of costs than is consistent with current economic practice. A recent decision by the U.S. Court of Appeals struck down the part of the DOI rule that generally limited natural resource recovery to "the use values" (State of Ohio et al. v. U.S. Department of the Interior, 880 F.2d 432. 1989). The Court found that Congress had established a distinct preference for restoration costs as a measure of recovery. The Court also found that lost use and non-use values (such as intrinsic values) generally must be compensated, in addition to restoration costs. The decision is expected to raise the costs of spills for liable parties.

As the foregoing implies, the benefits from design improvements have two primary components (apart from increased employee safety, for which estimates do not exist, and from the commercial value of the oil not lost). Limited data on each component exist.

Clean-up costs avoided. This is the reduction in spending for clean-up costs, including spending by both public and private sectors. Available estimates for spills prior to the EXXON VALDEZ put the average clean-up cost savings in the range of $1,500 to $38,000 per ton not spilled (adjusted to 1989 dollars).[6] The committee believes that past and present estimates of costs probably understate possible future costs of oil spill clean-up.

Environmental damages avoided. This is the decrease in environmental damages that either occur between the date of a spill and the date on which clean-up efforts cease, or continue after the completion of clean-up. Included within this category are commercial (especially fishing) and recreational benefits. While both "use values" and "non-use values" (existence, option, and bequest values) are relevant, measurement of these values is complex and difficult. Two studies (Cohen, 1986; Farrow, 1990), again for spills prior to EXXON VALDEZ, put the environmental benefits (other than clean-up savings) at $1,500 to $10,500 per ton (adjusted to 1989 dollars); the lower estimate comes from M. A. Cohen of Vanderbilt University, who did not attempt to take all environmental effects into account. Cohen postulates that a full accounting of environmental effects now would produce estimates in the range of $10,000 to $20,000 per ton, particularly in light of increasing public concern about environmental degradation and non-use values; indeed, one government analyst believes a $30,000 per ton upper-bound figure now would be credible. In addition, some speculation exists that the figure for the EXXON VALDEZ could range from $30,000 to $90,000 per ton spilled.[7]

As emphasized earlier, estimates of these benefits are quite difficult to pin down, even in one given situation after all the details are known. Even more speculative are judgments about decreases in environmental impact due to design changes. It is not possible to characterize in the abstract the environmental benefits from, say, reducing by half (compared to by 3/4) the size of spills that now are occurring. Similarly, there is no obvious consistent link between the amount of clean-up costs and the size of an oil spill. The development of a more reliable benefit-cost analysis requires further research to assess and value natural resource damages. Specifically, a better understanding of the environmental effects of oil spills and the feasibility and cost of restoration, as well as the development of an accepted methodology for valuing environmental benefits, is required. Therefore, at this time, cost-effectiveness analysis appears more feasible and promising than monetized benefit-cost analysis.

Spill Reduction Analysis

The extent to which various design alternatives would lessen the risk of oil spills cannot be known with certainty. However, as discussed in Chapter 5, the committee used one available methodology to assess the probable relative effectiveness of eight designs and design combinations. The resulting estimates of oil outflow reduction were shown in Tables 5-5 and 5-6. In each case, the reduction in oil spilled is relative to the base case of a tanker meeting current MARPOL requirements (with protectively located segregated ballast tanks, or PL/SBT). While the data are a plausible and carefully derived set of estimates, they are by no means definitive. Further work is needed to explore alternative assumptions about accident profiles.

TABLE 6-2 Oil Spill Performance

Design alternatives	Oil outflow relative to MARPOL tanker for composite accident**	
	Small tanker (40,000 DWT)	Large tanker (VLCC)
Double bottom	66 %	40 %
Double sides	133	109
Double hull	56	29
Hydrostatic control-passive	43*	51
Smaller tanks	72*	64
Int. oil-tight deck (IOTD) w/DS	47*	28*
Double sides w/hydro-passive	57*	27*
Double hull w/hydro-passive	50*	26*

*These figures represent committee judgments; the other figures were derived by DnV. See Tables 5-5 and 5-6.

**The composite accident is a weighted average of experience with collisions (40%) and with groundings (60%). The groundings are assumed to be evenly divided between low-energy (5 knots) and high-speed (10 knots) incidents.

The data from those two tables are merged in Table 6-2. To simplify the following analysis, this table assumes that half of the total tonnage of oil spilled from collisions and groundings occurs in low-energy accidents (5 knots) and half in high-energy accidents (10 knots). The numbers indicate, for each new design, how much oil is likely to spill in a standardized casualty, compared to the quantity spilled by a MARPOL ship. The assumption about accident energy levels is just that—an assumption—because data on actual casualty patterns do not exist.

The oil spill data from Table 6-1 now can be combined with the oil outflow reduction percentages from Table 6-2, to obtain an indication of the magnitudes of spillage that these design alternatives could avert. It should be stressed that data have many shortcomings, and the results must be viewed with skepticism.

The average amount of oil spilled annually in U.S. waters is roughly 9,000 tons of oil. However, there is a large variation from year to year, from a low of 300 tons to a high of 40,000 tons to date. This suggests the need for an analysis of a major spill year, in addition to a "typical year" analysis, at least for U.S. waters. The committee has not undertaken a "worst case" analysis, which would involve the total loss of cargo from one or more ships.

Several simplifying assumptions are made. The spill reduction data reflect only groundings and collisions; the committee does not have comparable data for explosions and structural accidents. Some of the design alternatives might have differing effects on, and perhaps even exacerbate, these other casualties. This caveat is important. However, in U.S. waters, over 80 per cent of oil spillage is due to groundings and collisions (65 percent due to groundings and 17 percent to collisions), so it seems worth proceeding. The analysis assumes that the listed design alternatives will have no net effect on spills from explosions or structural casualties—that only groundings and collisions will benefit.

The outflow reduction estimates assume the accident incident pattern is 60 percent groundings and 40 percent collisions, the ratio in the worldwide database used to develop the methodology. This world ratio of grounding to collision incidents (60 to 40) is approximately the same for major spills in U.S. waters; however; the ratio of spillage from groundings is closer to 80 and collisions 20, as noted in Table 6-1. Therefore, the following results unavoidably are skewed, giving somewhat greater weight to collision avoidance than appears warranted.

Finally, the designs and design combinations are assessed as potential improvements to a tanker fleet that is coming gradually into compliance with MARPOL requirements. Of course, much of the existing fleet predates the MARPOL convention and so does not have PL/SBT. To the extent that gradual replacement of pre-MARPOL tankers (even if no further design changes had been mandated) lessens future oil spills, these data overstate the problem. On the other hand, the data understate the problem to that extent that oil shipping increases in the future, and to the extent that MARPOL tankers increase risks associated with explosions and structural failures (all possibilities established earlier in this report).

The annual achievable reductions in oil spills are shown in Tables 6-3 and 6-4, for a typical year and for a major spill year, respectively. For each design, a spill-reduction range is indicated. The first number assumes that all accidents will occur in small tankers, and the second number assumes that all accidents will occur in large tankers.

These numbers indicate the extent of risk reduction delivered by the various designs. This is one aspect of risk assessment. Another key aspect is the acceptability of the residual risks, which turns in part on cost considerations, to be discussed in the following pages. (An additional factor is attitudes toward risk, discussed in general terms in Chapter 1.)

Cost Analysis: Economic Effects of Design Alternatives

A key task faced by the committee was to estimate the cost of pollution-resistant tanker designs. The economic effects can be measured as the increase in oil transportation costs due to changes in tanker capital and operating charges stemming from the structural design enhancements.

TABLE 6-3 Typical Year Performance

Design alternative	Oil spills averted in typical year
	Tons of oil saved (small tanker-large tanker)
Double bottom	2,600 - 4,500
Double sides	none-none
Double hull	3,300 - 5,300
Hydrostatic control-passive	4,300 - 3,700
Smaller tanks	2,100 - 2,700
Int. oil-tight deck w/DS	4,000 - 5,400
Double sides w/hydro	3,800 - 5,500
Double hull w/hydro	3,800 - 5,600

SOURCE: Tables 6-1 and 6-2. Rounded to nearest 100 tons. The typical year collision-grounding spill total is 7,500 tons. The first number in each pair shown assumes that all accidents will occur in small tankers, and the second assumes that all accidents will occur in large tankers.

Methodology

A simple microcomputer spreadsheet model was developed to help estimate these economic effects. The approach involves relating annualized vessel capital and operating costs to the annual volume of oil transported over a given trade lane. This results in an estimated unit transportation cost, such as $10 per ton of oil carried. Once the framework of the model was established for several sizes of "base case" single-hull tankers, variations were developed to reflect the cost impacts of the alternative designs.

The model groups tanker cost elements, with key assumptions as follows:

- Capital Cost. The acquisition cost of a newly built tanker is spread over 20 years at an assumed 12 percent cost of money, to arrive at an annualized capital charge or amortization amount.[8]
- Non-Voyage Operating Costs. These include all operating costs that do not vary with the particular voyage undertaken. These costs are for manning, stores, engine lubricants, insurance (both on the hull and engine, and for liability of all kinds), maintenance and repairs, and administration.
- Voyage Costs. These costs—essentially fuel and port charges—are developed on the basis of typical vessels' daily fuel consumption, speed, voyage distance, and port calls.

The sum of these three cost categories is the total cost for a tanker on the selected route over a year's time. Cargo delivered annually then is computed on the basis of voyage information. Total cost divided by oil volume transported yields the tanker transportation cost, measured in dollars per metric ton or cents per gallon.

TABLE 6-4 Major Spill Year Performance

Design alternative	Oil spills averted in major spill year
	Tons of oil saved (small tanker-large tanker)
Double bottom	13,600 - 24,000
Double sides	none-none
Double hull	17,600 - 28,400
Hydrostatic control-passive	22,800 - 19,600
Smaller tanks	11,200 - 14,400
Int. oil-tight deck w/DS	21,200 - 28,800
Double sides w/hydro	17,200 - 29,200
Double hull w/hydro	20,000 - 29,600

SOURCE: Tables 6-1 and 6-2. Rounded to nearest 100 tons. The largest spill year that occurred over the last decade is 40,000 tons. The first number in each pair shown assumes that all accidents will occur in small tankers, and the second assumes that all accidents will occur in large tankers.

The data used were chosen to realistically represent the costs of moving oil at the present time, using a newly built, foreign-flag tanker. Representative capital costs were drawn from a recent study by a major oil company and estimates developed by the committee. No attempt was made to optimize each design from a cost perspective. The committee also assumed there were no dimensional constraints on tanker design that would affect specific alternatives. Operating costs were based primarily on the results of a broad survey of tanker industry experience. In general, conservative and generally accepted values were employed.

Key assumptions that differentiate the alternative tanker designs are as follows:

- Capital costs are specified for each design. The designs are for equal cargo carrying capacity (deadweight tonnage), so those with greater ballast capacity may be somewhat larger in external dimensions or involve more internal steel, thus giving rise to differing initial costs.
- Maintenance and repair (M&R) costs are higher for alternative designs due to two reasons: (1) the higher capital cost will require M&R costs to be proportionally higher as well; and (2) the greater extent of (salt water) ballast tanks will mean increased corrosion, inspection, and steel replacement costs.
- Insurance will vary slightly in that hull and machinery (H&M) insurance is proportional to capital cost, while liability (P&I) insurance may decline slightly due to more favorable pollution experience with alternative designs.
- Fuel consumption is slightly higher for designs involving hydrostatic balance, due to the greater ship size involved.

TABLE 6-5 Oil Transport Costs, Per Ton

Size DWT × 10^3 Voyage nm	40 2,000	80 8,000	240 4,000
Alternative			
MARPOL ship	$4.33	$10.30	$3.65
Double bottom	4.77	11.31	4.06
Double sides	4.69	11.02	3.97
Double hull	5.01	11.87	4.26
MARPOL + hydro	5.34	12.71	4.51
Small tanks	4.74	11.25	4.01
IOTD w/DS	5.19	12.23	4.35
DS w/hydro	5.34	12.76	4.56
DH w/hydro	5.76	15.19	4.93

SOURCE: See calculations in Appendix G.

Other major operating factors, such as crew size, port costs, and load/discharge times, are assumed constant across design alternatives for each deadweight tonnage category. In fact, the operating cost elements subject to variation generally represent only 20 to 25 percent of total annualized cost. As the variations in these operating costs do not exceed 5 percent of total cost, it is fair to say that the cost differential associated with pollution-resistant tankers lies mainly in capital costs. The most expensive design costs an estimated 60 percent more to build than the base case MARPOL tanker.

Transport cost data were developed for the MARPOL tanker and eight alternative designs and design combinations. For each design, costs were developed for three vessel sizes. Table 6-5 shows that costs range from 7 to 47 percent more than the base case transport cost, depending on vessel size and alternative design. (Detailed calculations for this table and other calculations using the model are provided in Appendix G.)

From these data, the double hull with hydrostatic balance appears consistently to be the most costly solution. Double sides, double bottoms, and smaller tanks result in the lowest additional transport cost above the MARPOL design.

Vessel sizes and voyage distances then were combined to arrive at an overall indication of the transport cost implications of design alternatives. Assumptions were as follows:

U.S. seaborne oil increments of approximately 600 million tons per year were divided into three categories:

Imports:	350 m or 58 % of total
Coastal:	150 m or 25 %
Alaska:	100 m or 17 %
	100 %

Average vessel size and voyage lengths were assigned to each trade segment:

Imports:	80,000 DWT,	8,000 nautical miles round trip
Coastal:	40,000 DWT,	2,000 nautical miles round trip
Alaska:	240,000 DWT,	4,000 nautical miles round trip

RESULTS

Weighing the transport costs for each segment, and computing the results on the basis of 600 million tons of oil carried per year, generates a dollar figure for the increased transport cost associated with each design alternative. (See Table 6-6.) The incremental costs range from $340 to $2,050 million per year. The following conclusions can be drawn:

- Double sides, smaller tanks, and double bottoms are at the low end of the scale, in the $340 to $460 million range.
- The double hull with hydrostatic balance is most expensive, adding $2.05 billion to the U.S. annual oil transportation bill.
- Double hulls, double sides with either mid-height deck or hydrostatic balance, and MARPOL with hydrostatic balance are of intermediate expense, ranging from $710 to $1,100 million per year, or increments of 15 to 24 percent over the base case.

This analysis permits an estimate of the impact of design changes on consumers. The most expensive option, if universally adopted, would add less than two cents to the cost of a gallon of delivered crude oil. The impact on the price of gasoline at the pump might be roughly the same, although it could be more or less, depending on market factors. The moder-

TABLE 6-6 Incremental Transport Cost for Design Alternatives

Design alternative	Incremental Cost, in millions of dollars per year
MARPOL ship	$0
Double bottom	462
Double sides	339
Double hull	712
MARPOL w/hydro	1,080
Small tanks	430
IOTD w/DS	872
DS w/hydro	1,102
DH w/hydro	2,047

SOURCE: Table 6-5 and calculations in Appendix G.

ately expensive design options would add, at most, a cent to the cost of a gallon of delivered oil or gasoline.

These results should be viewed as indicative examples rather than as precise, due to several analytical limitations. First, foreign-flag vessel costs have been assumed throughout. This tends to underestimate the incremental capital and operating costs of U.S.-flag vessels. This assumption is one of convenience, based on the fact that no U.S.-flag tankers currently are under construction or on order. Second, the vessel sizes and route distances employed are representative only. A more precise analysis would utilize a more detailed mix of vessel sizes on a greater number of actual tanker routes. For example, the 4,000-mile round-trip distance assumed for all VLCCs underestimates the average round-trip distance for all VLCC traffic to the United States.

Third, the analysis deals with new construction only and ignores existing vessels, for which the transport economics are more complicated. Essentially, over the long term required to construct a new fleet, freight rates presumably will be based on economic costs. For design alternatives affecting the current fleet, such as a requirement for hydrostatic loading, the freight rate reaction may not be parallel to the cost change, for the following reasons. (1) If there is an oversupply of vessels, then shipowners may absorb some or all of the increased costs over the short run without passing them on in the form of higher freight rates. (2) If requirements result in a supply shortage, then transport charges may skyrocket, because transport demand is fixed in the short run. The history of tanker shipping is replete with short-term rate spikes that subside as supply and demand move into balance, with supply typically slightly exceeding demand. (3) Existing freight rates are based to a considerable degree on fully depreciated (older) vessels, so that base rates are lower than the economic values calculated here on the basis of new building costs.

In summary, the transport cost increases associated with each vessel design are indicative only. They relate to the long-term impacts of changes in new vessel construction and operation, and they are based on foreign-flag vessels. The economic effects of changes directed at the existing fleet have not been estimated; they could be either greater than or less than the new building impacts.

Cost-Effectiveness Analysis

At this point, the oil spill reduction data can be combined with the cost information to produce a cost-effectiveness analysis of design alternatives. This entails a comparison of the alternatives in terms of cost per ton of oil saved. The results are summarized in Table 6-7 and in more detail in Table 6-8.

TABLE 6-7 Summary of Cost-Effectiveness Results

	Shipping cost increase per ton of oil spill avoided in thousands of dollars	
	Typical year	Major spill year
Highest cost designs to lessen oil spills		
Double sides	(no oil saved)	
Double hull with hydrostatic control	$366-539	$69-102
Medium cost designs to lessen oil spills		
Double sides with hydrostatic control	200-344	38-64
MARPOL with hydrostatic control	250-292	47-55
Lowest cost designs to lessen oil spills		
IOTD with double sides	161-218	30-41
Double hull	134-216	25-40
Small tanks	159-205	30-38
Double bottom	103-178	19-34

SOURCE: Table 6-8.

In brief, the eight alternative designs and design combinations fall into three groups:

- The most costly ways to prevent oil spills are double sides, and double hulls with hydrostatic balance.
- Two alternatives could be described, in relative terms, as medium cost: Double sides with hydrostatic balance, and MARPOL ships using hydrostatic balance.
- The other four alternatives save oil in less costly fashion and cannot be distinguished on a cost basis: double hulls, small tanks, double sides with intermediate oil-tight deck, and double bottoms.

As shown earlier, in a typical year 7,500 tons of oil is spilled from groundings and collisions; costs for this case are shown in Column A of Table 6-8. Column B shows comparable costs for a year with 40,000 tons spilled, the major spill scenario. The first number in each column assumes all spills were from small tankers; the second number assumes all spills were from large tankers (VLCCs).

Selection of the most practical design—or even evaluation of any one design—depends on the yardstick chosen. Estimates of environmental/clean-up costs or claims data provide two such measures. If a design could save oil at a lower cost than either cleanup and environmental costs

TABLE 6-8 Added Transportation Costs Per Ton of Oil Saved

	Added yearly transport cost per ton of oil saved	
Design alternative	A Typical year*	B Year of major spills*
Double bottom	$178,000 - 103,000	$34,000 - 19,000
Double sides	(no oil saved)	(no oil saved)
Double hull	216,000 - 134,000	40,000 - 25,000
MARPOL w/hydro	251,000 - 292,000	47,000 - 55,000
Small tanks	205,000 - 159,000	38,000 - 30,000
IOTD w/DS	218,000 - 161,000	41,000 - 30,000
DS w/hydro	344,000 - 200,000	64,000 - 38,000
DH w/hydro	539,000 - 366,000	102,000 - 69,000

*The first number in each column assumes all spills are from small tankers; the second number assumes all spills are from large tankers (VLCCs). The incremental shipping costs shown in Table 6-6 for each design are divided by the amount of oil each design prevents from being spilled (shown in Tables 6-3 and 6-4). The smaller tanker number exceeds the large tanker number in all cases except MARPOL with hydrostatic control, which reflects the pattern of effectiveness results shown in Table 6-3.

or claims payments, this would constitute a strong argument for adoption of the design. However, data gathered by the committee suggest that none of the alternatives stand out as low-cost by these measures, if the future pattern of accident spills resembles that of the past. As noted previously, environmental and clean-up benefits from oil spill reductions, in the limited studies now available, range from $12,000 to $68,000 per ton saved. Alternatively, claims experience suggest figures of around $30,000 per ton saved, with the EXXON VALDEZ case possibly hitting $90,000 per ton.[9]

As noted previously, another way of viewing the cost of design alternatives is the societal insurance approach. For this, one merely divides the aggregate compliance costs by the quantity of delivered oil. In 1988-89, total oil imports (crude and refined) amounted to roughly 7.5 million barrels per day,[10] or 115 billion gallons per year. The most costly design—double hull plus hydrostatic balance—adds $2.047 billion to shipping costs, or 1.8 cents per gallon. The cost is lower if it is spread over the entire 600 million tons of oil moving annually through U.S. waters, and still lower for all the other design alternatives under consideration. This approach makes virtually every design appear quite inexpensive. Yet each of these alternatives also could be described, just as accurately, as creating hundreds of millions of dollars of extra shipping costs annually. Whether these costs are reasonable, given the related potential reduction in oil spills is a matter of opinion.

EXPERT JUDGMENT TECHNIQUE

Given the incomplete state of technical knowledge related to most of the design options, and the imprecise nature of the economic analysis, the committee was faced with the classic problem of decision making in the presence of uncertainty. Recognizing the deficiencies of the present study, the committee nevertheless reached general agreement on the conclusions and recommendations drawn from this chapter and the technical evaluations.

As a check on the consensus, the committee decided to employ a rating method to aggregate individual "expert judgments" on the various tank vessel design options. The rating method used a matrix to relate alternative designs and design combinations to important performance attributes.

Each committee member (expert) contributed a set of ratings, assigning a value for each attribute of each design. Each expert also assigned a weighting factor to each attribute, according to its judged importance. For each expert, attribute ratings were combined to estimate an overall rating for each design option. All of the judgments then were aggregated by arithmetic averaging, facilitated by the use of spreadsheet software. (The method and results are described in more detail in Appendix H.)

The rating method, it should be emphasized, was *not* a decision-making tool, but rather a means for confirming or clarifying consensus ideas. The principal conclusions to be derived from this exercise are as follows:

- None of the designs was judged to be clearly superior.
- The lowest-rated design was the reference (MARPOL) design, indicating that it is possible to improve upon this standard.
- The design options tended to cluster into three groupings. The most desirable overall were the double hull with hydrostatic control, the double hull, double sides with intermediate oil-tight deck, and double sides with hydrostatic control.
- The differences within each grouping were marginal. However, when applying the weighting factors, a slight preference for the two double-hull designs was evident. Of these, the double hull with hydrostatic control was viewed as the most desirable.

In sum, the rating method confirmed the consensus view that the current design standard (MARPOL) could be surpassed, that the double hull is an attractive alternative, and that the intermediate oil-tight deck with double sides holds promise.

POSSIBLE DISLOCATIONS TO WORLD SYSTEM OF OIL TRANSPORT

In recommending new regulations for tank vessel design, the effects of the new designs are compared with those of the existing vessel. Implicit in this comparison is the assumption that the new designs will be built or retrofitted throughout the fleet. This section considers factors that might invalidate such an assumption. These factors can be thought of as "legal loopholes": that is, means by which commercial interests can continue to operate legally while avoiding part or all of the expected cost of the new regulations—thus nullifying the corresponding benefits.

Changing from Oil to an Alternative Energy Source

A dramatic change would occur if new design regulations increased tank vessel costs so much that it became less expensive to use coal or natural gas as energy sources rather than oil. But such a change would require energy users to have equipment that could burn coal or natural gas, and heavy use of these fuels would involve a range of different impacts, environmental and otherwise. Moreover, for the design alternatives considered by the committee, the added transportation costs are less than one to two cents per gallon. This additional cost would not be sufficient to alter significantly the fuel balance between oil and other alternatives.

Diversion of Cargo from Tankers to Barges

If the difference in transportation costs of tankers and barges changed dramatically, cargo might be diverted to the latter, less expensive type of vessel. Again, there might be an environmental impact associated with such a change. But as long as any regulatory changes apply similarly to tankers and barges, there should not be significant diversion of cargo between these two types of vessels.

Trend to Less-regulated Vessel Sizes

If new regulations dramatically changed capital or operating costs, there might be an incentive to change to vessels just under the regulated size. For example, if vessels of 10,000 DWT or more required costly design improvements, operators might favor vessels of 9,999 DWT or smaller, which could have a less environmentally favorable design. However, given the economies of scale in vessel construction and operations, no major diversion of cargo to smaller-sized vessels should occur.

Lightering, LOOP, and Transshipment in a Foreign Port

To avoid complying with design regulations in OPA 90, until 2015, an oil importer can lighter more than 60 miles from shore or use the Louisiana Offshore Oil Terminal (LOOP) in the U.S. Gulf of Mexico. Or, an oil importer could send the oil to a Caribbean island, where it could be transshipped directly or indirectly (via a storage tank) to a small tanker, which could unload at a U.S. port. Smaller vessels entering U.S. ports, whether from lightering or transshipment operations, would be required to comply with the new law.

Lightering Outside of U.S. Jurisdiction

Rather than buy new vessels or retrofit to meet double-hull requirements, after 2015 an operator might choose to lighter from single-hull tankers outside U.S. waters. Such a pattern might subject lightering operations to the rougher weather of open seas, possibly increasing the risk of a spill.

Replacement of Jones Act Movements with Imports

In domestic trade, oil must be carried in U.S.-built vessels manned by U.S. citizens. These "Jones Act" vessels have higher costs than foreign vessels. Because design changes will increase the difference between U.S. and foreign costs, a possible result would be greater use of foreign-flag tankers to import product (rather than moving product within the United States). However, there is a much more significant factor already driving such a trend: U.S. refinery capacity. Because it is difficult to obtain a permit to build a new U.S. refinery, once existing capacity is fully utilized, additional demand for refined products will be met by imports. An additional factor is the depletion of Alaskan oil fields, and, if additional arctic oil reserves are not developed, the consequent reduced need for U.S.-flag tankers.

Creation of a Two-Tier Market

New U.S. design standards could bring about the creation of two distinct segments in the foreign-flag oil tanker market. This would occur if other major nations do not follow the lead of the United States. In this case, safer but more expensive vessels would serve the U.S. oil import trades, while traditional ships would continue to serve other trade lanes at lower freight rates.

The significance of this market segmentation would be two-fold. First, U.S. consumers would pay more for oil transportation (though perhaps less for spill cleanup) than would their foreign counterparts. Second, the operating flexibility of international tanker owners would be reduced, as traditional tankers would not be allowed in U.S. waters. Nearly 50 percent of the world tanker fleet now calls at U.S. ports in any given year (see Chapter 1). A two-tier market would create scheduling and other inefficiencies that would raise the cost of transportation for U.S. oil imports further.

If this scenario actually develops, the impact will not be significant for several years, until OPA 90 drives a meaningful number of vessels from U.S. waters. Any future regulatory actions should take into account the possibility of exacerbating the potential two-tier market effect.

Extending the Life of Existing Tank Vessels

Owners of older tank vessels are reluctant to spend money on replacements if the old vessels can be altered to provide adequate service at minor investment. To the extent that new regulations would increase the cost difference between buying new vessels and maintaining older ones, owners would be more inclined to extend the life of the older vessels. A key issue is the ability of regulatory agencies and classification societies to keep unsafe vessels out of U.S. waters.

The tendency of owners to extend the life of older vessels requires continuous monitoring by regulatory agencies and classification societies. Further regulatory actions should take into account the effect on this situation. A multi-faceted program to improve pollution control for existing vessels should be evaluated; the program should guard against encouraging retention of older vessels, although vessels that are well maintained and operated should not be prejudged based solely on age.

SUMMARY

The practical merit of the eight design alternatives analyzed depends on the contribution each could make in reducing oil spills, relative to the impact on oil transportation costs. This chapter illustrates the type of analysis that should be undertaken by policymakers as part of the decision-making process. The numerical findings reported here are less important than the nature of the analysis; reasonable people differ in their assessment of the probable oil outflow from each design, and, similarly, projections of actual cost effects are a matter of judgment.

This chapter identifies ways to characterize available evidence on both outflow and costs from a variety of perspectives. Further research is needed to explore the sensitivity of the numerical results to assumptions about benefits and costs of the designs.

One perspective focuses on oil transport costs. Based on the limited data available to the committee, none of the alternatives would add more than two cents to the cost of transporting each gallon of oil; in the aggregate, however, this would increase annual U.S. oil transport costs from $170 million to about $2.0 billion. A double hull with hydrostatic balance would be by far the most costly of the eight alternatives. Double sides, smaller tanks, and double bottoms appear to be the lowest cost alternatives.

A second perspective is that of cost-effectiveness. Each design can be characterized by combining relevant transport cost estimates with available data on oil spillage the design could avert. Substantial uncertainty surrounds estimates of the latter; this chapter relies on one set of results based on certain assumptions, and the results are at best indicative. Different assumptions would produce different results; the committee encourages more extensive investigations of these effects.

In the absence of other risk reduction measures, and assuming that the historical record of accidental oil spills is indicative of what may lie ahead, a double-hulled fleet might save 3,300 to 5,300 tons of spillage in U.S. waters in an average year. The double hull is among the four most cost-effective designs; the others are smaller tanks, double bottoms, and the intermediate oil-tight deck with double sides.

Whether the cost of implementing any new design is money well spent depends on the yardstick chosen for comparison. The actual costs of oil spills, including natural resource damages, would provide a logical basis for judgment, but available data on such damages are even less adequate than estimates of design costs. The major difficulty is that the effects of oil spills are neither well understood nor well documented. In addition, translating effects or damages into monetary terms is limited by uncertainties concerning the validity of valuation techniques, and the difficulty in valuing intrinsic benefits. At the same time, as the EXXON VALDEZ experience suggests, the cost of cleanup and natural resource damages is increasing dramatically. Thus, the reasonableness of costs remains a matter of opinion.

The uncertainties associated with estimates both of costs and, even more so, the benefits associated with design changes, led the committee to explore an additional approach to appraising the evidence. Judgmental rankings on a variety of design features were gathered from committee members, and the results indicated considerably more favorable assessment of four designs (double hull, intermediate oil-tight deck with double sides, and hydrostatic balance with either double hull or double sides) than of the other designs; no single design appeared superior in all respects. However, the rating method confirmed that it should be possible to improve upon the current design standard (MARPOL), that the double hull is an attractive alternative, and that the intermediate oil-tight deck with double sides potentially holds promise.

Additional factors that warrant concern are the possibilities that unilateral U.S. requirements could create a two-tier tanker market, and that requirements focusing on new vessels only could result in extended life for older vessels with adverse environmental consequences.

The available data suggest that the best of the design changes examined could prevent perhaps half of accident-related oil spills, which in sum comprise about one-fifth of maritime oil pollution. That is, some 10 percent of the total oil pollution problem might be eliminated by such a design change. This would increase transport costs by some $700 million annually, which translates into one or two cents, or less, per gallon. How reasonable this cost may be, even for the best design change, cannot be determined in the absence of some judgment about its effectiveness and cost relative to alternative risk-reduction measures, such as improvements in operations or enforcement, which are beyond the purview of the present study.

NOTES

1. The total quantity of oil moving in world waters in 1989 was 1,478 million metric tons, including 1,176 million tons crude and 392 million tons product. The total quantity of oil moving in U.S. waters in 1989 was 585 million tons, including 403 million tons crude and 182 million tons product. The total figures for 1989 are about the same as they were in 1980; shipments were somewhat lower during the intervening years but never fell below 450 million tons. [Source: Temple, Barker and Sloane, Inc., and U.S. Army Corps of Engineers (coastal petroleum movement).]
2. Charles Ehler, National Oceanic and Atmospheric Administration, and Eric Olsen, National Wildlife Foundation, presentations to the committee, June 6, 1990.
3. For tank vessels, liability limits under OPA 90 are eight times higher than previous limits under FWPCA, and now stand at $1,200 per gross ton. Limits do not apply if the spill resulted from gross negligence or willful misconduct of the responsible party (vessel owner, operator, or demise charterer), violation of applicable federal safety, construction, or operating regulations by the responsible party, or his employee, agent, or contractor, or if he (or his agent) fails to report the spill, cooperate in its removal, or comply with an order under Section 311 of FWPCA.
4. Gains to those whose livelihoods and recreation depend on actual use of the resource at risk.
5. Intrinsic benefits include such values as the loss of non-commercial biomass or non-commercially valued marine mammals or seabirds, as well as the value of the potential that species (especially those yet undiscovered) may be found useful for human consumption as food, medicine, genetic material, or other raw material.
6. Sources: The International Maritime Organization (1988, chapter 11) reports an average of $4,000 per ton, with a range of $71 to $21,000 per ton, in 1985 dollars. R. S. Farrow (1990, p. 69, Table 4-2) reports DOI estimates of roughly $2,000 per ton, in 1987 dollars. M. A. Cohen (1986, p. 185) put the average at roughly $1,100 per ton, in 1981 dollars. Somewhat lower estimates are reported by J. E. Mielke (1990). A range of $5 to $30,000 per ton (1983 dollars) was reported by I. C. White and J. A. Nichols (1983), and cleanup costs for the EXXON VALDEZ have been put as high as $60,000.
7. Farrow reports DOI estimates of roughly $9,600 per ton, in 1987 dollars. Cohen put the average at roughly $900 per ton, in 1981 dollars (1986, p. 185). His current view was obtained by telephone on January 10, 1991. The government analyst's view was obtained by telephone on October 23, 1990; his named cannot be cited because of pending litigation. Some believe the environmental damages figure for the EXXON VALDEZ could substantially exceed this range.
8. Estimates vary, especially for the intermediate oil-tight deck with double sides (IOTD w/DS). Values in Appendix G, Table 6-5, reflect capital cost of roughly 17 percent over base for a double hull, but 22 percent over base for the IOTD w/DS. A definitive study, however, of the IOTD w/DS showed that the estimated capital cost differential over base

is the same, or slightly less, than for double-hull designs (Mitsubishi Heavy Industries, Ltd., 1990).

9. Of course, if spills on the scale of the EXXON VALDEZ were to occur every year, then any of the designs would appear to be attractive investments, particularly in light of speculation about that accident's cleanup costs of $60,000 per ton and environmental damages of $30,000 or more per ton.
10. Statistical Abstract of the United States, 1990, page 571.

REFERENCES

American Petroleum Institute. 1989. Measuring Natural Resource Damages: An Economic Appraisal. API Publication 4490, Health and Environmental Services Department, Washington, D.C.

Cohen, M. A. 1986. The Costs and Benefits of Oil Spill Prevention and Enforcement. Journal of Environmental Economics and Management 13(2):167-188.

Farrow, R. S. 1990. Managing the Outer Continental Shelf Lands: Oceans of Controversy. New York: Taylor & Francis.

Gundlach, E. R. 1989. Amoco Cadiz Litigation: Summary of the 1988 Court Decision. Proc. 1989 Oil Spill Conference (Prevention, Behavior, Control, Cleanup). Washington, D.C.: API. pp. 503-508.

International Maritime Organization. 1988. Manual on Oil Pollution, Section IV, Combating Oil Spills. London: IMO.

Mielke, J. E. 1990. Oil in the Ocean: The Short- and Long-Term Impacts of a Spill. Report prepared by Congressional Research Service, Library of Congress, Washington, D.C. 90-356 SPR.

Mitsubishi Heavy Industries, Ltd. 1990. Mid-Deck Tanker. Design study received by the Committee on Tank Vessel Design, NRC, Washington, D.C., November 1990. Tokyo.

National Research Council. 1985. Oil in the Sea. Washington, D.C.: National Academy Press.

Temple, Barker & Sloane, Inc. 1990. Analysis of a Sample of 38 Major Persistent Oil Spills. Internal study conducted by TBS, Lexington, Massachusetts.

White, I. C. and J. A. Nichols. 1983. The Cost of Oil Spills. Proc. 1983 Oil Spill Conference (Prevention, Behavior, Control, Cleanup). Washington, D.C.: API. pp. 541-544.

U.S. Bureau of the Census. 1990. Statistical Abstract of the United States. Washington, D.C.: U.S. Department of Commerce.

7

The Need for Research

As noted repeatedly in Chapters 5 and 6, the committee was unable to identify one preferred tank vessel design or even to define specific standards for how a vessel should perform in an accident. The difficulty was not due to an overly cautious approach to the problem, or a lack of willingness on the part of the committee, but rather to an insufficient knowledge and technology base to resolve the problem. During its deliberations, the committee posed numerous questions for which answers seemed inadequate, or lacking entirely. Based on these and related discussions, literature reviews, committee studies, and assessment of the needs of the maritime community and society as a whole, the committee identified several areas where research is needed. These will be detailed in the following pages.

First, it should be noted that a growing number of research projects on improving tank vessel design and operations already are under way or planned. Any new research should take these projects into account.

RESEARCH COMPLETED, UNDER WAY, OR PLANNED

The committee is aware of studies and research efforts, mostly in other countries, that clearly are the result of the EXXON VALDEZ accident and the U.S. government's interest in legislation aimed at improving tanker design, which culminated in the Oil Pollution Act of 1990, enacted on August 18, 1990. These include:

• A study on oil outflow from collisions and groundings was completed in May 1990 under the joint auspices of the Norwegian Maritime Directorate, the Norwegian Shipowners Association, and Det norske Veritas. The study was sponsored by the Royal Norwegian Council for Scientific and Industrial Research (NTNF). A report of this work prompted the committee to commission additional work by DnV extending the basic work done by NTNF. The Norwegian studies assessed oil outflow due to collisions and groundings in ships of various designs; they did not evaluate other accidents or consequences such as fire, explosion, structural failure, or personnel safety.

• A study focusing on three designs is being conducted by Committee RR761, formed by the Japanese government. Membership includes ship owners, shipbuilders, academicians, and government representatives. The study has concentrated on the evaluation of double hulls (particularly grounding outflow), the vacuum system for minimizing grounding outflow, and the intermediate oil-tight deck as a means of improving grounding and collision protection. Results of the studies were presented to the committee and were helpful in drawing conclusions and recommendations. Like the Norwegian study, the work of Committee RR761 addressed only oil outflow from groundings and, secondarily, collisions. It did not estimate effects from fire, explosions, structural failure, safety considerations, etc.

• Studies by ship owners. Various ship owners' associations have conducted international studies of all aspects of oil pollution from tankers (Lloyd's Register of Shipping for the American Petroleum Institute, 1990; International Chamber of Shipping and the Oil Companies International Marine Forum, 1990; the International Association of Independent Tanker Owners, 1990a). The results have been presented to the committee.

Projects Planned

A common theme in the reports just described has been a recommendation for prompt, in-depth research into improved tanker design. The committee is aware of several ongoing or planned research projects seeking to meet that need, as follows.

• The Royal Norwegian Council for Scientific and Industrial Research has announced an in-depth study to build on the work of its 1990 study. The new study, to continue through 1994 with a budget of around $28 million, will seek to design a ship that reduces air pollution, sea pollution from operations, and sea pollution from accidents. The study will seek: (1) a reduction in probability of accidents through maneuvering and navigation measures; (2) a reduction in the consequences of accidents through hull design and safety systems; and (3) improved handling of oil spill cleanup.

• A seven-year program is planned by the Japanese government. This project has a budget of about $20 million to investigate new tanker designs that will minimize oil spills from collisions and groundings, and a budget of about $15 million to study the purification of exhaust gases from ships.

In addition to these comprehensive, government-supported efforts, numerous other, smaller programs have been initiated or are under development.

A NEW FOCUS FOR RESEARCH

As noted in Chapter 2, naval architects traditionally have not designed tank vessels, at the detail level, to withstand groundings and collisions. While the technology base is adequate and there are internationally recognized standards for the design of vessels to assure their integrity during normal operating conditions, current design practice does not address in-process accident behavior, which aircraft and automotive industries refer to as "crashworthiness." The state of the art in understanding fundamental forces and structural reactions during tank vessel accidents is limited, particularly for groundings. Similarly, although computer models exist for estimating structural damage in particular scenarios, these models have not been validated with actual accident data or full-scale testing.

To better understand actual vessel performance, and to move in the direction of establishing performance standards, more basic knowledge is required. "Performance standards" have meaning only if there is a mechanism to test various proposed designs against these standards; there must be a way to quantitatively predict damage to any design, and the vessel's post-accident performance.

Present analytical methods rely on simplified assumptions and limited numbers of primitive control parameters, local to the site of impact. Consequently, damage projections are but gross approximations, lacking both the detail accuracy required to quantify results beyond directional ranking, and the sensitivity to reflect any but the most major changes in assumptions and/or control factors.

Establishing Performance Standards

The committee has taken a strong position that performance rather than design criteria should be developed for a spill-free tanker. Design criteria tend to "fix" technology at a point in time, thus inhibiting innovation and removing the incentive to advance ship technology and design. Performance standards tend to promote new development in terms of structural and operational innovations that will result in meeting or surpassing the standards.

The significance of performance standards can best be explained using the example of the automotive industry. Safety standards were adopted in the United States in the early 1970s. All passenger cars currently are required not to exceed 20g deceleration (g is 9.81 m/sec^2 at the Earth's surface) in a 30 mile-per-hour head-on collision with a rigid barrier. There are also other requirements concerning, for example, intrusion of the steering wheel into the passenger compartment. Each auto manufacturer took a slightly different approach to meeting these standards, and the highly competitive market resulted in many innovative and ingenious approaches to crash-worthiness. While the performance standards for cars are checked and enforced through mandatory compliance tests, no such full-scale tests are possible for tank vessels. Therefore, an alternative, purely analytical method should be developed to accept or reject a given tank vessel design.

Because full-scale testing of tankers is prohibitively expensive, it is more practical to establish relative performance standards with respect to a chosen design, rather than absolute criteria. Such a reference vessel could be, for example, a conventional MARPOL tanker (the reference for the present study), or a suitable double-hull ship (the new standard for new tankships traveling in U.S. waters).

In the automobile industry, establishment of worldwide performance criteria was preceded by a decade of research and development by regulatory agencies and the industry. Interested parties within the maritime community could launch a similar research program. One focus of such a program would be to establish equivalence criteria for choosing alternative designs, based, for example, on the amount of cargo lost. The effectiveness of alternative designs could be examined in a wide range of collision and grounding scenarios. Additional small-scale and possibly full-scale static and dynamic tests would be necessary to calibrate various theoretical models and to obtain the scaling laws.

Specific Needs

Greater Understanding of Structural Behavior

Tank vessels vary significantly in structure, even within similar design types. Differences in basic structural arrangements, scantlings, and detail, both individually and cumulatively, will result in different failure sequences and post-accident vessel conditions. Experience proves this, but present technology cannot project it beyond gross generalities. Basic theoretical research is needed into relevant material behavior leading up to micro-element failure, and progressing through major structural energy dissipation into ship structures. Models of detail sequence behavior must be enlarged (1) to encompass the full variety of ship structures, (2) to reflect the global changes during the accident process, and (3) to account for the cargo-structure-water interface phenomena as the vessel's integrity is assaulted.

At present, dynamic progressions are defined through rudimentary static study of single-step movement from the intact state to local detail failure. This is an inadequate and misleading approach, if there is to be any attempt to meet dynamic performance standards of integrated structures. The path of structural failure follows the movement of destructive force as it stresses contiguous local structure into plastic deformation, first rupture and, ultimately, failure. As the dissipating force moves into peripheral structure, the process is repeated, moving like cracks in splitting sheet ice, seeking out the weakest path of resistance before dissipating in non-destructive stressing of steel. The projection of path and extent of destruction remain quantitative unknowns.

To achieve an understanding of structural behavior that is adequate to development of vessel performance standards, specific needs must be met, as follows:

- Improved analytical techniques are required to understand the hull rupture initiation process, a key controlling factor in whether cargo will be spilled. Present analyses use simplified assumptions and limited numbers of parameters. In properly designed hulls, rupture may not be inevitable; to that end, further research also should be directed toward developing innovative hull materials and structural configurations.
- Improved capabilities are needed for predicting the vertical, lateral, and horizontal extent of damage sustained to the ship bottom (of both single and double hulls). Current methods use simplified assumptions. Improved methods also would help determine outflow from breached tanks, determine stability of damaged ships, and plan salvage of stranded ships.

To make design tools available, a development program would need to integrate theory, modeling, structural testing, and verification with historical accident data. Other than the last aspect, all of these are within the ability of academia and industry. Unfortunately, verification is not possible with existing databases and is not part of industry practice.

More Detailed Casualty Databases

The preceding discussion points out another significant need. The committee found existing casualty databases incomplete and/or misleading: Of the dozens of accidents referenced in the study, not one is publicly documented to serve as an adequate resource for the most general accident description.

Records lack the detail documentation (vessel speed or description of grounding obstacle, for example) that would help relate a particular scenario to the exact damage done. Such information either is considered proprietary (frequently for legal reasons), or is recorded in an oversimplified form, or not at all. In addition, there is seldom any database link between general accident statistics (such as the number or volume of spills) and the supporting details (vessel description, initial cause of accident, extent of damage in structural terms). An accident analysis typically must rely on manual search through incomplete records from numerous sources (such as the Coast Guard, shipbuilders, operators, classification societies, and marine casualty organizations). Therefore, better procedures are needed for inspecting newly damaged ships, including detailed determination of collapse/tearing of structure and full photographic coverage. This type of inspection, if mandatory, would provide an engineering and statistical data bank that could enhance understanding of the phenomena involved and lead to improved designs.

Research leading to establishing a mandated damage-assessment protocol should be undertaken; complete and accurate cause-and-effect data would enable development of an accurate vessel performance model. This research would require the cooperation of engineers, ship repair facilities, ship owners and operators, and legal experts.

Research on the Influence of Tank Contents

Ship structural behavior following an accident is influenced by cargo and ballast status. Tank contents affect not only total vessel mass, but also the performance of the containment structures as they process damaging forces. The effect may be either to inhibit or to exacerbate structural failure—and resulting oil outflow. The interaction of tank contents with the sequence of structural failure is neither understood nor taken seriously into account; the issue is no more than an aside in post-accident investigations. Research into the impact phenomena of tank contents (or lack thereof) will be required to quantify post-accident vessel status accurately.

The computational models to be developed for predicted potential oil outflow from a damaged ship should link kinetic energy, structural configuration, and the extent of damage. The next step is to determine the relationship between extent of damage, relative hydrostatic pressure, tank size, and oil outflow. A first step has been taken (the Det norske Veritas method used in Chapter 5), using a very simplified model for describing the ship in conjunction with a probabilistic approach. A more accurate and complete analysis, based on past accidents in combination with a simplified description of probable events, may be appropriate.

In addition, more research is needed to understand the dynamic mixing of sea and cargo in a damaged cargo tank following an accident. Forces contributing to this process, which can lead to greater cargo outflow than explained by hydrostatic pressure, are poorly understood.

Models for Residual Strength

Calculation of the reserve strength of partially damaged hulls is an essential aspect of managing groundings, as the ship must be unloaded, salvaged, and brought to a repair dock. Procedures for evaluating progressive collapse of a hull girder, subjected to a combination of still water and wave loads, need further development and application. Such models have been developed for offshore oil platforms. Some non-linear finite element programs can be used, but present codes need further refinement if they are to be practically applied and more widely used.

Full-Scale Test to Destruction

A carefully planned and fully instrumented full-scale grounding test would contribute substantially to understanding of tank vessel structural response under traumatic stress. Each year a number of tankers reach the end of their useful life cycle. Possibly one or more of such tankers could be converted into experimental ships and subjected to a controlled grounding.

An analogous crash test of a fully instrumented Boeing 707 was sponsored by the Federal Aviation Agency with industrial involvement. The knowledge gained from the "test to destruction" of an instrumented surplus tanker would advance all of the structural research areas that have been identified.

Summary of Research Needs

The technology base must be enhanced across the spectrum, from exploratory research to proof-of-concept development.

The result of the research projects described should be a reliable mechanism that can, irrespective of a vessel's details, accurately project structural and cargo behavior as a function of vessel design, in any selected accident scenario. Whether the accident is a grounding or collision, the model must perform within the envelope of the given operating environment.

The needs include: (1) an integrated micro-understanding of the dynamics of ship structural failure, and related factors; (2) long-range research in failure theory (interactive behavior of over-stressed integrated structures in association with hydrodynamic loads); (3) protocols leading to mandatory engineering documentation of casualties; and (4) computational models resulting in outflow predictions.

RECOMMENDED U.S. RESEARCH STRATEGY

The U.S. government is sponsoring a variety of research efforts related to different aspects of oil spill prevention and cleanup. The present study is one example. Some research will be funded under the Oil Pollution Act of 1990. However, at the present time, no comprehensive multi-year plan has been announced that would have significant impact on ship design focused to reduce or prevent oil outflow during accidents.

The government should cooperate in a comprehensive, multi-year research and development program that would result in computational models for predicting vessel structural response during an accident and consequential cargo outflow. These are the engineering tools basic to development of vessel performance standards.

The program should (1) define the documentation, procedures, and protocols that would remedy the absence of quantified engineering casualty data, (2) ensure adequate theoretical knowledge and application technology to design tank vessels to meet performance standards, and (3) achieve optimal pollution control by integrating use of design alternatives with operational and cleanup options.

The scope of such a program would require the cooperation of government, the engineering and computer science communities, legal experts, shipbuilders, shipowners, and classification societies. The project should be coordinated with foreign efforts through the IMO.

REFERENCES

Det norske Veritas. 1990. Potential Oil Spill from Tankers in Case of Collision and/or Grounding: A comparative study of different VLCC designs. Report conducted for the Royal Norwegian Council for Scientific and Industrial Research, Oslo. DnV 90-0074.

International Association of Independent Tanker Owners. 1990a. Measures to Prevent Accidental Pollution. Oslo, Norway: The Association.

International Chamber of Shipping, Oil Companies International Marine Forum, and International Association of Independent Tanker Owners. 1990. Oil Tanker Design and Pollution Prevention. Study prepared for the Committee on Tank Vessel Design, Washington, D.C.

Lloyd's Register of Shipping. 1990. Statistical Study of Outflow from Oil and Chemical Tanker Casualties. Report conducted for American Petroleum Institute, Washington, D.C. Technical Report STD R2-0590.

Summary of References

American Petroleum Institute. 1989. Measuring Natural Resource Damages: An Economic Appraisal. Prepared by Health and Environmental Sciences Dept., API, Washington, D.C. API Publ. 4490.

Card, J. C. 1975. Effectiveness of Double Bottoms in Preventing Oil Outflow from Bottom Damage Incidents. Marine Technology 12(1):60-64.

Chevron Shipping Company. 1990. Double Hull Tanker Design. Paper prepared for Society of Naval Architects and Marine Engineers convention, sent to Committee on Tank Vessel Design, NRC, Washington, D.C., October, 1990. San Francisco.

Cohen, M. A. 1986. The Costs and Benefits of Oil Spill Prevention and Enforcement. Journal of Environmental Economics and Management. 13(2):167-188.

Cutter Information Corp. 1990. Oil Spill Intelligence Report. Newsletter published by Cutter, Arlington, Massachusetts, June 14, 1990.

Det norske Veritas. 1990. Potential Oil Spill from Tankers in Case of Collision and/or Grounding: A comparative study of different VLCC designs. Report conducted for the Royal Norwegian Council for Scientific and Industrial Research, Oslo. DnV 90-0074.

Energy Information Administration. 1989. Petroleum Supply Annual. Washington, D.C.: Government Printing Office.

Exxon Corp. 1982. Large Tanker Structural Survey Experience. Paper published by Exxon, New York.

Farrow, R. S. 1990. Managing the Outer Continental Shelf Lands: Oceans of Controversy. New York: Taylor & Francis.

Ferguson, J. M. 1990. Structural Integrity and Life Expectancy. Paper published by Lloyd's Register of Shipping, London.

Gray, W. O. 1979. Requirements for Inert Gas Systems. Paper presented at IMCO Tokyo Seminar on Tanker Safety and Pollution Prevention, Tokyo, February 19-23, 1979.

Gundlach, E. R. 1989. Amoco Cadiz Litigation: Summary of the 1988 Court Decision. Proc. 1989 Oil Spill Conference (Prevention, Behavior, Control, Cleanup). Washington, D.C.: API. pp. 503-508.

Iarossi, F. (president, American Bureau of Shipping). 1990. Getting Ready for a Shortening Life. Fairplay International September 20:37.

International Association of Independent Tanker Owners. 1990a. Measures to Prevent Accidental Pollution. Oslo, Norway: the Association.

International Association of Independent Tanker Owners. 1990b. Quest for the Environmental Ship. Draft of paper to be published by the Association, Oslo, Norway.

International Chamber of Shipping. 1978. Ship to Ship Transfer Guide. London: Witherby.

International Chamber of Shipping, Oil Companies International Marine Forum, and International Association of Independent Tanker Owners. 1990. Oil Tanker Design and Pollution Prevention. Study prepared for the Committee on Tank Vessel Design, NRC, Washington, D.C.

International Maritime Organization. 1988. Manual on Oil Pollution, Section IV, Combating Oil Spills. London: IMO.

International Maritime Organization. 1989. Analysis of Serious Casualties to Sea-going Tankers, 6,000 gross tonnage and above, 1974-1988. London:IMO.

Lloyd's Register of Shipping. 1989. Maritime Overseas Corporation 64,000 DWT Product Oil Carrier Sloshing Investigation. Report prepared for MOC, New York, October, 1989. CSD 89/33.

Lloyd's Register of Shipping. 1990. Statistical Study of Outflow from Oil and Chemical Tanker Casualties. Report conducted for the American Petroleum Institute, Washington, D.C. Technical Report STD R2-0590.

Mielke, J.E. 1990. CRS Report for Congress, Oil in the Ocean: The Short- and Long-Term Impacts of a Spill. Report prepared by Congressional Research Service, Library of Congress, Washington, D.C., July 24, 1990. 90-356 SPR.

Ministry of Transport-Japan. 1990. Prevention of Oil Pollution. Report prepared for IMO Marine Environment Protection Committee, received by Committee on Tank Vessel Design, NRC, Washington, D.C., November, 1990. Toyko.

Minorsky, V.U. 1959. An Analysis of Ship Collisions with Reference to Protection of Nuclear Power Plants. Journal of Ship Research 3:1-4.

Mitsubishi Heavy Industries, Ltd. 1990. Mid-Deck Tanker. Design study received by the Committee on Tank Vessel Design, NRC, Washington, D.C., November 1990. Tokyo.

National Research Council. 1975. Petroleum in the Marine Environment. Report based on a workshop held by the Ocean Affairs Board, Airlie, Virginia, May 21-25, 1973.

National Research Council. 1981. Reducing Tankbarge Pollution. Washington, D.C.: National Academy Press.

National Research Council. 1985. Oil in the Sea. Washington, D.C.: National Academy Press.

Ponce, P. 1990. An Analysis of Total Losses Worldwide and for Selected Flags. Marine Technology 27(2):114-116.

Royal Institute of Naval Architects. 1990. The Naval Architect. September:E357,E381,E385.

Society of Naval Architects and Marine Engineers. 1980. Ship Design and Construction. New York: SNAME.

Tanker Advisory Center. 1990. Guide for the Selection of Tankers. New York: TAC.

U.S. Bureau of the Census. 1990. Statistical Abstract of the United States. Washington, D.C.: U.S. Department of Commerce.

U.S. Coast Guard. 1973. Note by the United States—Report on Study I Segregated Ballast Tanker. Prepared for IMCO International Conference on Marine Pollution, London, October 8-November 2, 1973.

U.S. Coast Guard. 1989a. Navigation and Inspection Circular 10-82. Published by the Coast Guard, Washington, D.C., September 18, 1989.

U. S. Coast Guard. 1990a. Navigation and Inspection Circular 2-90. Published by the Coast Guard, Washington, D.C., September 21, 1990.

U. S. Coast Guard. 1990b. Report of the Tanker Safety Study Group, Chairman H. H. Bell (rear admiral, USCG, retired). Washington, D.C.: U. S. Department of Transportation.

U.S. Coast Guard. 1990c. Report on the Trans-Alaska Pipeline Service (TAPS): Tanker Structural Failure Study (draft). Washington, D.C.: U. S. Department of Transportation.

U. S. Coast Guard. 1990d. Update of Inputs of Petroleum Hydrocarbons Into the Oceans Due to Marine Transporation Activities. Paper submitted to IMO Marine Environment Protection Committee 30, September 17, 1990.

U.S. Maritime Administration. 1990. Foreign Flag Merchant Ships Owned by U.S. Parent Companies. Report prepared by Office of Trade Analysis and Insurance, Washington, D.C., January 1, 1990.

Vaughan, H. 1977. Damage to Ships Due to Collision and Grounding. Paper published by Det norske Veritas, Oslo, Norway, August 1977. DnV 77-345.

White, I.C. and J.A Nichols. 1983. The Cost of Oil Spills. Proc. 1983 Oil Spill Conference (Prevention, Behavior, Control, Cleanup). Washington, D.C.: API. pp. 541-544.

Other References—Unpublished and Internal

American Waterways Operators. Letter report to the Committee on Tank Vessel Design, NRC, Washington, D.C., April 6, 1990.

Clarkson Research Studies, Ltd. FAX to D. Perkins, National Research Council, Washington, D.C., August 31, 1990.

Jones, N. 1990. Some Comments on the Collision Protection of Ships with Double Hulls. Paper presented at meeting of the Committee on Tank Vessel Design, NRC, Washington, D.C., June 6, 1990.

Liu, D. 1990. Letter to the Committee on Tank Vessel Design, NRC, Washington, D.C., May 10, 1990.

Liu, D. Structural Design of Tankers. Paper presented to meeting of the Committee on Tank Vessel Design, NRC, Washington, D.C., March 26, 1990.

Temple, Barker & Sloane, Inc. 1990. Analysis of a Sample of 38 Major Persistent Oil Spills. Internal study conducted by TBS, Lexington, Massachusetts.

U. S. Coast Guard. 1989b. Marine Safety Center internal memorandum, May 25, 1989.

U. S. Coast Guard. 1990e. Assessment of Success of Tankships with Double Bottoms and PL/SBT in Mitigating Pollution Due to Casualties. Internal analysis by the Coast Guard, Washington, D.C., March 12, 1990.

Wierzbicki, T., E. Rady, and J.G. Shin. 1990. Damage Estimates in High Energy Grounding of Ships. Paper presented at meeting of the Committee on Tank Vessel Design, NRC, Washington, D.C., June 6-7, 1990.

Additional Background Material

Alaska Oil Spill Commission. 1990. Spill: The Wreck of the Exxon Valdez, Implications for Safe Marine Transportation. Anchorage: The Commission.

American Institute of Merchant Shipping. 1974. Tanker Double Bottoms: Yes or No? Washington, D.C.: The Institute.

American Petroleum Institute. 1988. Oil Spill Studies: Measurement of Environmental Effects and Recovery. Guidelines based on API workshops Feb. 23-25, 1982 and May 19-20, 1983. Prepared by Health and Environmental Sciences Dept., API, Washington, D.C.

American Petroleum Institute. 1989. Task Force Report on Oil Spills. Washington, D.C.: API.

Beaumont, J. G. 1990. Ship Construction and Safety (with particular reference to Oil Tankers). Paper published by Lloyd's Register of Shipping, London, March, 1990.

Björkman, A. 1990. Letter to D. Perkins, National Research Council, Washington, D.C., August 23, 1990. Data comparing Coulombi Egg hull design to other designs.

Caiafa, C. A., and L. M. Neri. 1986. The Controlled Impact Demonstration (CID Structural Experiments, Philosophy, and Results). Paper presented at the annual winter meeting of the American Society of Mechanical Engineers, Anaheim, California, December 7-12, 1986.

Czimmek, D. W., and C. R. Jordan. 1981. Optimization of Segregated Ballast Distribution and its Impact on Tanker Economics. Marine Technology 18(2):127-148.

Electric Power Research Institute. 1986. Seismic Hazard Methodology for the Central and Eastern United States, Vol. I: Methodology. Palo Alto, California: EPRI. NP-4726.

Energy Information Administration. Annual Energy Outlook 1990. Washington, D.C.: U.S. Department of Energy.

Finnish Tanker Safety Committee. 1987. Safety Requirements for Tankers. Report to the Ministry of Trade and Industry, Helsinki.

Gray, W. O. 1979. Accidental Spills from Tankers and Other Vessels. In The Prevention of Oil Pollution. London: Graham and Trotman. pp. 79-114.

Herbert Engineering Corp. 1990. VLCC Light-Loading Damage Stability Calculation. Report prepared for the American Petroleum Institute, Washington, D.C. HEC File 9004.

Hussain, M. 1990. An Active Inert Gas Controlled Method to Reduce Spillage Resulting From Hull Rupture in Oil Tankers. Paper prepared for API Forum on Alternative Tank Vessel Design, Washington, D.C., June 5, 1990.

Inter-governmental Maritime Consultative Organization. 1977. Tanker Safety and Tanker Inspection and Certification. Presentation to IMCO Maritime Safety Committee by the United States of America, April 12, 1977.

International Chamber of Shipping and Oil Companies International Marine Forum. 1986. Guidance Manual for the Inspection and Condition Assessment of Tanker Structures. Prepared on behalf of Tanker Structure Co-operative Forum. London: Witherby.

International Maritime Organization. 1988. Fire Protection Aspects Related to Vacuum Systems to Minimize The Effect of Pollution by Oil After Damage. Paper submitted to IMO Fire Protection Subcommittee by Sweden, December 22, 1988.

International Maritime Organization. 1990. Prevention of Oil Pollution. Paper submitted to IMO Marine Environment Protection Committee by Denmark, Finland, Iceland, Norway, and Sweden, February 9, 1990.

International Oil Pollution Compensation Fund. 1989. Annual Report. London: IOPC.

Jacobsson, M. 1989. The International Oil Pollution Compensation Fund: Ten Years of Claims Settlement Experience. Proc. 1989 Oil Spill Conference (Prevention, Behavior, Control, Cleanup). Washington, D.C.: API. pp. 509-511.

Keeney, R. I., and D. Von Winterfeldt. 1988. Probabilites are Useful to Quantify Expert Judgments. Reliability Engineering and System Safety 23(4):293-298.

Keith, V. F., and J. D. Porricelli. 1990. A Double Hull Oil Tanker Without a Loss of Cargo Carrying Capacity. Paper prepared for API Forum on Alternative Tank Vessel Design, Washington, D.C., June 5, 1990.

Kimon, P. M., R. K. Kiss, and J. D. Porricelli. 1973. Segregated Ballast VLCCs: An Economic and Pollution Abatement Analysis. Marine Technology 10(4):334-363.

Kopp, R. J., and V. K. Smith. 1989. Benefit Estimation Goes to Court: The Case of Natural Resource Damage Assessments. Journal of Policy Analysis and Management 8(4):593-612.

Leighou, R. B. 1942. Chemistry of Engineering Materials. New York: McGraw-Hill.

McKenzie, A. 1990. Double Hulls and the Prevention of Oil Spills. Paper prepared for API Forum on Alternative Tank Vessel Design, Washington, D.C., June 5, 1990.

McNatt, T. R., and O. F. Hughes. 1990. Tank Vessel Structural Design Using the Computer Program Maestro. Paper prepared for API Forum on Alternative Tank Vessel Design, Washington, D.C., June 5, 1990.

McNeely, J. A. 1988. Economics and Biological Diversity: Developing and Using Economic Incentives to Conserve Biological Resources. Gland, Switzerland: International Union for the Conservation of Nature and Natural Resources.

Nalder, E. 1989. Tankers Full of Trouble: Crude Deliveries Push Safety to the Limit. Reprinted from The Seattle Times, November 12-17, 1989.

National Research Council. 1990. Crew Size and Maritime Safety. Washington, D.C.: National Academy Press.

National Resources Defense Council. 1990. No Safe Harbor: Tanker Safety in America's Ports. New York: NRDC.

National Wildlife Federation. 1990. The Day the Water Died: A compilation of the November 1989 Citizens Commission Hearings on the Exxon Valdez Oil Spill. Washington, D.C.: NWF.

Noble, P. G. Safe Transport of Oil at Sea. Paper prepared for API Forum on Alternative Tank Vessel Design, Washington, D.C., June 5, 1990.

Norwegian Shipowners Association. 1989. Report on Accidental Pollution from Crude Oil Tankers, Measures for Prevention and Abatement. Oslo, Norway: The Association.

Ohyagi, M. 1987. Statistical Survey on Wear of Ship's Structural Members. Technical Bulletin of Nippon Kaiji Kyokai 5:75-85.

Okamoto, T., T. Hori, M. Tateishi, R. Masaru, S. M. H. Rashed, and S. Miwa. 1985. Strength Evaluation of Novel Unidirectional-Girder-System Product Oil Carrier by Reliability Analysis. Trans. SNAME 93:55-78.

Organisation for Economic Co-operation and Development. 1989. Environmental Policy Benefits: Monetary Valuation. Paris: OECD.

Ortiz, N. R., T. A. Wheeler, R. L. Keeney, and M. A. Meyer. 1989. Use of Expert Judgment in NUREG-1150. Paper presented at American Nuclear Society/European Nuclear Society International Topical Meeting on Probability, Reliability, and Safety Assessment, Pittsburgh, Pennsylvania, April 1989.

Porricelli, J. D., V. F. Keith, and R. L. Storch. 1971. Tankers and the Ecology. Trans SNAME 79:169-221.

Skinner, S. K., and W. K. Reilly. 1989. The Exxon Valdez Oil Spill, a report to the President. Prepared by the National Response Team. Washington, D.C.: U.S. Department of Transportation and U.S. Environmental Protection Agency.

Smith, V. K. 1987. Nonuse Values in Benefit Cost Analysis. Southern Economic Journal 54(1):19-26.

Stopford, M. 1990. Forecasts for the International Shipbuilding Market: Demand, Pricing, and Capacity. Paper prepared for Shipbuilders Council of America seminar, Washington, D.C., January 31, 1990.

Tanker Structure Co-operative Forum. 1990. Effect of Structural Design Aspects on Mitigating Pollution, draft. Work Group Project prepared with participation of the American Bureau of Shipping, Paramus, New Jersey.

Temple, Barker & Sloane, Inc. 1985. Summary of Analysis of Oceangoing Tank Barge Navigation Incidents in Inland Waters. Prepared for Sonat Marine Inc., Philadelphia.

Townsend, R. 1990. Shipping Safety and America's Coasts. Report prepared by Townsend Environmental for the Center for Marine Conservation, Washington, D.C.

Tornay, E. G. 1990. Limiting Cargo Loss from Damaged Tankers. Paper presented at Ship Operations, Management, and Economics Symposium, Kings Point, New York, April 16, 1990.

U.S. Coast Guard. 1974. Double Hull Effectiveness Analysis. Washington, D.C.: U.S. Department of Transportation.

U.S. Coast Guard. 1978. Tanker Safety and Pollution Prevention—Requirements for U.S. Tankers in Domestic Trade. Washington, D.C.: U.S. Department of Transportation.

U.S. Coast Guard. 1980. Commercial Vessel Safety: Economic Benefits—Appendix A, Estimating Procedures for Benefits of Marine Safety Regulations. Washington, D.C.: U.S. Departent of Transportation. DOT-CG-351-A.

U.S. Coast Guard. 1986, 1989. Marine Safety Manual, Volume I—Administration and Management (M16000.6), Volume II—Material Inspection (M16000.7), and Volume IV—Technical (COMDTINST m16000.9). Washington, D.C.: U.S. Government Printing Office.

U.S. Coast Guard. 1989. Development and Assessment of Measures to Reduce Accidental Oil Outflow from Tank Ships. Washington, D.C.: U.S. Department of Transportation.

U.S. Coast Guard. 1989. Polluting Incidents In and Around U.S. Waters. Washington, D.C.: U.S. Department of Transportation. COMDTINST M16450.2H

U.S. Congress, Senate. 1977. Recent Tanker Accidents: Legislation for Improve Tanker Safety. Hearings before the Committee on Commerce, Science, and Transportation, Washington, D.C., March 8, 15, 16, and 18, 1977. Washington, D.C.: U.S. Government Printing Office.

U.S. Congress, Senate. 1977. The Tanker and Vessel Safety Act of 1977: Report of the Committee on Commerce, Society and Transportation. Washington, D.C.: U.S. Government Printing Office.

U.S. Department of Commerce. 1977. Tanker Pollution Abatement Report: A Study of Tanker Construction Design, Equipment, and Operating Features Related to Improved Pollution Abatement. Washington, D.C.: Maritime Administration. MA-SC7302-78012.

U.S. Department of Commerce. 1983. Assessing the Social Costs of Oil Spills: the Amoco Cadiz Case Study. Rockville, Maryland: National Ocean Service. NOAA-83092912.

U.S. Environmental Protection Agency. 1990. Reducing Risk: Setting Priorities and Strategies for Environmental Protection. Prepared by Science Advisory Board, Washington, D.C. SAB-EC-90-021.

U.S. General Accounting Office. 1990. Federal Costs Resulting from the Exxon Valdez Oil Spill. Washington, D.C.: GAO. GAO/RCED-90-91FS.

U.S. Minerals Management Service. 1990. Offshore Oil Terminals: Potential Role in U.S. Petroleum Distribution, draft. Washington, D.C.: U.S. Department of the Interior.

U.S. Office of Technology Assessment. 1978. An Analysis of Oil Tanker Casualties 1969-1974. Prepared for the U.S. Senate Committee on Commerce, Science, and Transportation and the National Ocean Policy Study. Washington, D.C.: U.S. Government Printing Office.

APPENDIXES

APPENDIX

A

MARPOL 73/78 Annex I*

*Source: International Maritime Organization (excerpted from U.S. Coast Guard Commandant Instruction M16455, August 30, 1985).

APPENDIX A*

MARPOL 73/78 Annex I (excerpt)

CHAPTER III — REQUIREMENTS FOR MINIMIZING OIL POLLUTION FROM OIL TANKERS DUE TO SIDE AND BOTTOM DAMAGES

Regulation 22

Damage Assumptions

(1) For the purpose of calculating hypothetical oil outflow from oil tankers, three dimensions of the extent of damage of a parallelepiped on the side and bottom of the ship are assumed as follows. In the case of bottom damages two conditions are set forth to be applied individually to the stated portions of the oil tanker.

(a) *Side damage*

(i)	Longitudinal extent (ℓ_c):	$\frac{1}{3}L^{\frac{2}{3}}$ or 14.5 metres, whichever is less
(ii)	Transverse extent (t_c): (inboard from the ship's side at right angles to the centreline at the level corresponding to the assigned summer freeboard)	$\frac{B}{5}$ or 11.5 metres, whichever is less
(iii)	Vertical extent (v_c):	from the base line upwards without limit

(b) *Bottom damage*

		For 0.3L from the forward perpendicular of the ship	Any other part of the ship
(i)	Longitudinal extent (ℓ_s):	$\frac{L}{10}$	$\frac{L}{10}$ or 5 metres, whichever is less
(ii)	Transverse extent (t_s):	$\frac{B}{6}$ or 10 metres, whichever is less but not less than 5 metres	5 metres
(iii)	Vertical extent from the base line (v_s):	$\frac{B}{15}$ or 6 metres, whichever is less	

(2) Wherever the symbols given in this Regulation appear in this Chapter, they have the meaning as defined in this Regulation.

Regulation 23

Hypothetical Outflow of Oil

(1) The hypothetical outflow of oil in the case of side damage (O_c) and bottom damage (O_s) shall be calculated by the following formulae with respect to compartments breached by damage to all conceivable locations along the length of the ship to the extent as defined in Regulation 22 of this Annex.

(a) for side damages:

$$O_c = \Sigma W_i + \Sigma K_i C_i \qquad (I)$$

(b) for bottom damages:

$$O_s = \tfrac{1}{3}(\Sigma Z_i W_i + \Sigma Z_i C_i) \qquad (II)$$

where: W_i = volume of a wing tank in cubic metres assumed to be breached by the damage as specified in Regulation 22 of this Annex; W_i for a segregated ballast tank may be taken equal to zero,

C_i = volume of a centre tank in cubic metres assumed to be breached by the damage as specified in Regulation 22 of this Annex; C_i for a segregated ballast tank may be taken equal to zero.

$K_i = 1 - \frac{b_i}{t_c}$ when b_i is equal to or greater than t_c, K_i shall be taken equal to zero,

$Z_i = 1 - \frac{h_i}{v_s}$ when h_i is equal to or greater than v_s, Z_i shall be taken equal to zero,

b_i = width of wing tank in metres under consideration measured inboard from the ship's side at right angles to the centreline at the level corresponding to the assigned summer freeboard,

h_i = minimum depth of the double bottom in metres under consideration; where no double bottom is fitted h_i shall be taken equal to zero.

Whenever symbols given in this paragraph appear in this Chapter, they have the meaning as defined in this Regulation.

(2) If a void space or segregated ballast tank of a length less than ℓ_c as defined in Regulation 22 of this Annex is located between wing oil tanks, O_c in formula (I) may be calculated on the basis of volume W_i being the actual volume of one such tank (where they are of equal capacity) or the smaller of the two tanks (if they differ in capacity) adjacent to such space, multiplied by S_i as defined below and taking for all other wing tanks involved in such a collision the value of the actual full volume.

$$S_i = 1 - \frac{\ell_i}{\ell_c}$$

where ℓ_i = length in metres of void space or segregated ballast tank under consideration.

(3) (a) Credit shall only be given in respect of double bottom tanks which are either empty or carrying clean water when cargo is carried in the tanks above.

(b) Where the double bottom does not extend for the full length and width of the tank involved, the double bottom is considered non-existent and the volume of the tanks above the area of the bottom damage shall be included in formula (II) even if the tank is not considered breached because of the installation of such a partial double bottom.

(c) Suction wells may be neglected in the determination of the value h_i provided such wells are not excessive in area and extend below the tank for a minimum distance and in no case more than half the height of the double bottom. If the depth of such a well exceeds half the height of the double bottom, h_i shall be taken equal to the double bottom height minus the well height.

Piping serving such wells if installed within the double bottom shall be fitted with valves or other closing arrangements located at the point of connexion to the tank served to prevent oil outflow in the event of damage to the piping. Such piping shall be installed as high from the bottom shell as possible. These valves shall be kept closed at sea at any time when the tank contains oil cargo, except that they may be opened only for cargo transfer needed for the purpose of trimming of the ship.

(4) In the case where bottom damage simultaneously involves four centre tanks, the value of O_s may be calculated according to the formula

$$O_s = \frac{1}{4}(\Sigma Z_i W_i + \Sigma Z_i C_i) \qquad \text{(III)}$$

(5) An Administration may credit as reducing oil outflow in case of bottom damage, an installed cargo transfer system having an emergency high suction in each cargo oil tank, capable of transferring from a breached tank or tanks to segregated ballast tanks or to available cargo tankage if it can be assured that such tanks will have sufficient ullage. Credit for such a system would be governed by ability to transfer in two hours of operation oil equal to one half of the largest of the breached tanks involved and by availability of equivalent receiving capacity in ballast or cargo tanks. The credit shall be confined to permitting calculation of O_s according to formula (III). The pipes for such suctions shall be installed at least at a height not less than the vertical extent of the bottom damage v_s. The Administration shall supply the Organization with the information concerning the arrangements accepted by it, for circulation to other Parties to the Convention.

Regulation 24

Limitation of Size and Arrangement of Cargo Tanks

(1) Every new oil tanker shall comply with the provisions of this Regulation. Every existing oil tanker shall be required, within two years after the date of entry into force of the present Convention, to comply with the provisions of this Regulation if such a tanker falls into either of the following categories:

(a) a tanker, the delivery of which is after 1 January 1977; or

(b) a tanker to which both the following conditions apply:

(i) delivery is not later than 1 January 1977; and

(ii) the building contract is placed after 1 January 1974, or in cases where no building contract has previously been placed, the keel is laid or the tanker is at a similar stage of construction after 30 June 1974.

(2) Cargo tanks of oil tankers shall be of such size and arrangements that the hypothetical outflow O_c or O_s calculated in accordance with the provisions of Regulation 23 of this Annex anywhere in the length of the ship does not exceed 30,000 cubic metres or $400\sqrt[3]{DW}$, whichever is the greater, but subject to a maximum of 40,000 cubic metres.

(3) The volume of any one wing cargo oil tank of an oil tanker shall not exceed seventy-five per cent of the limits of the hypothetical oil outflow referred to in paragraph (2) of this Regulation. The volume of any one centre cargo oil tank shall not exceed 50,000 cubic metres. However, in segregated ballast oil tankers as defined in Regulation 13 of this Annex, the permitted volume of a wing cargo oil tank situated between two segregated ballast tanks, each exceeding ℓ_c in length, may be increased to the maximum limit of hypothetical oil outflow provided that the width of the wing tanks exceeds t_c.

(4) The length of each cargo tank shall not exceed 10 metres or one of the following values, whichever is the greater:

(a) where no longitudinal bulkhead is provided:

$0.1L$

(b) where a longitudinal bulkhead is provided at the centreline only:

$0.15L$

(c) Where two or more longitudinal bulkheads are provided:

(i) for wing tanks:

$0.2L$

(ii) for centre tanks:

(1) if $\frac{b_i}{B}$ is equal to or greater than $\frac{1}{5}$:

$0.2L$

(2) if $\frac{b_i}{B}$ is less than $\frac{1}{5}$:

— where no centreline longitudinal bulkhead is provided:

$$\left(0.5\frac{b_i}{B} + 0.1\right)L$$

— where a centreline longitudinal bulkhead is provided:

$$\left(0.25\frac{b_i}{B} + 0.15\right)L$$

(5) In order not to exceed the volume limits established by paragraphs (2), (3) and (4) of this Regulation and irrespective of the accepted type of cargo transfer system installed, when such system interconnects two or more cargo tanks, valves or other similar closing devices shall be provided for separating the tanks from each other. These valves or devices shall be closed when the tanker is at sea.

(6) Lines of piping which run through cargo tanks in a position less than t_c from the ship's side or less than v_c from the ship's bottom shall be fitted with valves or similar closing devices at the point at which they open into any cargo tank. These valves shall be kept closed at sea at any time when the tanks contain cargo oil, except that they may be opened only for cargo transfer needed for the purpose of trimming of the ship.

Regulation 25

Subdivision and Stability

(1) Every new oil tanker shall comply with the subdivision and damage stability criteria as specified in paragraph (3) of this Regulation, after the assumed side or bottom damage as specified in paragraph (2) of this Regulation, for any operating draught reflecting actual partial or full load conditions consistent with trim and strength of the ship as well as specific gravities of the cargo. Such damage shall be applied to all conceivable locations along the length of the ship as follows:

(a) in tankers of more than 225 metres in length, anywhere in the ship's length;

(b) in tankers of more than 150 metres, but not exceeding 225 metres in length, anywhere in the ship's length except involving either after or forward bulkhead bounding the machinery space located aft. The machinery space shall be treated as a single floodable compartment;

(c) in tankers not exceeding 150 metres in length, anywhere in the ship's length between adjacent transverse bulkheads with the exception of the machinery space. For tankers of 100 metres or less in length where all requirements of paragraph (3) of this Regulation cannot be fulfilled without materially impairing the operational qualities of the ship, Administrations may allow relaxations from these requirements.

Ballast conditions where the tanker is not carrying oil in cargo tanks excluding any oil residue, shall not be considered.

†(2) The following provisions regarding the extent and the character of the assumed damage shall apply:

(a) Side damage

(i) Longitudinal extent — $\frac{1}{3}(L^{\frac{2}{3}})$ or 14.5 metres, whichever is less

(ii)	Transverse extent (Inboard from the ship's side at right angles to the centreline at the level of the summer load line)	$\frac{B}{5}$ or 11.5 metres, whichever is less
(iii)	Vertical extent	From the moulded line of the bottom shell plating at centreline, upwards without limit

(b)	Bottom damage		For 0.3L from the forward perpendicular of the ship	Any other part of the ship
	(i)	Longitudinal extent	$\frac{1}{3}(L^{\frac{2}{3}})$ or 14.5 metres, whichever is less	$\frac{1}{3}(L^{\frac{2}{3}})$ or 5 metres, whichever is less
	(ii)	Transverse extent	$\frac{B}{6}$ or 10 metres, whichever is less	$\frac{B}{6}$ or 5 metres, whichever is less
	(iii)	Vertical extent	$\frac{B}{15}$ or 6 metres, whichever is less, measured from the moulded line of the bottom shell plating at centreline	$\frac{B}{15}$ or 6 metres, whichever is less, measured from the moulded line of the bottom shell plating at centreline

(c) If any damage of a lesser extent than the maximum extent of damage specified in sub-paragraphs (a) and (b) of this paragraph would result in a more severe condition, such damage shall be considered.

(d) Where the damage involving transverse bulkheads is envisaged as specified in sub-paragraphs (1)(a) and (b) of this Regulation, transverse watertight bulkheads shall be spaced at least at a distance equal to the longitudinal extent of assumed damage specified in sub-paragraph (a) of this paragraph in order to be considered effective. Where transverse bulkheads are spaced at a lesser distance, one or more of these bulkheads within such extent of damage shall be assumed as non-existent for the purpose of determining flooded compartments.

(e) Where the damage between adjacent transverse watertight bulkheads is envisaged as specified in sub-paragraph (1)(c) of this Regulation, no main transverse bulkhead or a transverse bulkhead bounding side tanks or double bottom tanks shall be assumed damaged, unless:

(i) the spacing of the adjacent bulkheads is less than the longitudinal extent of assumed damage specified in sub-paragraph (a) of this paragraph; or

(ii) there is a step or a recess in a transverse bulkhead of more than 3.05 metres in length, located within the extent of penetration of assumed damage. The step formed by the after peak bulkhead and after peak tank top shall not be regarded as a step for the purpose of this Regulation.

(f) If pipes, ducts or tunnels are situated within the assumed extent of damage, arrangements shall be made so that progressive flooding cannot thereby extend to compartments other than those assumed to be floodable for each case of damage.

(3) Oil tankers shall be regarded as complying with the damage stability criteria if the following requirements are met:

(a) The final waterline, taking into account sinkage, heel and trim, shall be below the lower edge of any opening through which progressive flooding may take place. Such openings shall include air pipes and those which are closed by means of weathertight doors or hatch covers and may exclude those openings closed by means of watertight manhole covers and flush scuttles, small watertight cargo tank hatch covers which maintain the high integrity of the deck, remotely operated watertight sliding doors, and side scuttles of the non-opening type.

(b) In the final stage of flooding, the angle of heel due to unsymmetrical flooding shall not exceed 25 degrees, provided that this angle may be increased up to 30 degrees if no deck edge immersion occurs.

(c) The stability in the final stage of flooding shall be investigated and may be regarded as sufficient if the righting lever curve has at least a range of 20 degrees beyond the position of equilibrium in association with a maximum residual righting lever of at least 0.1 metre within the 20 degrees range; the area under the curve within this range shall not be less than 0.0175 metre radians. Unprotected openings shall not be immersed within this range unless the space concerned is assumed to be flooded. Within this range, the immersion of any of the openings listed in sub-paragraph (a) of this paragraph and other openings capable of being closed weathertight may be permitted.

(d) The Administration shall be satisfied that the stability is sufficient during intermediate stages of flooding.

(e) Equalization arrangements requiring mechanical aids such as valves or cross-levelling pipes, if fitted, shall not be considered for the purpose of reducing an angle of heel or attaining the minimum range of residual stability to meet the requirements of sub-paragraphs (a), (b) and (c) of this paragraph and sufficient residual stability shall be maintained during all stages where equalization is used. Spaces which are linked by ducts of a large cross-sectional area may be considered to be common.

(4) The requirements of paragraph (1) of this Regulation shall be confirmed by calculations which take into consideration the design characteristics of the ship, the arrangements, configuration and contents of the damaged compartments; and the distribution, specific gravities and the free surface effect of liquids. The calculations shall be based on the following:

(a) Account shall be taken of any empty or partially filled tank, the specific gravity of cargoes carried, as well as any outflow of liquids from damaged compartments.

(b) The permeabilities assumed for spaces flooded as a result of damage shall be as follows:

Spaces	*Permeabilities*
Appropriated to stores	0.60
Occupied by accommodation	0.95
Occupied by machinery	0.85
Voids	0.95
Intended for consumable liquids	0 to 0.95*
Intended for other liquids	0 to 0.95*

(c) The buoyancy of any superstructure directly above the side damage shall be disregarded. The unflooded parts of superstructures beyond the extent of damage, however, may be taken into consideration provided that they are separated from the damaged space by watertight bulkheads and the requirements of sub-paragraph (3)(a) of this Regulation in respect of these intact spaces are complied with. Hinged watertight doors may be acceptable in watertight bulkheads in the superstructure.

(d) The free surface effect shall be calculated at an angle of heel of 5 degrees for each individual compartment. The Administration may require or allow the free surface corrections to be calculated at an angle of heel greater than 5 degrees for partially filled tanks.

(e) In calculating the effect of free surfaces of consumable liquids it shall be assumed that, for each type of liquid at least one transverse pair or a single centreline tank has a free surface and the tank or combination of tanks to be taken into account shall be those where the effect of free surfaces is the greatest.

"(5) The Master of every new oil tanker and the person in charge of a new non-self-propelled oil tanker to which this Annex applies shall be supplied in an approved form with:

(a) information relative to loading and distribution of cargo necessary to ensure compliance with the provisions of this Regulation; and

(b) data on the ability of the ship to comply with damage stability criteria as determined by this Regulation, including the effect of relaxations that may have been allowed under sub-paragraph (1)(c) of this Regulation.

---

* The permeability of partially filled compartments shall be consistent with the amount of liquid carried in the compartment. Whenever damage penetrates a tank containing liquids, it shall be assumed that the contents are completely lost from that compartment and replaced by salt water up to the level of the final plane of equilibrium.

APPENDIX

B

Structural Design of Tankers*

by
DONALD LIU
American Bureau of Shipping

DESIGN LOADS

The loads to be considered in the structural design of a tanker consist of: ship and cargo weight, hydrostatic internal and buoyancy pressure in still-water, wave loads and dynamic components resulting from wave induced motions. These loads are of variable nature and may be acting simultaneously. They can be broken down into the following components:

Internal/External Liquid Pressure Differential

The liquid pressure differential for any combination of internal/external liquid levels, with the ship on even keel or in rolled position such as:

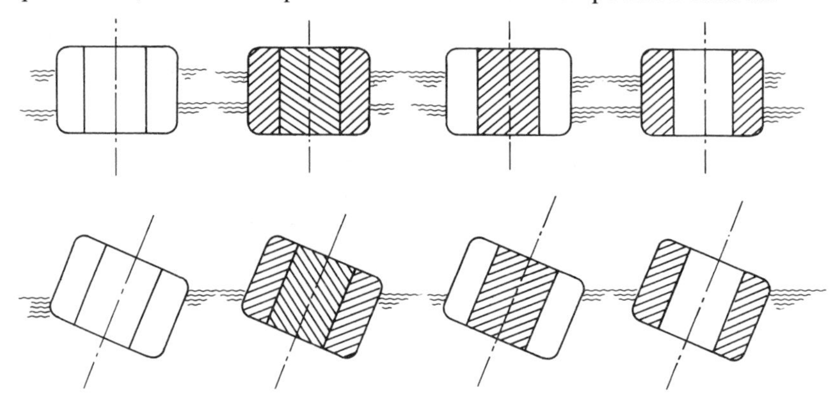

*Excerpts from Liu, D. Structural Design of Tankers. Paper presented to a meeting of the Committee on Tank Vessel Design, National Research Council, Washington, D.C., March 26, 1990.

Shear Forces

The shear forces generated by the non-uniform distribution of cargo ship weight and buoyancy along the length of the vessel, in still water and waves.

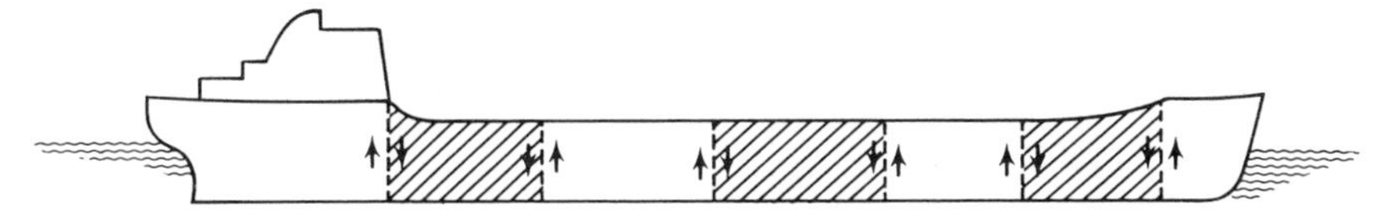

Hull Girder Longitudinal Bending Moment

The hull girder bending moment* resulting from the distribution of cargo, ship weight and buoyancy along the length of the vessel, in still water and wave.

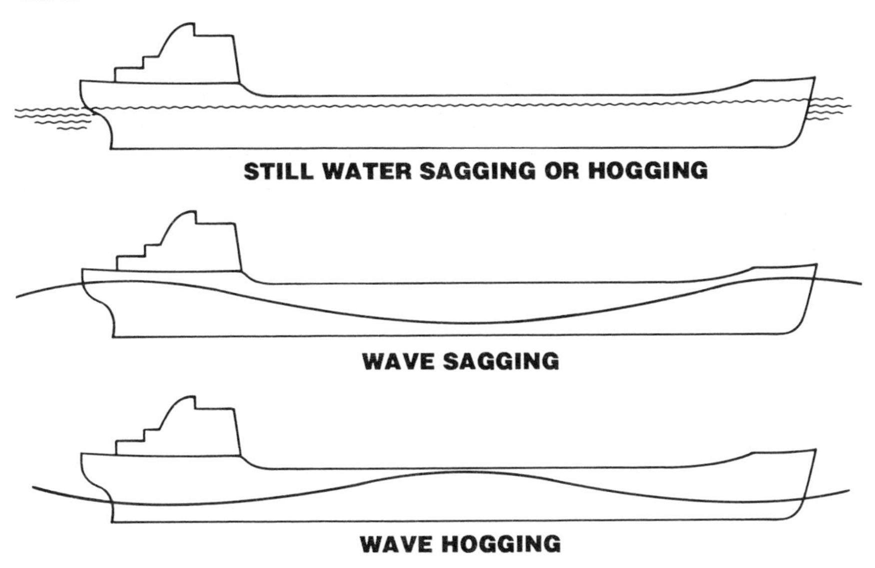

Slamming

The impact loads on the flat bottom forward due to pitching as shown below:

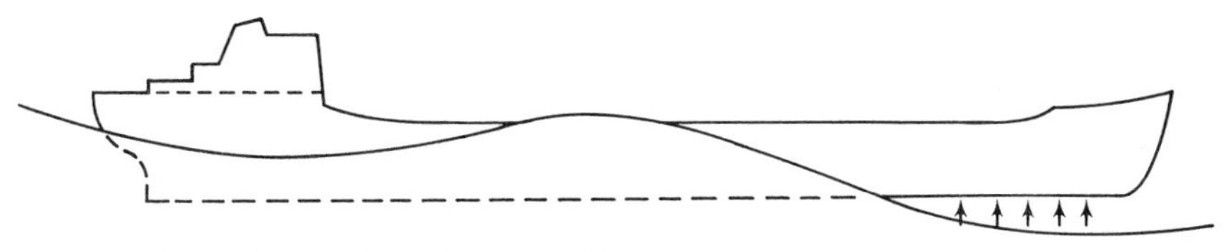

*The summation of forces that tend to bend the hull.

Green Water on Deck

The impact loads due to bow slamming and the effect of green water* on deck.

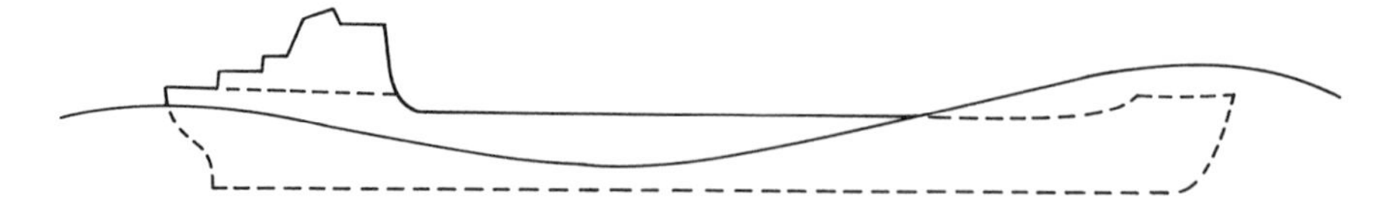

Liquid Cargo Sloshing

In partially filled tanks, the sloshing and impact loads of liquid cargo on bulkhead, shell and deck tank boundaries, resulting from the inertial reactions between the accelerated mass of the liquid and the ship structure when the ship is rolling and pitching in a seaway can be very large. These loads vary with the number of swash bulkheads and level of liquid in the tank. The next page shows the result of a recent ABS study. At 35% full, the sloshing pressure without swash bulkheads is [Figure B-1(A)] 33,640 pressure per square foot (psf), [Figure B-1(B)] 8,060 psf with one swash bulkhead, and [Figure B-1(C)] 4,362 psf with two swash bulkheads. Without swash bulkheads, the sloshing pressure jumps from 33,640 psf to 68,073 psf when the liquid level is increased from 35% to 75% full [from Figure B-1(A) to Figure B-1(D)].

Thermal Loads

Thermal loads result from non-linear temperature differentials in ship structure, and have to be considered in the design of vessels carrying hot cargoes.

SCANTLINGS SELECTION

The design objectives of the structural optimization for the selection of scantlings and dimensions of structural members are the following:

- Reduce steel weight—to save on cost of building material and loss of cargo
- Simplify details and fabrication procedures—to save labor costs
- Keep stresses and deflections within acceptable limits—to avoid structural problems later in service.

---

*High seas washed onto the deck.

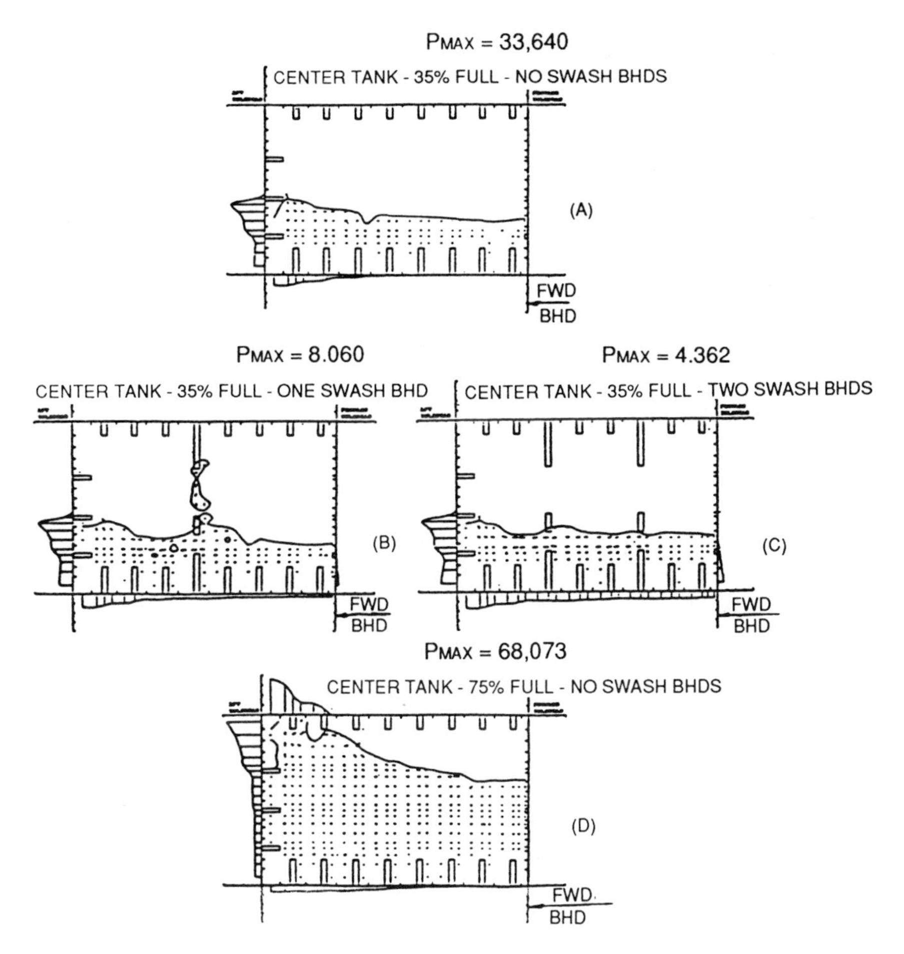

FIGURE B-1 Liquid Cargo Sloshing. Dynamic loads on tank boundaries due to rolling and pitching.

ENGINEERING ANALYSIS

As an alternative to using rule equations, the classification society rules permit use of scantlings obtained by systematic analysis based on sound engineering principles, provided they meet the overall safety and strength standards predicated by the rules.

This rule provision is normally utilized for the design of large tankers, whose structure and loads are too massive and complex to rely entirely on rule empirical equations. The normal practice is to base the tanker's design on engineering analysis, and to use rule equations only as a check, for guidance and control. The potential benefit from utilizing engineering analysis is stress verification and rational distribution of steel, which results in steel weight optimization without sacrificing strength.

Ship Moment

The first step in the analysis consists of calculating the hull girder longitudinal vertical bending moment, for all anticipated loaded and ballast conditions. This calculation is based on the offsets describing the hull geometry, and on the lightship and cargo weight distribution. A computer program is utilized for this purpose, which in addition to the hull girder bending moment yields also the equilibrium draft and trim, as well as the hull girder shear forces and vertical deflections.

Ship Motion

The next step in the analysis is to calculate the vertical, lateral and torsional dynamic components of the hull girder bending moments, as well as the motions, point accelerations and pressure distribution along the vessel's length. A computer program based on two-dimensional, six-degree of freedom strip-theory, is used to calculate motions and loads for the vessel in arbitrary headings in long and short crested seas, combined with probabilistic considerations of statistical wave data normally measured in the North Atlantic.

3-D Global 3-D Finite Element Analysis (FEM)

The data obtained from ship moment and ship motion above are used to carry out a three-dimensional global FEM analysis for the entire vessel, or for a portion of the hull girder. This analysis gives the overall structural response in the form of element stresses and displacements.

3-D Local Finite Element Analysis (FEM)

The three-dimensional fine mesh FEM analysis is used to obtain the detailed stress distribution in way of localized structural details of interest, such as the typical connection of side shell longitudinal to the web frame and transverse bulkhead shown in Figure B-2. The displacements from the 3-D global model, at the appropriate coincident nodes, are used as boundary conditions for the 3-D fine mesh models.

2-D FEM Local Analysis

The two-dimensional fine mesh FEM analysis is carried out for selective web frames, or other structural members, where precise distribution of stresses

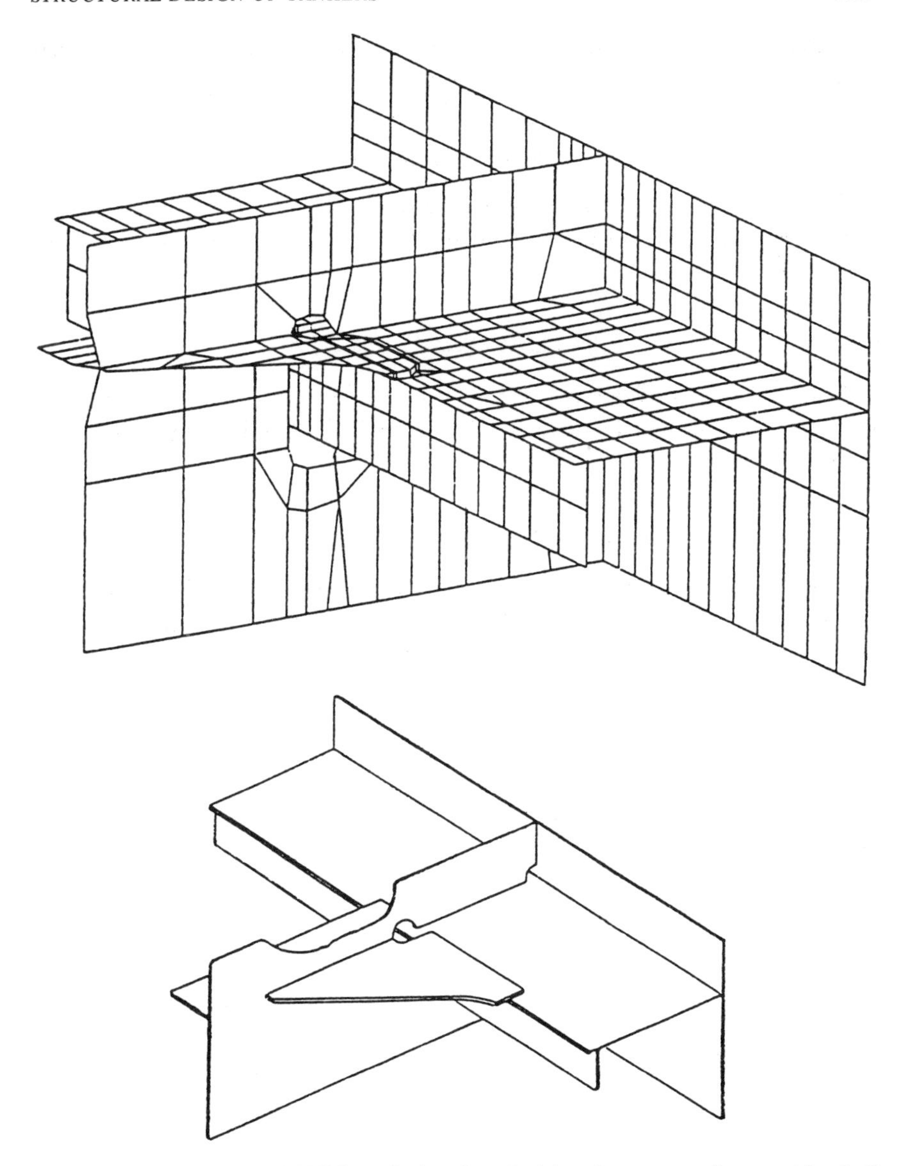

FIGURE B-2 3-D local FEM analysis of typical bracket connection to side shell longitudinal.

is required. The use of fine mesh models permits the avoidance of localized high stresses and a better resolution of details.

Sloshing Analysis

This analysis is used to determine the dynamic loads on the tank boundaries from the motion of the fluid within the tank due to ship movements in a seaway.

Thermal Stress Analysis

This analysis provides the distortions and stresses in the hull structure induced by non-linear temperature differentials in vessels carrying hot cargoes.

Fatigue and Fracture Mechanic Analysis

Based on the combined effect of loading, material properties and flaw characteristics, this analysis predicts the service life of the structure, and determines the most effective inspection plan.

Vibration Analysis

This analysis is used to determine the extent of vibrations in the ship structure induced by the interaction of fluids, structure, machinery and propellers.

APPENDIX

C

Large Tanker Structural Survey Experience*

The major difficulty in adequately surveying VLCCs is the physical size of the task. Table 4-1 illustrated the scope of work required to give complete coverage to a typical 250,000 DWT vessel. Historically, owners inspected vessels during routine dry-dockings, thus limiting survey to structure that was readily accessible. Special staging rarely was used due to its cost and the expense of extending the dry-dock periods.

Owners now generally feel that the only practical time for conducting complete inspection of a vessel is during ballast voyages. At that time, tanks can be flooded progressively to different heights, and structure can be accessed by raft. This method has limitations in that safety considerations do not permit placing inspectors in close proximity to the deck head. However, significant areas of structure can be inspected successfully.

The most critical aspect of the at-sea survey, and the principal limitation on its success, is adequate cleaning of tanks. Safe access for extended periods requires thorough removal of residual hydrocarbon, in order to ensure a gas-free environment. Heavy accumulations of wax, sludge, sediment, and scale have to be removed to expose bare steel for inspection and to ensure that structure can be climbed safely. The efficiency of this operation is dependent on the type and amount of tank cleaning equipment, its level of maintenance, and the previous services of the vessel (length of voyage, time spent in floating storage, and the type of crude oil carried).

Cleaning requires extensive planning and effort by management and crew. In many cases, the process may begin several voyages prior to the survey, with schedulers positioning the vessel to maximize cleaning opportunities

*Excerpts from paper: Exxon Corporation, 1982.

and the crew making preliminary inspections of tanks to ascertain where increased cleaning procedures are necessary. It is sometimes necessary to provide additional manpower to remove sediment from critical survey areas.

The direct cost of a survey in 1981 was approximately $75,000 (1990 dollars),* including charges for a steel inspector, ultrasonic team, and analysis and reporting by a naval architect. Additional costs were incurred for incremental fuel consumed in shifting ballast water during the survey, as well as for cleaning the vessel to an adequate standard. Ballast shifting alone typically costs $12,000.* This level of expense emphasizes the need for thorough preparation by the vessel prior to embarking a survey team to minimize the necessity of re-inspection of inaccessible tanks.

A normal ballast voyage provides about 20 days to complete an inspection. This is a tight schedule and requires careful coordination between the survey team, ship's staff and shore management. A planning meeting is held with ship's staff after embarkation to review safety procedures associated with tank entry, to acquaint the crew with the procedures used by the team, and to plan the sequence of full inspection. Emphasis is placed on conformance with approved company and industry safety standards. The survey team is required to reject any tank for inspection that does not meet these standards.

An inspection sequence is selected that minimizes ballast shifts to conserve time and fuel. The first tank structure inspected is normally the bottom plating and adjoining structure in the tanks that the chief officer requires to be immediately available for minimal ballast movement. These tanks then are ballasted and bottom structure in remaining tanks is inspected. The team then returns to the partially ballasted tanks to complete the inspection of the upper portion of the tanks. Inflatable rubber rafts are used to move about the tank, although a rigid boat which can be split for access through a tank hatch and reassembled in the tank is undergoing operation evaluation. Within each bay, free climbing of the structure is employed to improve the speed and efficiency of the inspection. The remaining tanks are completed in a similar manner.

The inspector checks the structure for corrosion wastage, buckling, and cracking. Crack detection is accomplished visually, as suitable equipment is not available. The side shell attachments, bottom fore and aft girder bracket attachments, and stringer platform to bulkhead attachments are among locations examined where experience has indicated a high probability of local failure.

While the inspector is making his visual inspection, the ultrasonic techni-

*Costs are reported in 1990 dollars, based on U.S. Department of Labor, Bureau of Labor Statistics, Employment Cost Index—Private Industry, total compensation. Costs are rounded to the nearest thousand dollars.

cians take readings at predetermined locations in the tanks. These locations are selected ship-by-ship prior to the survey to give representative readings of the structure and coverage of known suspect areas. Additional ultrasonic readings are taken as necessary by the inspector during the visual inspection of the tank. On average, about 8,000 readings are taken, although in excess of 11,000 readings have been taken in cases of severe corrosion.

Due to the large areas to be covered, the inspector has to focus on historically suspect areas to optimize the effectiveness of the survey. In-house training, based on past experience with these ships, is necessary to ensure satisfactory results from the inspector.

Complete coverage of the structure generally has not been achieved. However, on adequately cleaned vessels 85-90 percent of the tank section has been inspected. The most difficult area to inspect is the deckhead due to a lack of access other than by elaborate staging methods. Rafting is not used for the deckhead since they survey team would have to be temporarily trapped between deep transverse web frames as the water level was raised and lowered to inspect each bay. As a result, the deckhead is inspected visually from the highest practical water level, or upper walkways (if fitted) and ultrasonic readings are taken from the main deck of the plating and of any deck longitudinals available through deck openings.

Although tank cleanliness has been the major factor limiting the effectiveness of the surveys, other factors have been pertinent. Heavy seas causing roll of 5° or more prohibit safe tank work. High temperatures and humidity encountered in places such as the Red Sea and the Arabian Gulf can result in extremely difficult work conditions. For instance, a tank ambient temperature of 35°C with 95 percent relative humidity restricts effective working time to as little as 15 minutes per hour. Higher tank temperatures and humidity can easily occur and very little can be done to reduce them.

Upon completion of the survey all data is returned to the office for analysis and report preparation. The survey report includes estimates of necessary steel renewals, coatings, and anodes. The nature and extent of any structural defects are detailed and repairs, modifications or changes in operating procedure are specified.

APPENDIX

D

35,000 DWT Tanker Cargo Capacity and Damage Stability Study

Seaworthy Systems
November 1990

35,000 DWT TANKER CARGO CAPACITY AND DAMAGE STABILITY

A generic 35,000 DWT tanker was developed to investigate the effects of various segregated ballast tank (SBT) arrangements on the vessel's cargo carrying capacity and damaged stability characteristics. The arrangements were provided with sufficient ballast capacity to meet the draft and trim requirements for tankage was divided into 10 transverse compartments. Eight arrangements were investigated. A midship section of the single skin design is illustrated in Figure D-1. Double bottoms of 2m (B/15) and B/5 in depth are presented in Figures D-2 and D-3. Double sides of 2m and B/5 in width are presented in Figures D-4 and D-5. Double hulls of 2m and B/5 in depth are presented in Figures D-6 and D-7. Centerline bulkheads were added to the 2m double side and double hull arrangements for damage survival considerations. Principal dimensions of the generic design are as follows:

L^{BP} = 638'
B = 89'
D = 46.75'
t = Varies based upon cargo carrying capacity

NOTE: Most parametric studies of alternate tanker designs deal with VLCC-size tankers. Yet, much of the tanker traffic of interest to the United States is and will continue to consist of smaller tankers. Accordingly, Seaworthy Systems, Inc., was requested to examine alternate designs for a "handy-sized" tanker of 35,000 DWT. The study is attached. It is emphasized that the various designs were not optimized. Nonetheless, they are valid to demonstrate that double hull tankers in the lower size range can be designed with damage resistance far in excess of present requirements.

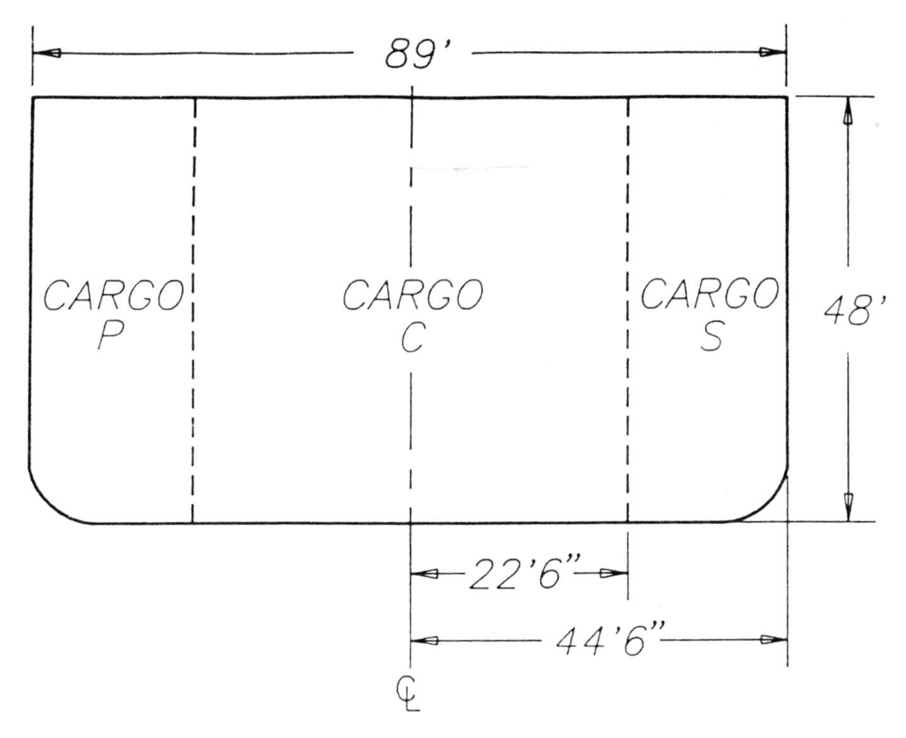

FIGURE D-1 Typical midbody 35,000 DWT tanker.

Protectively Located Ballast Requirements

The SBT arrangement must meet capacity and location requirements of 33 CFR 157.10. The ballast capacity must be sufficient to allow the vessel to meet ballast draft and maximum trim requirements. The ballast must be arranged in wing or double bottom tanks to minimize cargo outflow in the event of damage to the hull. Also, the ballast arrangement must not result in a stress numeral of over 100. (Stress numerals were not investigated in this study.)

The ballast capacity and arrangement must keep the propeller 100 percent immersed. Trim in the ballast condition must not exceed 1.5 percent of the vessel's length. The draft amidships must be at least 2 meters plus 2 percent of the vessel's length.

The projected area of PL/SBT tankage must exceed the amount required by an equation which is based upon the vessel's L, B, D and deadweight. For wing tanks to be considered effective they must be at least 2 meters wide. The minimum depth of the double bottom is B/15 or 2 meters, whichever is less. In this study we used 2 meter sides as required and 2 meter double bottoms in order to allow for tank inspection.

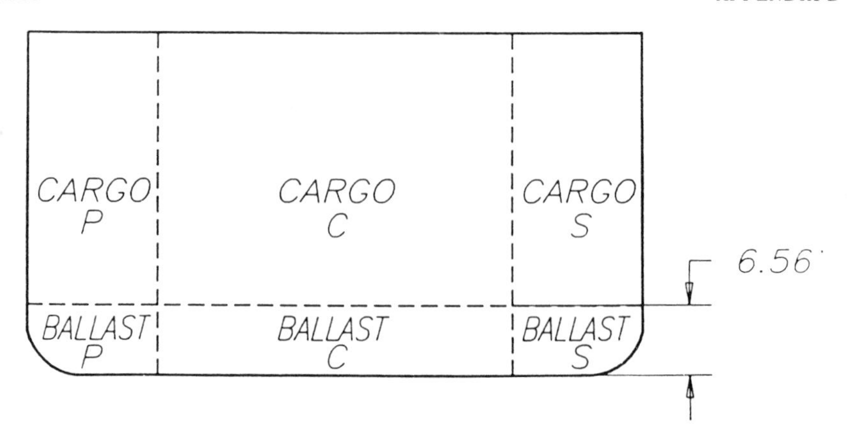

FIGURE D-2 35,000 DWT tanker with 2 meter (B/15) double bottom.

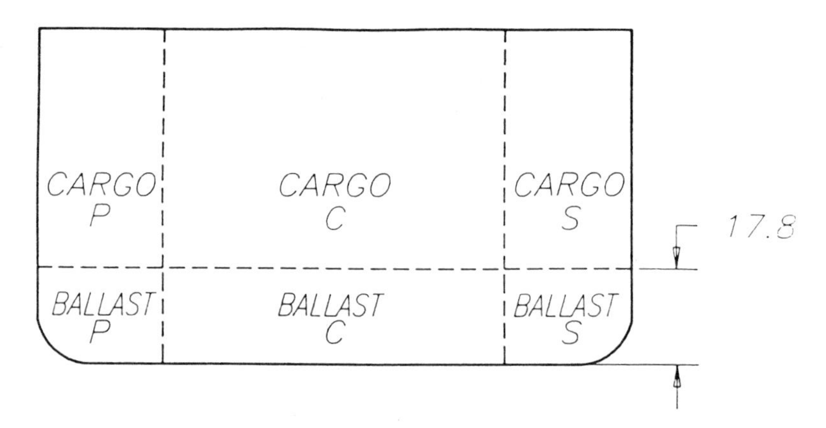

FIGURE D-3 35,000 DWT tanker with B/5 double bottom.

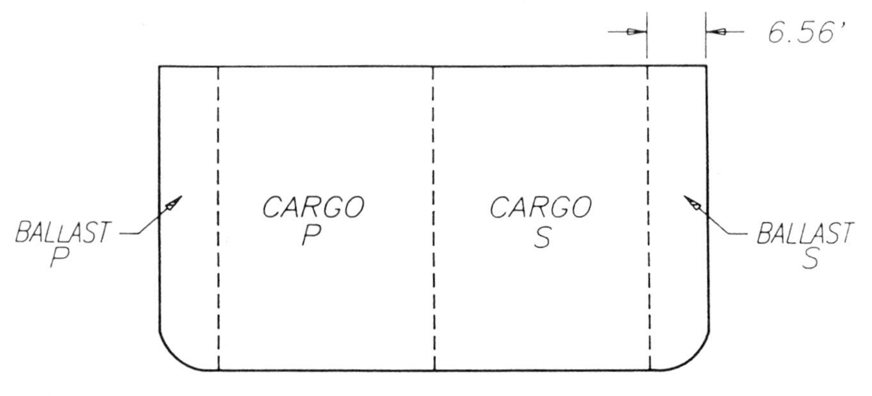

FIGURE D-4 35,000 DWT tanker with 2 meter (B/15) double side.

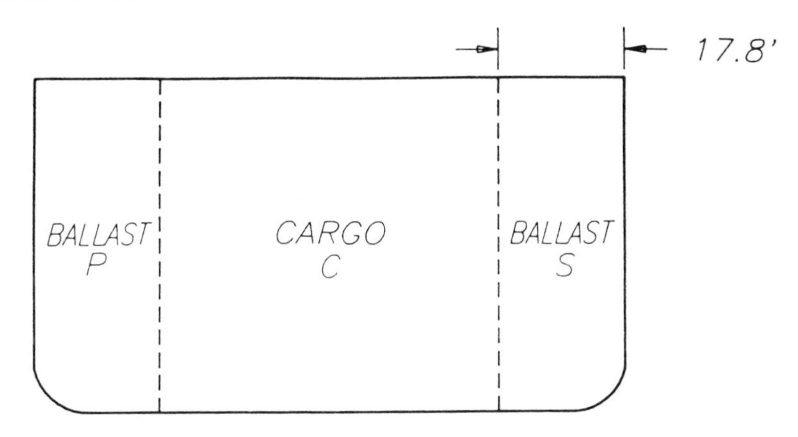

FIGURE D-5 35,000 DWT tanker with B/5 double side.

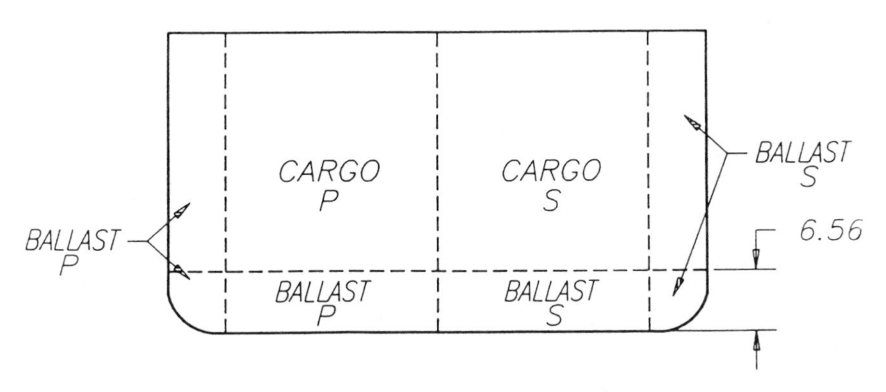

FIGURE D-6 35,000 DWT tanker with 2 meter (B/15) double hull.

Draft, trim, and ballast distribution were calculated for the eight arrangements. The arrangements with double sides, bottom and hull exceed the area requirements. The 2 meter double bottom, sides and hull arrangements require additional ballast capacity to meet the draft and trim requirements. The ballast arrangements for the various configurations are provided in Figures D-2 through D-7.

Cargo Capacity

Table D-1 compares the cargo carrying capacities of the various SBT configurations. All of the SBT tank arrangements result in vessels which are volume rather than deadweight limited. The single skin design loses 29

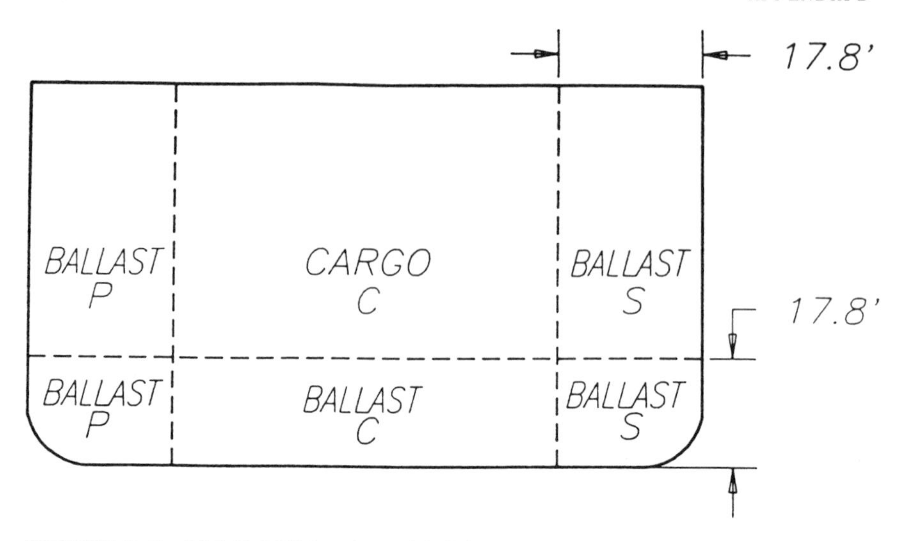

FIGURE D-7 35,000 DWT tanker with B/5 double hull.

percent of its volume to segregated ballast. The 2 meter double bottom loses 26 percent. The 2 meter double sides loses 27 percent. The 2 meter double hull loses 28 percent. The double skin arrangements lose less due to the effective use of wing and bottom tankage. The B/5 double bottom and sides lose 10 percent more cargo. The single skin design would lose a bit less if the bulkhead arrangements were optimized for the ballast tankage. The double hull, bottom or side arrangements meet MARPOL draft and trim requirements but were not optimized; therefore, internal tanks had to be used for ballast.

Damage Stability

The damage stability analysis was performed with the vessels fully loaded with cargo at a density of 6.63 BBL/LT (Bunker C). The loaded cargo tanks were assumed to be five percent permeable to account for the density of the cargo and the amount of cargo carried in the tank. The segregated ballast tanks were assumed to be 99 percent permeable which corresponds to an empty tank. Cases investigated include 1, 2, and 4 compartment damage.

For vessels to meet damage stability requirements, they must comply with the following. The equilibrium heel due to unsymmetrical flooding must not exceed 25 degrees. The vessel's righting lever curve must have a range of 20 degrees beyond the equilibrium heel angle. The maximum righting arm must be at least 4 inches (0.1 meter). All the configurations were able to meet these requirements for 1 and 2 compartment damage.

The worst case scenario for 4 compartment damage stability is presented in Table D-2. This table illustrates the results of the analysis. The single skin hull has the least equilibrium heel angle. The 2m (B/15) double hull provides the greatest maximum righting arm. The greatest range of stability is 70 degrees. The following four configurations meet this range; PL/SBT single skin, 2m double bottom, 2m double sides, 2m double hull. The B/5 double side and double hull have excessive heel and a reduced range of stability. The B/5 double bottom arrangement results in the worst damage stability characteristics. The maximum righting arm is only 0.4 feet which meets the requirement, but its range of positive stability is only 17°, which does not meet the requirements.

Conclusions

The use of SBT result in designs that are volume-(cubic) rather than deadweight-limited. Although further optimization of tank arrangements may result in less volume loss in a single skin design than is shown in this analysis, the double bottom, side or hull arrangement can also be optimized to result in similar cargo capacities. The study also indicates that the double skin can be taken too far resulting in excessive loss of cargo carrying capacity and degradation of damage stability characteristics.

35,000 DWT TANKER - LOSS OF CARGO CARRYING CAPACITY

ALTERNATIVE	CARGO VOLUME AVAILABLE (FT3)	CARGO DEADWEIGHT			DEAD-WEIGHT OR VOLUME LIMITED	% VOLUME AVAILABLE FOR CARGO
		1) BUNKER C 6.63 BBL/LT	2) DIESEL OIL 7.38 BBL/LT	3) GASOLINE 8.70 BBL/LT		
SINGLE SKIN (NO PL/SBT)	1,730,000	35,580	35,580	33,673	1) DWT 2) DWT 3) VOLUME	100%
SINGLE SKIN* W/ PL/SBT	1,232,684	31,485	28,285	23,994	1) VOLUME 2) VOLUME 3) VOLUME	71 %
2M DOUBLE BOTTOM*	1,272,198	32,494	29,191	24,763	1) DWT 2) VOLUME 3) VOLUME	74 %
2M DOUBLE SIDES*	1,264,197	32,290	29,008	24,607	1) DWT 2) VOLUME 3) VOLUME	73 %
2M DOUBLE HULL*	1,239,397	31,656	28,439	24,124	1) VOLUME 2) VOLUME 3) VOLUME	72 %
B/5 DOUBLE BOTTOM	1,098,883	28,067	25,215	21,389	1) VOLUME 2) VOLUME 3) VOLUME	64 %
B/5 DOUBLE SIDES	1,083,955	27,686	24,872	21,099	1) VOLUME 2) VOLUME 3) VOLUME	63 %
B/5 DOUBLE HULL	672,407	17,174	15,429	13,088	1) VOLUME 2) VOLUME 3) VOLUME	39 %

*These designs were not optimized to maximize cargo capacity

TABLE D-1 35,000 DWT Tanker—Loss of Cargo Carrying Capacity

35,000 DWT TANKER

4 COMPARTMENT DAMAGED STABILITY
WORST CASE SCENARIO

ALTERNATIVE	DAMAGED SCENARIO	EQUILIBRIUM HEEL (DEGREES)	MAXIMUM RIGHTING ARM (FEET)	RANGE OF STABILITY (DEGREES)
SINGLE SKIN	DAMAGE TO NO. 6,7,8&9 STBD CARGO TANKS	1	4.34	70
B/15 (2M) DOUBLE BOTTOM	DAMAGE TO NO. 5,6,7&8 STBD DOUBLE BOTTOM	3	2.95	70
B/15 (2M) DOUBLE SIDE	DAMAGE TO NO. 4,5,6&7 STBD WING TANKS	6	3.26	70
B/15 (2M) DOUBLE HULL	DAMAGE TO NO. 3,4,5&6 STBD WING TANKS	7	9.50	70
B/5 DOUBLE BOTTOM	DAMAGE TO NO. 3,4,5&6 STBD DOUBLE BOTTOM	13	0.40	17
B/5 DOUBLE SIDE	DAMAGE TO NO. 3,4,5&6 STBD WING TANKS	16	3.05	50
B/5 DOUBLE HULL	DAMAGE TO NO. 3,4,5&6 STBD WING TANKS	12	5.2	50

TABLE D-2 35,000 DWT Tanker—Four Compartment Damaged Stability, Worst Case Scenario

APPENDIX

E

Double Hull Tanker Design*

DOUBLE BOTTOM HEIGHT REQUIREMENTS

The existing IMO requirements are 1/15th the beam of the ship or two meters, whichever is less. We believe a better criteria would be 1/15th the beam of the ship but in no case less than two meters nor be required to be greater than three meters. Two meters is adequate for smaller vessels; but as the vessel size is increased and structural stiffeners increase in size, personnel access and air circulation become problems if the height of the double bottom is simply limited to two meters. There are many ways to design a double bottom, but in larger size vessels either longitudinals encroach into the available space to make access very difficult or longitudinal girders and transverse frames produce compartments that have the same or worse effect. On the other hand, we believe that three meters is sufficient height, even in the largest size vessels (VLCC's). If the height of the double bottom is increased beyond that, the vessel's stability is further impaired—the greater the double bottom height, the harder it is to design a vessel that has good damage stability characteristics. Furthermore, maintenance of the double bottom overhead becomes more difficult due to accessibility if the height is increased much beyond three meters. To illustrate the situation, compare the access of a "2M" double hull design currently being built in the Far East (Figure E-1) to that of a current Chevron design, Figure E-2. Note that reasonable access is closely linked to structural design, not simply the height/width of void spaces.

---

*Chevron Shipping Company (adapted), October 1990.

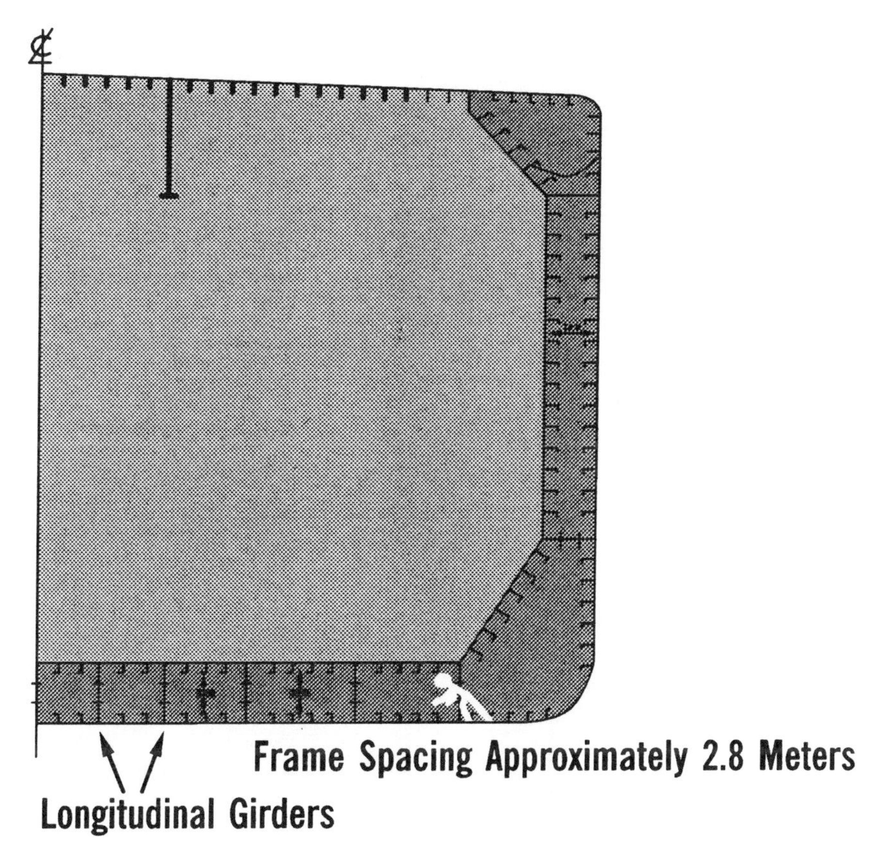

FIGURE E-1 Vessel currently being built in the Far East: ballast tank (midships) 2 meters wide.

DOUBLE SIDE TANK WIDTH REQUIREMENTS

Double side tanks require a minimum width of two meters. We believe the criteria should be 1/15th the beam but in no case less than two meters nor be required to be greater than three meters. No reasonable double side tank will be able to prevent a high impact collision from penetrating through the side tank. The width criteria should be established based on a width sufficient to prevent damage to the inner side in a glancing-type contact.

In addition to damage prevention to the inner side, access and maintenance issues must be addressed. This is somewhat less of a problem in the side tanks than in the double bottom tanks because there is less structure in the upper portions of these tanks; but in the bottom part, the problem is the same.

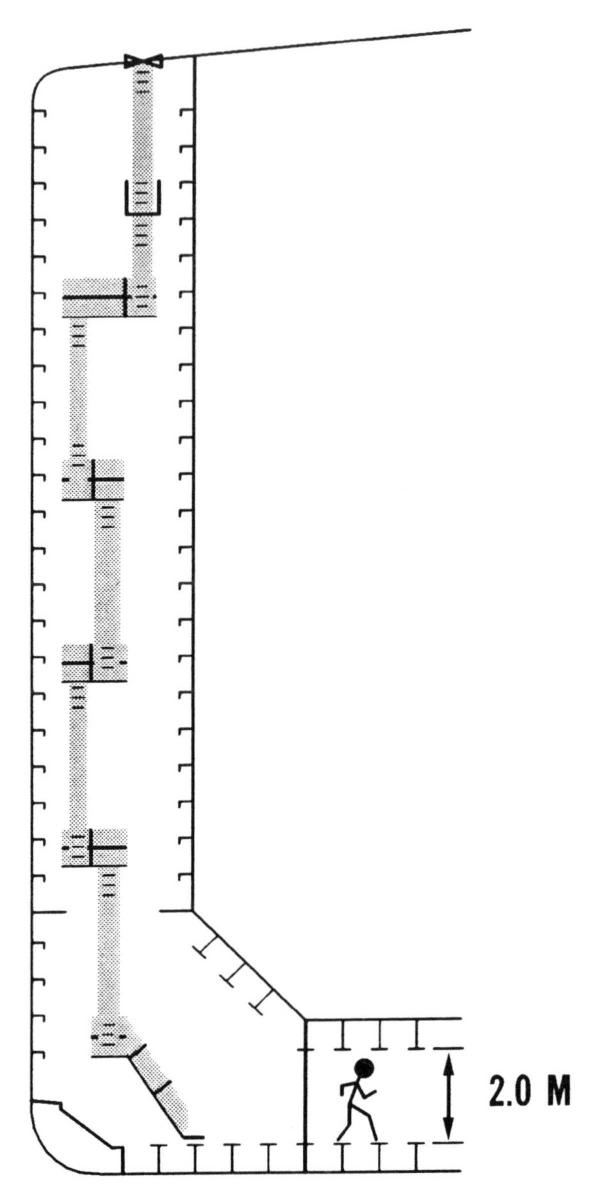

FIGURE E-2 Chevron 130,000 DWT ballast tank (midships).

MAINTENANCE AND SAFETY

The double hull tanker will require frequent inspection of the double hull spaces to ensure that coating failures and corrosion are detected and dealt with on a timely basis. Therefore, access for maintenance must be sufficient to allow for easy inspection of the spaces and must allow for the effective removal of injured personnel. Openings in the structure at least one meter × 0.6 meter should be provided in every compartment for personnel movement in both the fore and aft and the athwartships directions. Provision for adequate air circulation in all double hull tanks is essential and purge pipes may be required in some locations. Figure E-3 shows the purge pipe arrangement that we made in order to get air circulation through the ballast tanks that have a dead leg at the centerline bulkhead. For entrance of men, it will be necessary to know that there is a satisfactory atmosphere throughout the tank. The more subdivided the tank is, the more difficult this becomes. We believe that a purge line is essential to get proper air circulation in the tank.

COATINGS

The massive amount of ballast tank surface area that is present in a double hull tanker makes corrosion protection a vital issue. These tanks should be coated and protected with cathodic protection methods to provide protection backup when coating failures occur. Failure to do so will result in corrosion and eventual leakage of cargo into the double bottom space with the resulting buildup of explosive vapors. Application of high quality coatings, with cathodic protection systems to give good corrosion protection in the ballast tank spaces will be critical to long-term operation without problems.

DAMAGE STABILITY CRITERIA

The current IMO requirements for damage stability which, in the case of most large tankers, means damage to two tanks. Our experience in grounding incidents is that the longitudinal extent of damage is frequently greater than is provided for under current criteria. We believe a double hull tanker should be able to survive damage which extends the full longitudinal extent of the cargo area of the vessel if the damage does not penetrate the vessel's inner bottom. For a double hull ship, intact stability is reduced by a combination of the higher center of gravity of the cargo and the possible large free surface in the center cargo tanks if a center line bulkhead is not fitted. If a double-bottom centerline bulkhead is fitted and asymmetric flooding can occur when the outer hull is breached, the combination of reduced

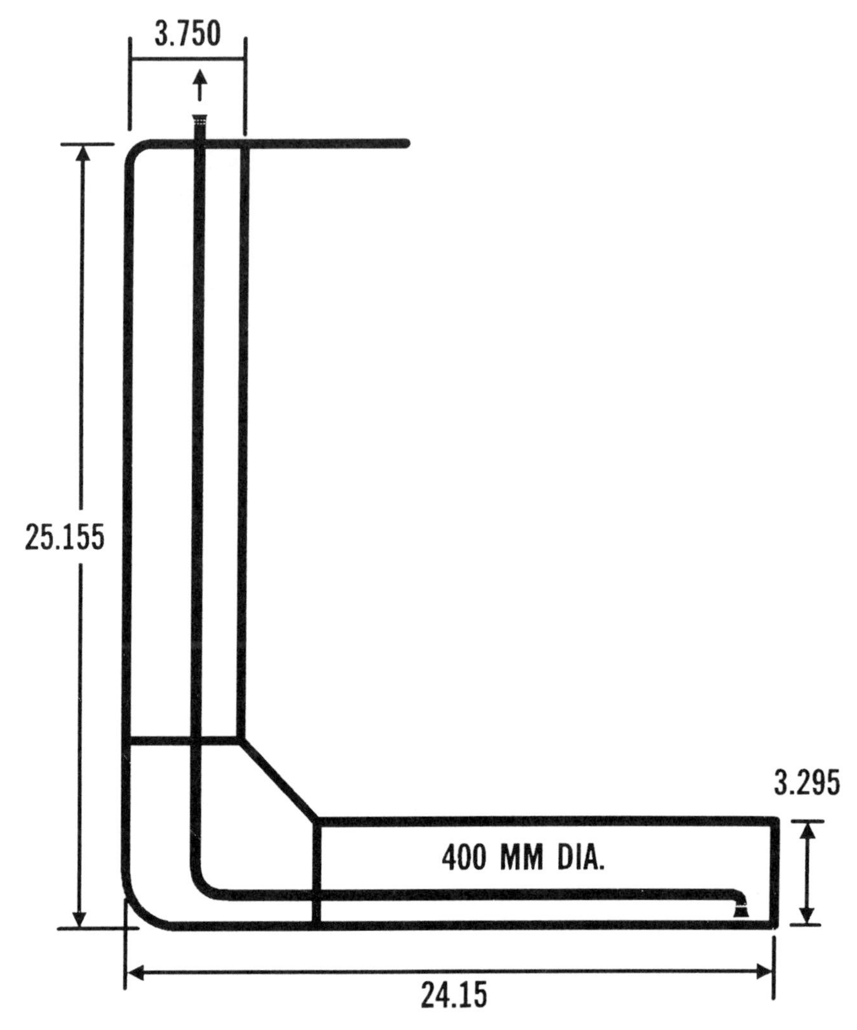

FIGURE E-3 Double bottom "L" tank purging arrangement.

initial stability and the creation of large heeling moments can cause the vessel to capsize when damage extends over less than half the vessel's length. We should strive to require a standard of damage stability that provides for the vessel's survivability at least equal to the survivability of a typical single hull vessel. Figure E-4 shows the free-floating condition of a single hull after being damaged nearly its full length (damaged area shaded). As can be seen, the vessel is heeled to 33.5 degrees but remains afloat.

In Figure E-5, we depict an industry standard 130,000 DWT vessel with 2.0 meter double hull and "L" shaped ballast tanks. If it is loaded 98% full

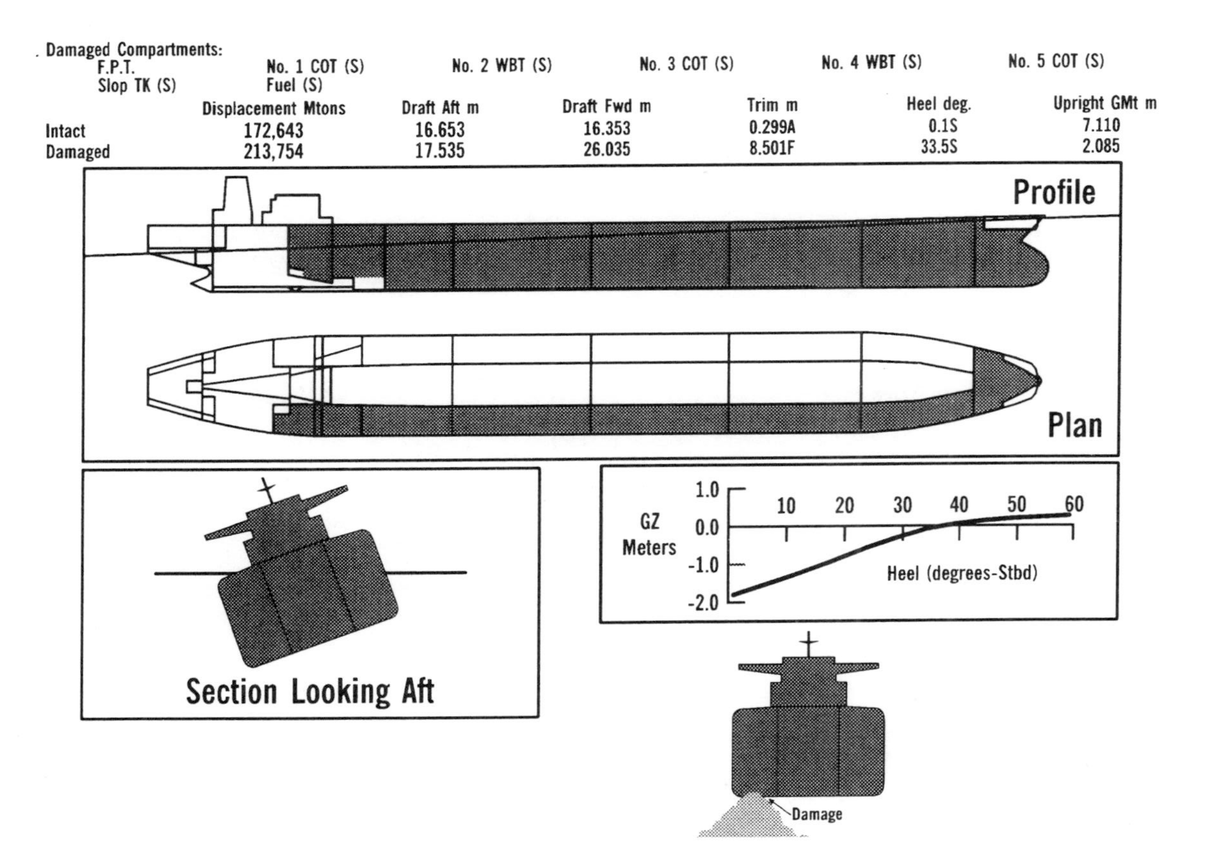

FIGURE E-4 Free-floating damaged condition.

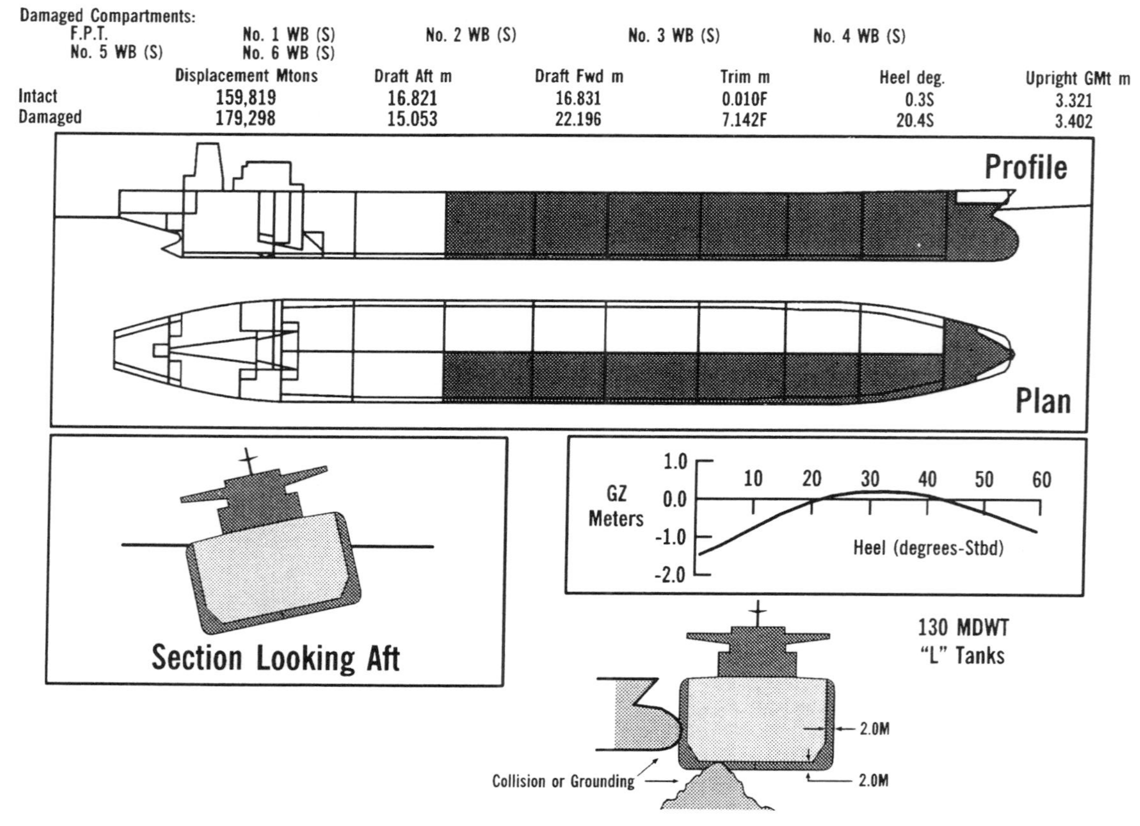

FIGURE E-5 Industrial standard 130,000 DWT free-floating damaged condition—98% full.

and damaged to the extent indicated in the figures by the shaded areas, it will survive. If it is damaged further (into No. 7 cargo tank), it will capsize. If the same vessel is loaded with a heavier cargo so that, although fully loaded, the cargo tanks are only 96% full, then the vessel capsizes with far less damage (see Figure E-6). While it is prudent good marine practice to fill tanks to 98%, whether it is done in actual practice, is a question.

THE CHEVRON DESIGN

The Chevron 130,000 DWT vessels have double-hull dimensions that meet B-15 criteria and an arrangement of double-hull that allow it to survive substantial bottom damage. It uses a combination of "U", "J", and "S" tanks, as described in Figures E-7 and E-8. It also has access and maintenance arrangements that allow relatively easy inspection and maintenance. We believe, with this design, we've mitigated some of the problems inherent in double-hull designs. The tank arrangement and damage stability characteristics of the Chevron vessel is shown in Figure E-9. The Chevron vessel utilizes many "U" ballast tanks to minimize the capsizing moments that are created by "L" tanks. The "U" tanks are fine, if full or empty, but create problems during loading and discharge of the vessel if too many are partially full simultaneously. Two sets of "J" tanks were arranged to provide for heel control while working cargo. The "S" tanks were included to provide buoyancy in a damaged condition, as the "U" tanks will flood across the vessel and not cause heeling, but sinkage will be greater. To minimize the free surface effects which could be created inadvertently by flooding too many "U" tanks simultaneously, three center cargo tanks have centerline subdivision bulkheads.

This arrangement has very good damage stability characteristics as shown in Figure E-9. The vessel can be damaged from the bow through to the engine room and remain stable and afloat. The computer-generated sketch of the vessel in this figure doesn't show the starboard side "S" tanks to be undamaged, but that is the case. If loaded to 96%, the stability is reduced, but the vessel survives this condition, as well, as shown in Figure E-10.

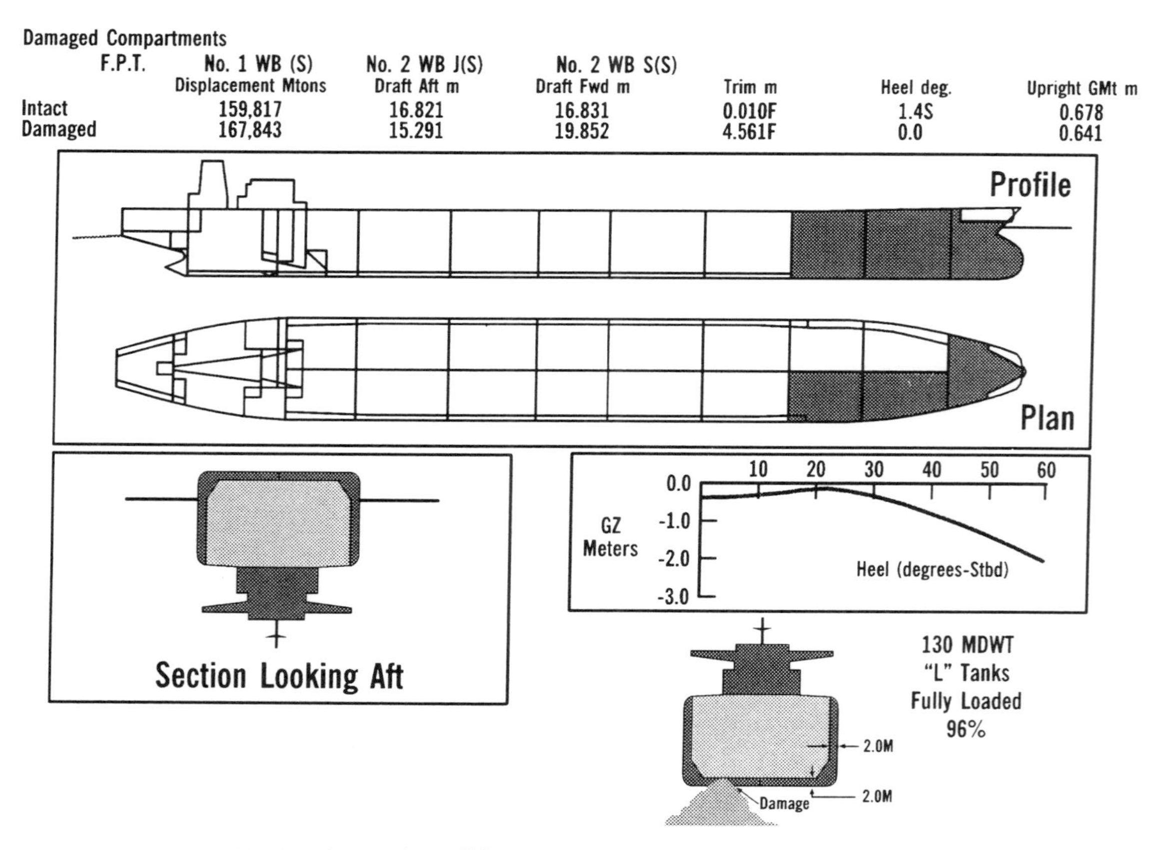

FIGURE E-6 Free-floating damaged condition.

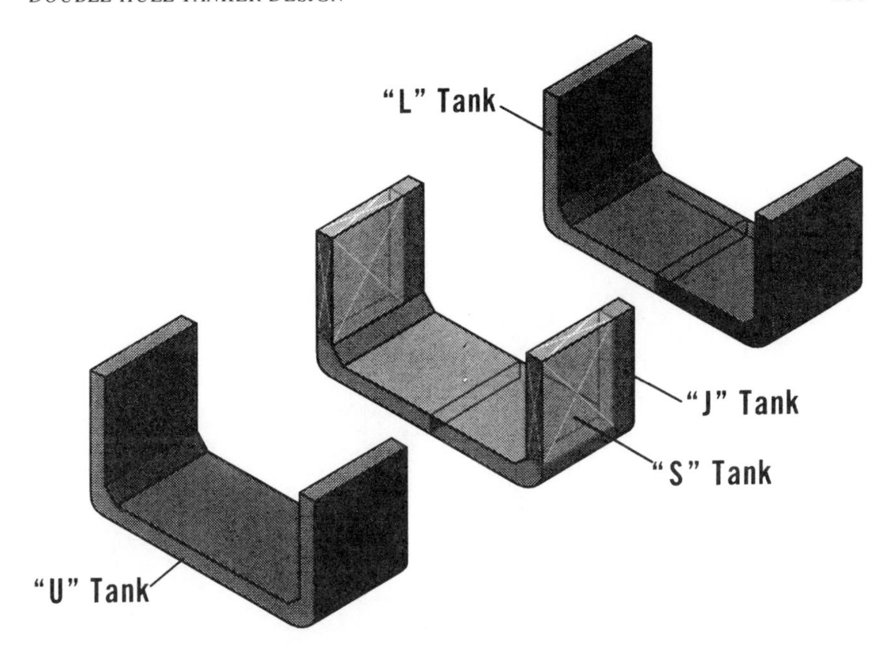

FIGURE E-7 Double-hull tank configurations.

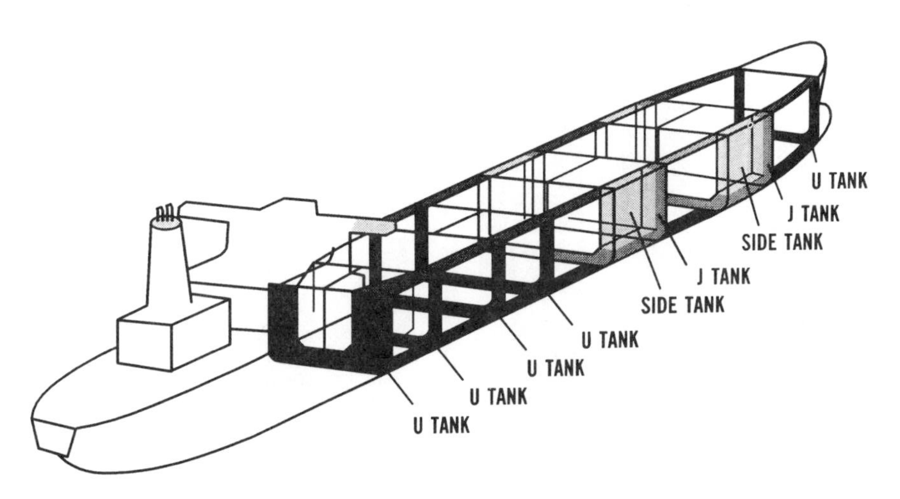

FIGURE E-8 Chevron 130,000 DWT double hull tank arrangement.

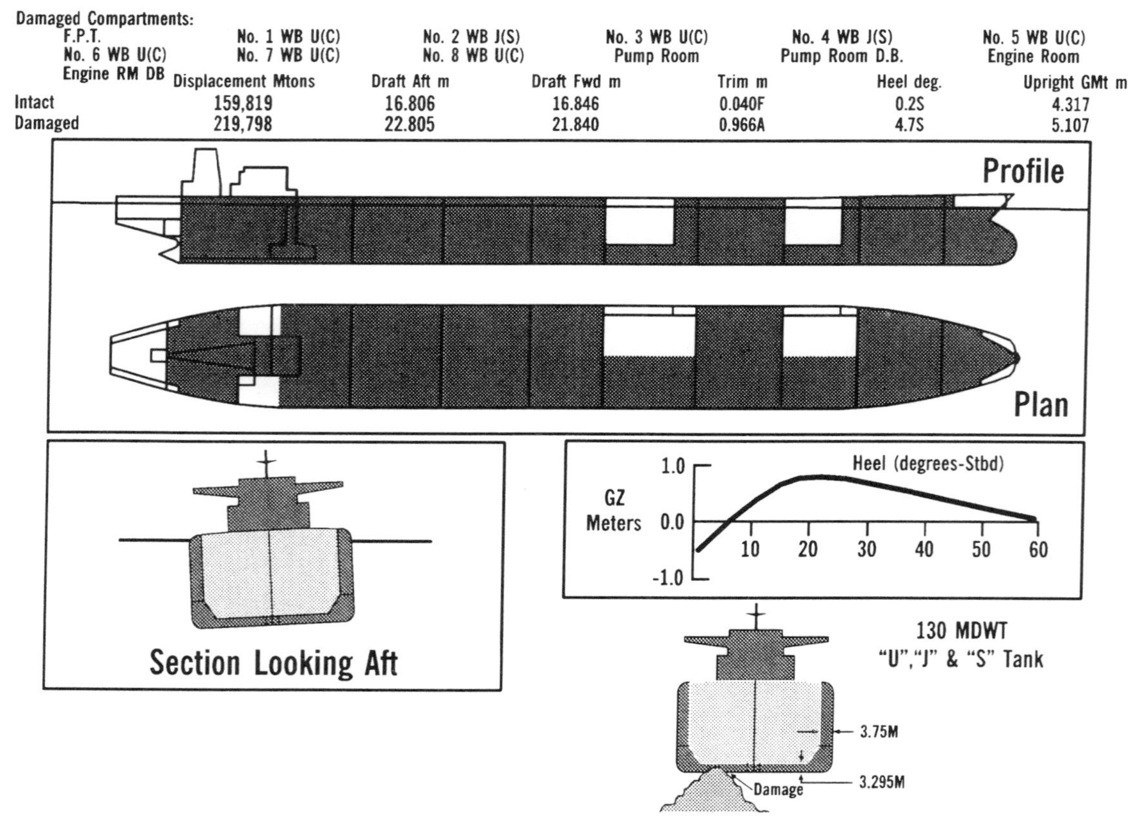

FIGURE E-9 Chevron 130,000 DWT double hull free-floating damaged condition—98% full.

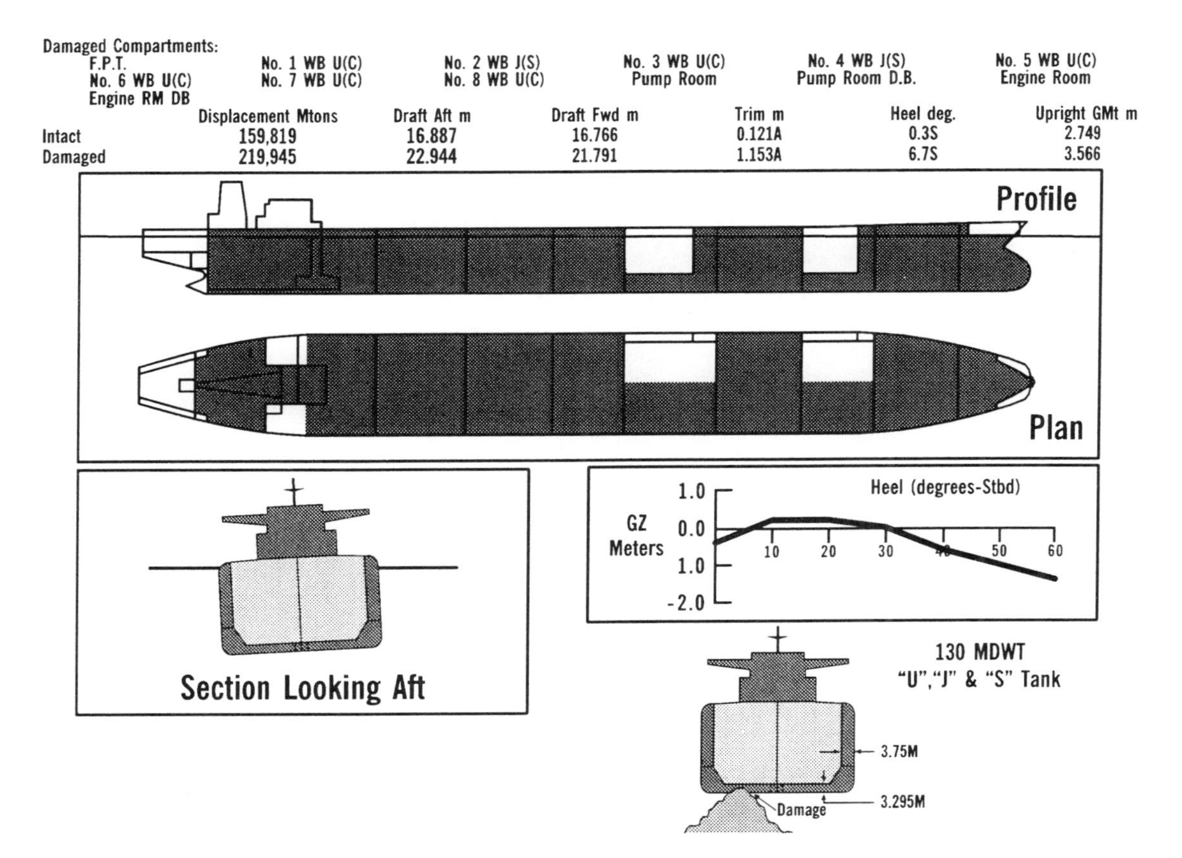

FIGURE E-10 Chevron 130,000 DWT double hull free-floating damaged condition—96% full.

APPENDIX

F

Comparative Study on Potential Oil Spill in Collision and/or Grounding—Different Tanker Designs

DET NORSKE VERITAS

The appendixes to this study are not included. Committee comments on assumptions and conclusions can be found at the end of the Det norske Veritas report on page 299 and following.

DET NORSKE
VERITAS

Report

Date 15 August 1990	Dept. DS0261	Project No. 261098

Approved by
for Det norske Veritas Classification A/S - DSO

W. Magelssen, Head of Technical Department

DIVISION SHIP AND OFFSHORE
Address: Veritasveien 1, Høvik, Norway
Postal Address: P.O Box 300, N-1322 Høvik, Norway
Telephone: (47-2) 47 99 00
Telex: 76192 verit n
Facsimile: (47-2) 47 99 11

Client, Sponsor	Client's ref.	Type of Report
National Research Council	D.W. Perkins	TECHNICAL

Summary

The potential oil outflow has been calculated using a probabilistic program (PROBAN) for 13 different VLCC designs, and 8 different 40,000 dwt tanker designs in case of collision and/or grounding. The influence of hydrostatically balanced loading of cargo tanks, and of vacuum systems on oil outflow has been analysed.

A double hull provides best protection against overall oil spill. A double bottom provides a relatively better protection against overall oil spill than a double side. A passive vacuum system reduces oil outflow much in grounding. Long and narrow tanks leak less than short, wide tanks.

The results show that a VLCC with double sides (> B/20) and a double bottom (> B/15) is likely to spill only ~ 33% of the quantity a conventional modern VLCC SBT spills. The corresponding reduction for a 40,000 dwt tanker with double bottom and double sides is ~ 37%.

DnVC Report No. 90-0161	Subject Group H1

Title of Report

COMPARATIVE STUDY ON POTENTIAL OIL SPILL IN COLLISION AND/OR GROUNDING - DIFFERENT TANKER DESIGNS

Work carried out by	Work verified by
P.E. Köhler L. Jørgensen	S. Valsgård

Date of last revision 25.01.91	Rev. No. 1	Number of pages 92

4 indexing terms

- TANKERS
- COLLISION
- GROUNDING
- OIL POLLUTION

Distribution statement:

- [x] No distribution without permission from the responsible department
- [ ] Limited distribution within Det norske Veritas Classification A/S
- [ ] Unrestricted

PREAMBLE

The recent casualties of larger oil tankers resulting in severe oil spills, together with oil spills from a number of smaller tankers grounded, have caused growing concern in many countries.

A review of the casualties indicates that human and operational aspects have been decisive factors in most casualties. In view of the serious consequences of every single oil spill, however, improvements in existing measures related to ship design and equipment will also have to be considered with the objective of reducing the potential for oil spills.

In this report an assessment of the potential oil outflow from VLCCs and 40,000 dwt tankers in case of collision and grounding is presented. The objective of the study has been to investigate how double sides, double bottom, tank size and tank location influence the amount of oil escaping from a damaged VLCC.

The study is based on available statistical information on damage location and extent. A novel probabilistic approach has been adopted to account for uncertainties in damage location and extent.

Additional oil spill caused by a grounded tanker possibly breaking in parts has not been considered in this study as this would necessitate a rather detailed analysis on the ultimate strength margins in damaged condition. Possible additional oil spill when salvaging a grounded tanker has not been considered in the study.

Neither has structural damage caused by explosion and fire resulting in oil spill been considered.

The results presented in this report are applicable to tankers in general.

The VLCC study is an extension of a joint Scandinavian study carried out earlier in 1990/8/. At the request of the National Research Council, two additional VLCCs - the intermediate oil tight deck and the hydrostatically balance loaded VLCC respectively, have been analysed and the results included in the graphs together with the other VLCCs. The 40,000 dwt tanker study was carried out exclusively for National Research Council.

Det norske Veritas Classification A/S and the participants of the first VLCC study accept no liability for any loss, damage or expense allegedly caused directly or indirectly, when using or referring to the results presented in this report.

LIST OF CONTENTS

1. SCOPE OF WORK

The objective of this study is to shed some light on the merits of building tankers with double sides and/or double bottom in order to reduce the probability of oil outflow in case of collision and/or grounding.

The analysis is focused on the cargo area only; - no crude oil is assumed to be located forward of the collision bulkhead, nor aft of the forward engine room bulkhead. Possible fuel oil leakage has not been considered in the study; - the fuel oil tanks are located aft of the forward engine room bulkhead. As the fuel tank location is assumed similar in all designs studied, pollution potential from fuel tanks is equal and hence not included in the study.

In this study the potential oil outflow from 11 different VLCC designs with double sides and/or double bottom have been compared with the potential oil outflow from a modern conventional VLCC of 280,000 dwt having segregated ballast tanks. In addition, the potential oil outflow from a modern conventional VLCC SBT with hydrostatically balanced loading of cargo tanks has been evaluated.

All VLCC designs have the same main particulars and body plan as the modern conventional VLCC. The length of the cargo area is equal for all VLCCs; the total cargo and ballast capacity respectively is approximately equal to that of the VLCC SBT within the cargo area for all VLCCs analysed.

Eight 40,000 dwt designs have been analysed. All designs have the same main particulars and body plan. The length of the cargo area is equal for all designs; - the ballast capacity within the cargo area varies from approximately 14,000 to 17,500 tons depending on arrangement.

The oil outflow in collision and grounding has been estimated using available statistical information on damage extent, combined with basic laws of mechanics and PROBAN[1)], a probabilistic analysis program developed by Veritas Research/1/.

The calculated potential oil outflow in collision and grounding has been combined using general casualty frequency statistics to provide an overall ranking of the different designs.

The influence of vacuum systems on the oil outflow in grounding has been investigated for VLCCs by amending the theory and formulae developed in connection with the DNVC - Pollution Prevention Class Notation work into the PROBAN oil outflow model.

The results from the VLCC and 40,000 dwt tanker studies have been discussed and extrapolated for 80,000 dwt tankers.

---

1) This program has been installed on computers at following organisations/companies in the United States: NASA, Boeing, ALCOA, Chevron and Conoco. A short description of PROBAN is given in Appendix 1.

2. ESTIMATION OF OIL OUTFLOW IN COLLISION AND GROUNDING

The potential oil outflow is estimated separately for collision and grounding casualties, and combined using a weighing based on collision and grounding casualty frequencies for tankers.

2.1 Oil Outflow in Collision

2.1.1 Basic Assumptions

Several simplifying assumptions have been made w.r.t. the collision process. In order to avoid optimistic results the assumptions made are conservative.

It is assumed that oil begins to escape from a cargo tank when the bow of the striking ship touches and penetrates one of the sides or corners of the tank. No large deformations resulting in yielding of tank sides have been assumed. Hence, at contact the tank is penetrated.

In the analysis it is postulated that the bow is wedge shaped and remains wedge shaped after having penetrated the hull plating. As the vertical bow penetrates the tank, the whole tank is ruptured from bottom to top. This assumption has been made in order to conform with the MARPOL/2/ assumption that all oil will escape from a damaged tank over time.

A consequence of these assumptions is that the results presented for collision strictly apply for the final condition only i.e. when all oil has escaped from a holed tank. For more accurate estimates on oil outflow during the collision a detailed modeling of bow crushing behaviour and ship side structural response during the collision process, including the changes in ship speeds and headings, and added mass effects, would be necessary/4,5/. This has been beyond the scope for this study.

It should be stressed that only damage in way of the collision contact point has been considered. Damages far away from the contact point, such as possible tearing in weld seems due to excessive tension potentially resulting in oil leakage, have not been considered.

As for possible explosions due to friction heat and sparks generated in a collision, it is assumed that the inert gas system is effective; - should the system fail then structural damage may become extensive. These aspects have not been considered in the study.

SEE COMMITTEE INTERPRETATIONS OF DNV COLLISION ASSUMPTIONS ON PAGE 299.

2.1.2 Damage Analysis Procedure

The procedure for calculating the potential oil outflow in collision is shown in the flow chart below(Fig.2.1).

The interdependence between collision damage length and damage penetration used in this study is shown in Fig.2.2. No information is given on ship speeds and sizes, nor on the number of tankers included in the basic population of 296 ships.

The damage length and damage penetration values in Fig.2.2 include implicitly elastic and plastic deformation of the bow and side structure, the speeds of both ships, changes in ship motions and accelerations as well as any mass effects.

A multilognormal distribution for damage penetration and length has been calculated based on Fig.2.2. The distribution has been used in the PROBAN analysis for defining the extent of structural damage.

The distribution of collision points along the ship hull is shown in Fig.2.3 for 332 collisions. The distribution includes struck and striking ships. This explains the large percentage of collisions in the bow region. Considering only struck ships, the distribution becomes bell shaped with about 30% of collisions aft and in the bow. No information on ship types or sizes is given; - it is assumed that the distribution is valid also for tankers.

Fig.2.2 includes damages to struck ships as well as striking ships. No information is given on the number of damages to struck and striking ships respectively. Refering to the collision point distribution in Fig.2.3, damages to the relatively more rigid bow region are emphasized in Fig.2.2. The multilognormal distribution calculated may thus be somewhat optimistic for side damages i.e. the damage lengths and penetrations might be short.

The effect of a double side structure on damage penetration has been considered by introducing an equivalent penetration depth. This equivalent penetration depth is obtained as a function of the single side penetration depth, the stiffness of the inner side structure in relation to the outer side structure and the distance between the outer and inner sides. For simplicity, elastic deformation only is assumed. In the analysis it has been assumed that the inner side has a mean stiffness of 0.7 of the outer side with a standard deviation of 0.1.

The ship length has been divided into 50 sections of equal length. Using the damage extent distribution, the cargo volume punctured i.e. number and size of tanks, and the resulting oil outflow, has been calculated for the 51 locations(ends of each section) along the ship hull.

The collision point distribution has been used for weighing the oil outflow calculated at the 51 locations. The oil outflow is obtained as volume(m^3), and may be converted to tons using the specific gravity of oil.

In Appendix 2 the limit state functions used for calculation of oil outflow in collision and grounding are given for design 1A as illustration.

Fig.2.1. Flow Chart for Collision Analysis

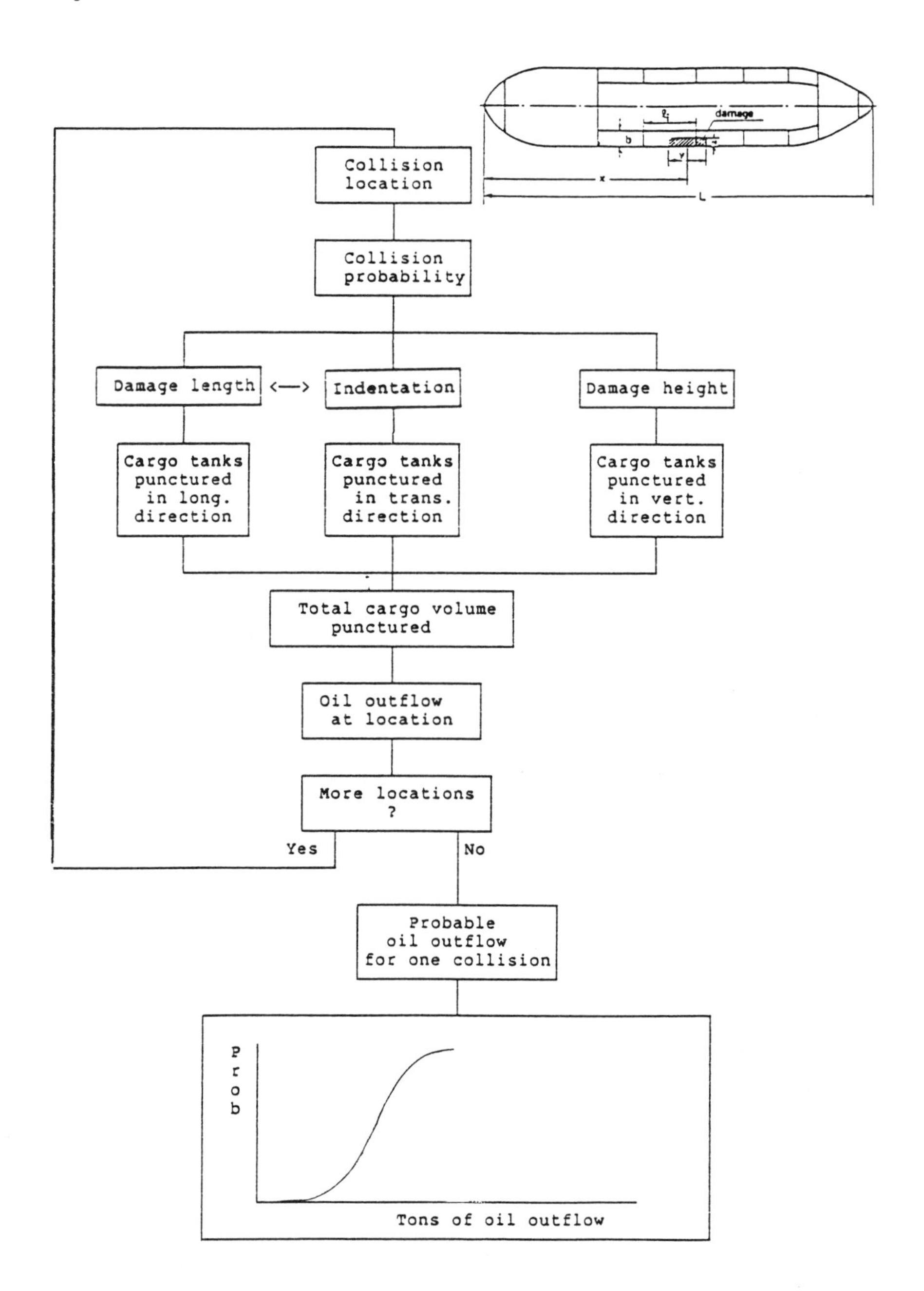

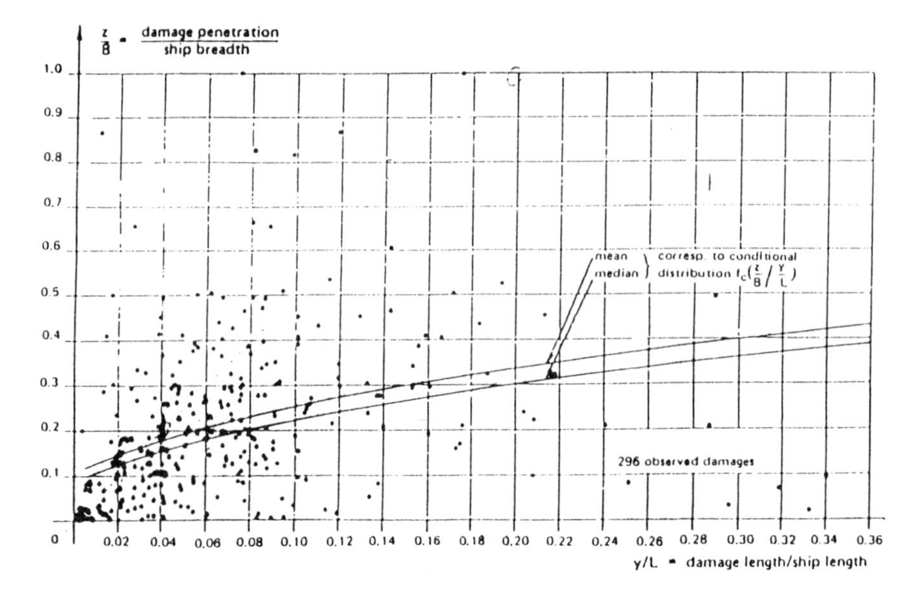

Fig.2.2. IMO Data on Stability and Subdivision - 296 Collisions/3/

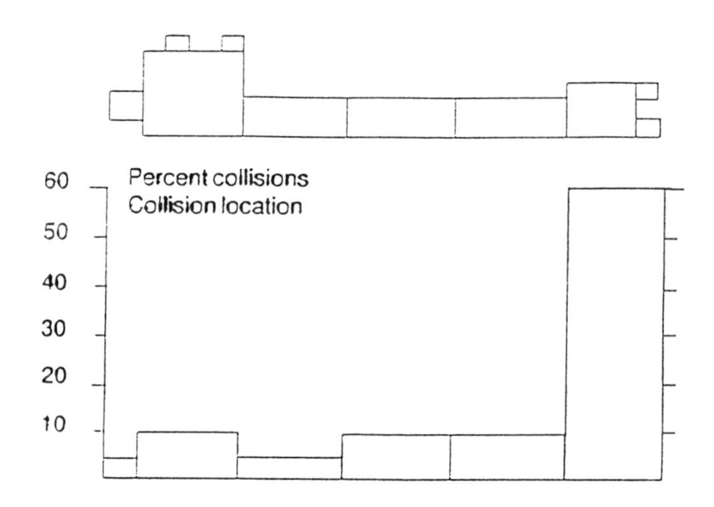

Fig.2.3. Location of Damage for 332 Collisions

2.2 Oil Outflow in Grounding

2.2.1 Basic assumptions

The simplifying assumptions made in the grounding analysis relate to the damage extent and oil outflow. Again, the assumptions made are considered to be conservative.

It is assumed that the ship has forward speed in grounding, i.e. the damage starts in the bow of the ship and develops towards the stern.

Damage caused by grounding while the ship is adrift, when the ship is turning or going astern, has not been considered.

The ground is assumed to be a solid rock which does not crush during the grounding process. The rock is assumed to be wedge shaped(i.e. triangular in the transverse yz-plane) having a constant breadth. A sensitivity analysis has been performed regarding the influence of the rock shape on damage extent(see below). Grounding on sand banks or mud bottom has not been considered as the damage in this case primarily would be local indentations, possibly combined with tearing of welds.

As the rock comes into contact with one of the sides of a tank, the bottom or a corner of a tank, the hull plating is assumed to be penetrated. Oil begins to escape until hydrostatic equilibrium is achieved in the damaged tank. Should the ship side be damaged in grounding, then it is assumed that all oil which is below the intact side plating, will escape from the tank and be substituted with water. In addition, due to the reduced hydrostatic pressure, the oil level will drop in the tank proportionally to the specific gravities of water and oil, and height of side damage.

In the analysis statistical information on maximum vertical extent of damage has been used for the whole damage length. This conservative assumption has been made due to lack of detailed information on the vertical damage extent, or penetration, as a function of damage length. Consequently, the damage lengths calculated may be too short as too much energy is absorbed vertically(see discussion below).

Possible tearing of welds well away from the damage location causing leakage of oil have not been considered(ref. similar assumption in the collision analysis above).

The influence of ship motions during the grounding process have not been considered in detail. During a grounding the ship may develop a forward trim due to shallow water i.e. downward suction of the bow, or due to sudden loss of buoyancy as the forward bottom plating is peeled off. The ship may also run up on the rock lifting the bow. In the analysis a variation in ship draught with $\pm$ 0.5 m has been included partly to cover these aspects.

Tidal water effects on oil outflow have not been considered in this study. At low tide more oil may escape from a ship sitting on the rock because of reduced draught. Tidal water may also result in excessive hogging moments when the ship sits on the rock. These moments might result the hull breaking in parts with uncontrolled pollution as an outcome; - this problem may be studied separately in detail using available programs for ultimate strength analysis of the hull girder.

SEE COMMITTEE INTERPRETATIONS OF GROUNDING ASSUMPTIONS ON PAGE 299.

2.2.2 Damage Analysis Procedure

The flow chart for the grounding damage analysis procedure is shown in Fig.2.4.

In analysing the extent of grounding damage, the energy absorbed by the bottom structure being crushed is estimated by following expression/6/:

$$W_s = 352\ V_s + 126\ A_s \tag{2.1}$$

where V_s is the volume of material displaced
A_s is the total area of fracture or tearing

The kinetic energy to be absorbed by the bottom structure in grounding may be expressed as

$$E_k = 0.5\ mv^2 \tag{2.2}$$

where $m = m_s(1+a_s)$
m_s = mass of the laden ship
a_s = added mass coefficient for surge
v = ship's speed at grounding

Assuming all kinetic is transformed into deformation energy, the extent of grounding damage may be expressed for a single bottom ship as follows:

$$E_k = W_s = L_d(352B_d t_{pe}+126t_{pa}) = 0.5\ m_s(1+a_s)v^2$$

or

$$L_d = 0.5\ m_s(1+a_s)v^2/(352B_d t_{pe}+126t_{pa}) \tag{2.3}$$

where L_d is length of damage in longitudinal direction
t_{pe} is equivalent thickness of bottom plating
t_{pa} is actual thickness of bottom plating
B_d is the breadth of damage

Fig. 2.4. Flow Chart for Grounding Analysis

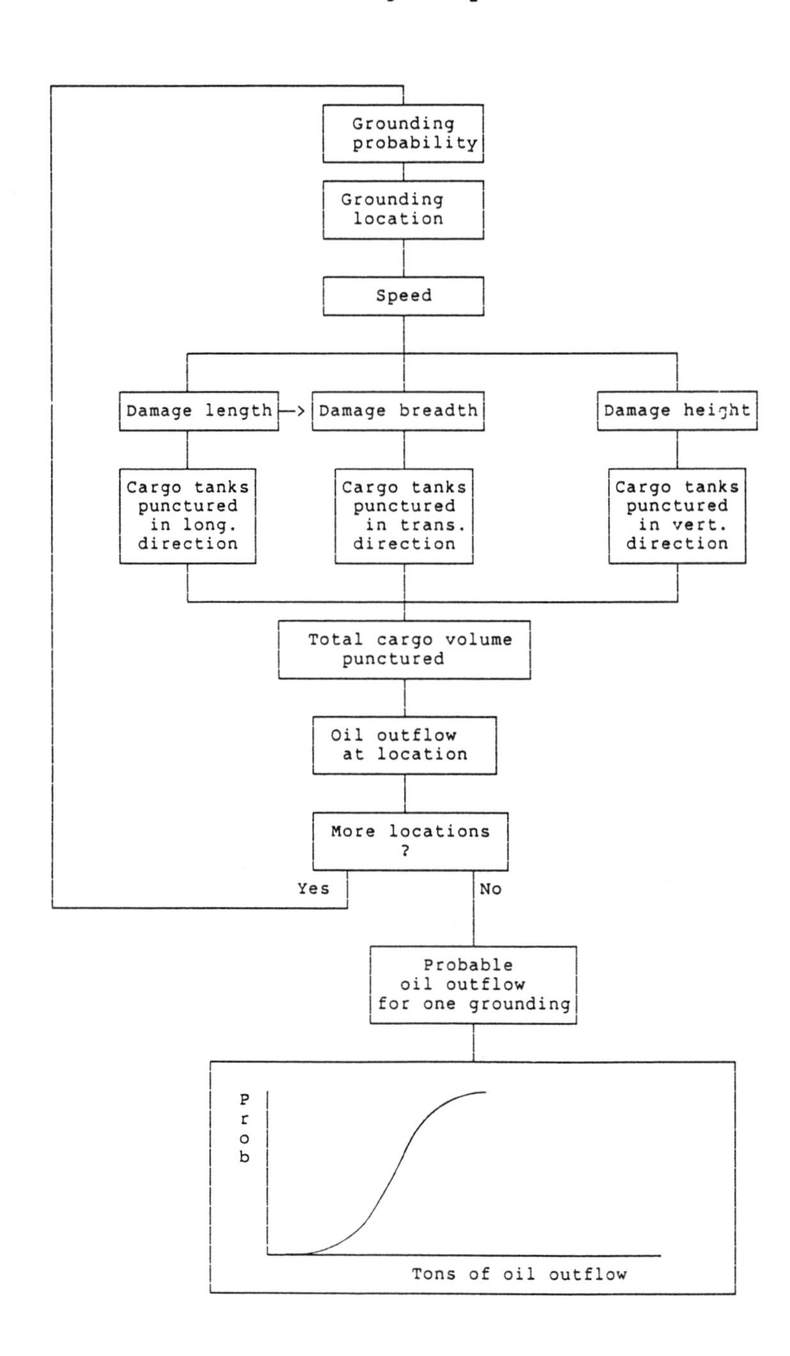

For a ship with a double bottom the corresponding expression is given by

$$L_d = 0.5m_s(1+a_s)v^2/[352(B_{d1}t_{pe1}+B_{d2}t_{pe2})+126(t_{pa1}+t_{pa2})] \quad (2.4)$$

where B_{d1} is breadth of damage to the outer bottom
B_{d2} is breadth of damage to the inner bottom
t_{pe1} is equivalent thickness of the outer bottom
t_{pe2} is equivalent thickness of the inner bottom
t_{pa1} is actual thickness of outer bottom
t_{pa2} is actual thickness of inner bottom

The damage height has been estimated from Fig.2.5 which gives the vertical extent of damage plotted against B/15. From this figure the mean 0.60512 has been calculated with a standard deviation of 0.3807. Most ships included in Fig.2.5 are ships with a deadweight less than 40,000 tons, only 4 being above. The damage height for the largest ship was less than one metre.

In order to study the effects of draught variation on grounding damage extent, the initial draught has been used with a standard deviation of 0.5 m. The draught deviation adopted partly covers variations in the specific gravities of crude oils carried.

The damage breadth has been assumed to be 2.5 times the damage height(IMO/MARPOL/2/ assumption).

In a recent report/7/ Prof.T.Wierzbicki of MIT has studied damage penetration vs. damage breadth, and the validity of the Vaughan formula. The conclusion was that the formula is valid for a breadth-penetration ratio of 6.7, and hence the MARPOL assumption above may result in too long damage lengths(by a factor of 1.8). On the other hand, the use of maximum vertical damage penetration for the whole damage length in the present study results in shorter damages and thus the effect of the above MARPOL breadth assumption on damage length is balanced to some extent.

The number of grounding points in transverse direction has been taken as 7. In case of the grounding point being on the port or starboard side, the damage will extend further in the longitudinal direction as the damage breadth will be reduced (see Eq. 2.3 and 2.4). When calculating the damage length at port and starboard sides, it is assumed that the side structure also will be damaged.

Available information on 24 tanker groundings shows that in 21 cases the damage began in the bow, and in 3 cases in the midship. Out of the damages in the bow, 8 extended to the midship and 3 to the stern. All damages beginning midships extended to stern. The tankers analysed were > 20,000 tdw.

The potential oil outflow in grounding is calculated at 51 locations along the ship hull(as in the collision analysis). The obtained oil outflow is weighed with the probability for damage at that specific location.

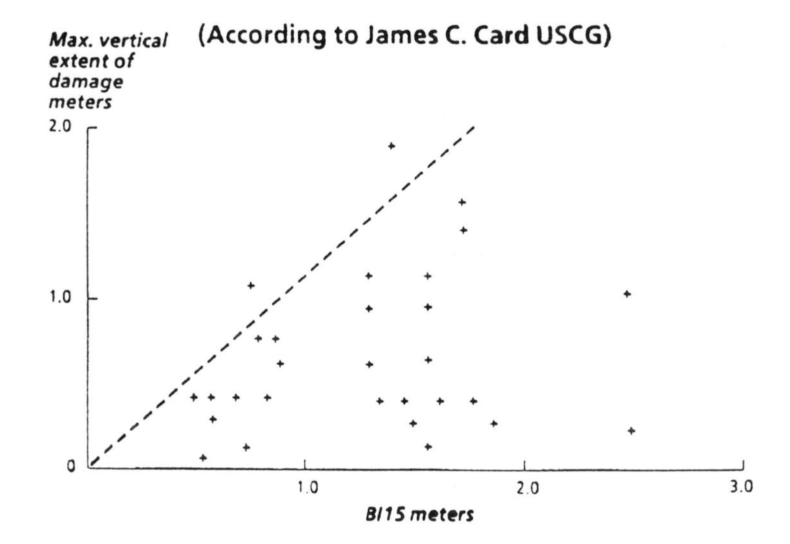

Fig.2.5. Vertical Extent of Grounding Damage

The estimated oil outflow is given in m³, and may be converted to tons using the specific gravity of oil.

3. PROBABILISTIC RANKING OF VLCC DESIGNS

3.1 VLCC Designs Analysed

The VLCC designs considered in this study are all designs that have surfaced recently with the prime objective of reducing the oil outflow in case of a severe casualty.

Common for these designs is that they are hybrids of modern conventional VLCCs with segregated ballast tanks. The oil outflow is expected to be reduced by introducing either double side or double bottom or both, and possibly in addition, utilizing the vacuum method for reducing oil outflow in grounding.

The influence of an intermediate oil tight deck on the potential oil spill has also been studied. The effect of hydrostatically balanced loading of cargo tanks on potential oil outflow from original VLCC SBT has been analysed.

3.1.1 General Features

All VLCC designs analysed have following main particulars:

L_{pp} = 315,0 m
B = 57,2 m
D = 30,4 m
T = 20,8 m
C_B = 0.83

For all designs, following data has been kept approximately equal:

- Total ballast capacity ≈ 107,000 m3
- Ballast capacity in cargo area ≈ 93,000 m^3
- Cargo capacity in cargo area ≈ 330,000 m^3
- Slop tank after

In addition, all designs have following draught and trim values:

- IMO minimum ballast draught = 8.3 m
- Design ballast draught = 9.95 m
- Trim ≈ -1.7...-2.0 m

MARPOL/2/ requirements to tank lengths have been considered as have requirements to

- Hypothetic Outflow max. 30,000 m^3
- Max centertank 50,000 m^3
- Max Wingtank 22,500 m^3

3.1.2 Particulars of the VLCC Designs

The numbers below refer to arrangement numbers in Fig.3.1- 3.13, and to numbers in graphs showing the estimated oil outflow for each design. For easy reference, a fold out page showing the VLCC designs compared in this study has been enclosed at the end of this report.

In Fig.3.1-3.13 the box in the low right hand corner shows the IMO/MARPOL/2/ requirements to protective area, the DnVC requirements for Environmental Class Notation PP1 and PP2 respectively, and the corresponding actual values.

Approximate tank volumes are given in Table 3.1 for each design.

0. Modern Conventional VLCC with segregated ballast tanks

This design is used for comparison in order to obtain an indication of how well the other designs perform in case of collision and grounding.

The modern conventional VLCC SBT barely meets the IMO requirements, and falls much short in meeting the DnVC PP1 and PP2 requirements.

The length over depth ratio for the modern conventional VLCC is 10.0-10.5 against 11.5-12.5 for older VLCCs. The increased freeboard of the modern conventional VLCC may result in more oil outflow in case of grounding as compared with an older VLCC.

1. Double side and double bottom

- 1A. With two longitudinal bulkheads
- 1B. With bulkhead at centreline

The long tanks in 1A have to be furnished with wash bulkheads in order to reduce sloshing loads. In 1B the tanks are approximately half the length of tanks in 1A.

This design exceeds the IMO, DnVC PP1 and PP2 requirements by a large margin.

2. Double bottom in whole cargo area

Assuming all ballast in the cargo area is carried in the double bottom only, a double bottom of appr. 6.6 m height is required.

This design exceeds the IMO and DnVC PP1 requirements, but is short of the DnVC PP2 requirement.

3. Double side in whole cargo area, single bottom

- 3A. Short tanks
- 3B. Centreline bulkhead in cargo tanks 1 and 2

Design 3A has not been analysed further as a centreline bulkhead is required due to MARPOL requirements; - hence 3A approaches 3B.

3B exceeds the IMO and DnVC PP1 requirements, but falls short of meeting the DnVC PP2 requirement.

4. Double bottom with ballast in side tanks

- 4. Ballast in two side tanks midship
- 4A. Ballast in two side tank aft midship
- 4B. Ballast in two side tanks bow and stern

For designs 4 and 4A the double bottom height and the width of side tanks are equal, and all tanks have the same corresponding length.

Design 4B has a lower double bottom height and narrower side tanks than 4 and 4A.

All three alternative designs exceed the IMO as well as the DnVC PP1 and PP2 requirements respectively.

5. Double side in whole cargo area and partial double bottom

In this design the double bottom extends over tanks 1 and 2 only, centre tanks 3-11 have single bottom. There is a centreline bulkhead in tanks 1 and 2.

Please observe the rather low double bottom height and that the side tanks are narrower than in 3B.

IMO and DnVC PP2 requirements are exceeded.

6. Double bottom in side tanks, single bottom in centre tank

This design carries no ballast in centre tanks as opposed to designs 2 and 4. To compensate, there are 5 short ballast side tanks.

The IMO and DnVC PP1 requirements are met, but not DnVC PP2.

7. Tank size half of MARPOL requirements

The number of tanks is significantly higher in this design than for designs above. The ballast is carried in side tanks only, every second side tank carries crude oil. The width of centre tank and side tanks is the same as for the modern conventional VLCC.

Please note that the ballast capacity in cargo area is slightly less than for other designs(~ 102,700 m^3), and the cargo capacity of ~ 320,700 m^3 is also lower than the cargo capacity for the other designs above.

This design meets the IMO requirements but not the DnVC requirements for PP1 and PP2.

8. Intermediate Oil Tight Deck

An intermediate deck is fitted 18.5 m above the baseline. All ballast is carried in the side tanks above the intermediate deck. The ballast tanks extend over the whole cargo area.

This design does not meet with the MARPOL requirements for protective tank location, nor DnVC PP requirements.

9. Hydrostatically Balanced Loading

The draught for the original VLCC is increased from 20.8 m to 24.4 m (A-freeboard) by ballasting in order to increase the cargo capacity when filling the cargo tanks to hydrostatic balance.

The VLCC with hydrostatically balanced loading can take 262,500 t of crude oil, or about 96% of the cargo in the original VLCC.

Table 3.1. Cargo Tank Volumes for the VLCCs(m^3)

Tank	ORIGINAL	1A	1B	2	3B	4	4A	4B	5	6	7	8
1 SB,BB	27800	19000	15000	13000	23000	11600	11600		18000	19400	10200	15400
C	38400	40000		27000		47000	47000	40000		43700	16500	38400
2 SB,BB		20000	22100	14900	14900	14900	11000	11000	19400	19400		9200
C	25600	41300		27000		44400	44400	37600		43700	17200	25400
3 SB,BB		20000	18400	14900			14900	12600		19400	10200	9200
C	25600	41300		27000	36900	44400	44400	43000	30200	43700	17200	25400
4 SB,BB	22800	20000	22100	14900		14900		11000		19400		13800
C	39000	41300		27000	29500	44400	44400	37600	30200	43700	17200	38100
5 SB,BB		3600	18400	14900		15000	15000	12600			10200	9200
C	25600			27000	36900	39500	39500	43000	30200	7900	17200	25400
6 SB,BB	22000		22100	14900								13600
C	30000			20300	29500	4800	4800	37600	30200		17200	29600
7 SB,BB			18400					2600			10200	
C					36900				30200		17200	
8 SB,BB			22100									
C					29500				30200		17200	
9 SB,BB			7300								10200	
C					36900				30200		17200	
10 SB,BB					3600							
C									30200		17200	
11 SB,BB							7500				10200	
C											17200	
12 SB,BB											5100	
C												

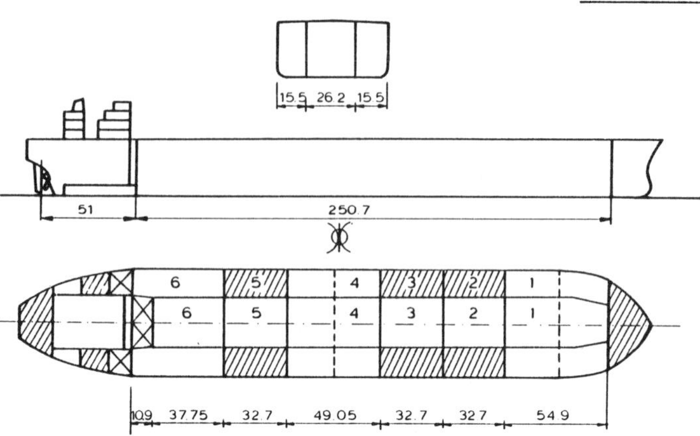

	IMO	PP1	PP2
REQUIRED	5917	11833	19229
ACTUAL	9005	8661	8661

Fig.3.1. Modern Conventional VLCC SBT(Original)

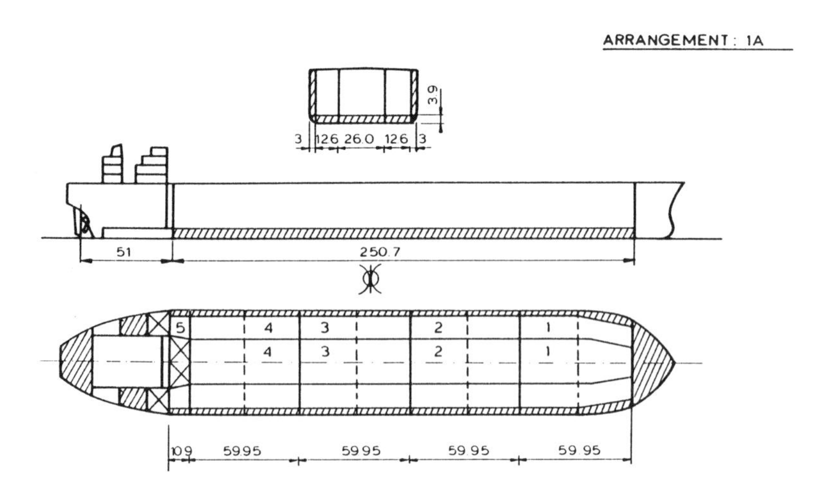

	IMO	PP1	PP2
REQUIRED	5917	11833	19229
ACTUAL	29583	29583	29583

Fig.3.2. Double side/double bottom (1A)

ARRANGEMENT: 1B

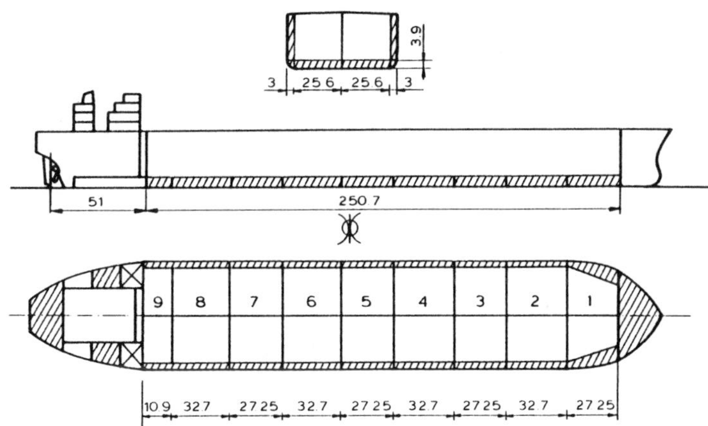

	IMO	PP1	PP2
REQUIRED	5917	11833	19229
ACTUAL	29583	29583	29583

Fig.3.3. Double side/double bottom (1B)

ARRANGEMENT: 2

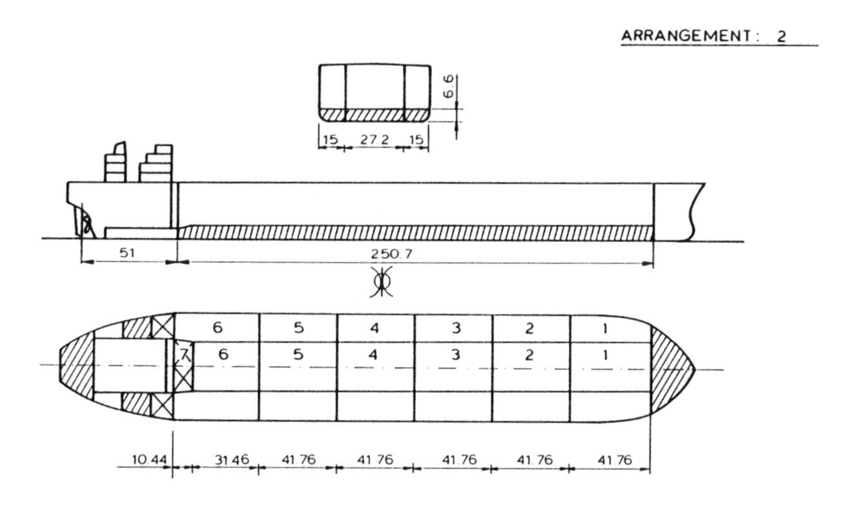

	IMO	PP1	PP2
REQUIRED	5917	11833	19229
ACTUAL	17649	17649	17649

Fig.3.4. Double bottom (2)

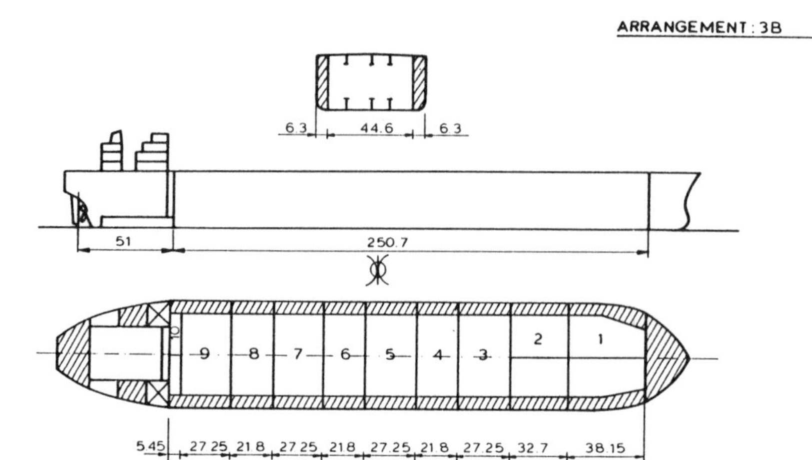

	IMO	PP1	PP2
REQUIRED	8875	13312	20708
ACTUAL	18401	18401	18401

Fig.3.5. Double side, single bottom (3B)

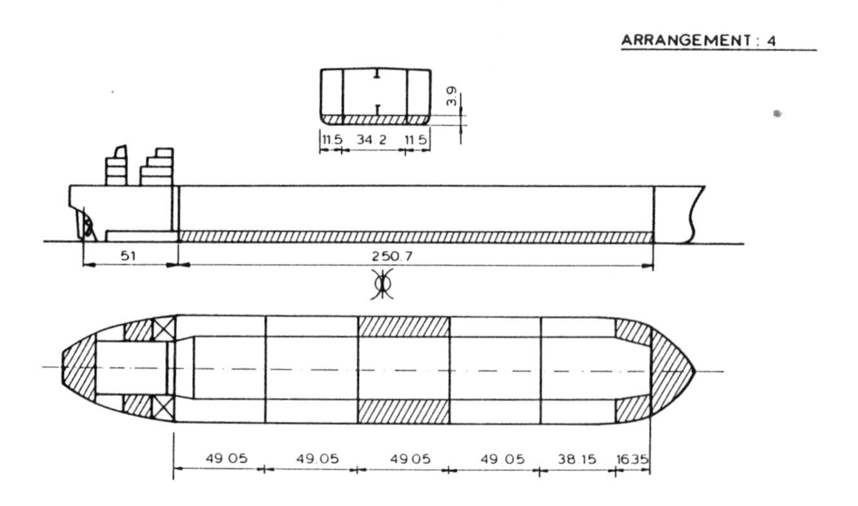

	IMO	PP1	PP2
REQUIRED	5917	11833	19229
ACTUAL	19761	19628	19628

Fig.3.6. Double bottom, single side (4)

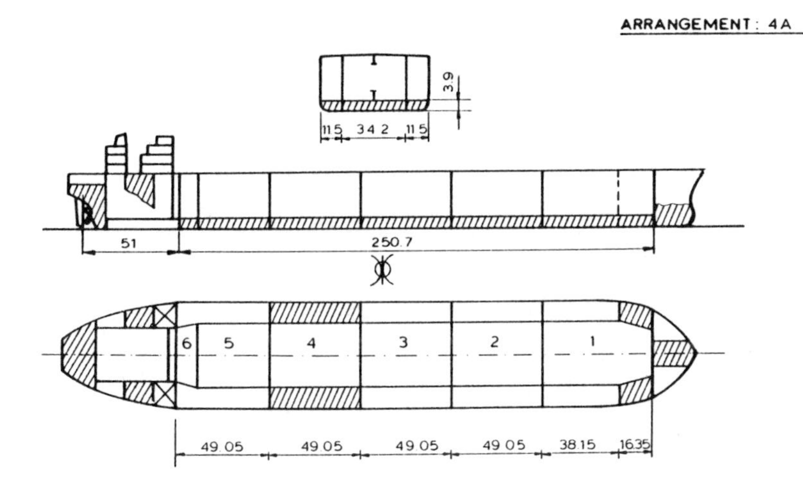

	IMO	PP1	PP2
REQUIRED	5917	11833	19227
ACTUAL	19761	19628	19628

Fig.3.7. Double bottom, single side (4A)

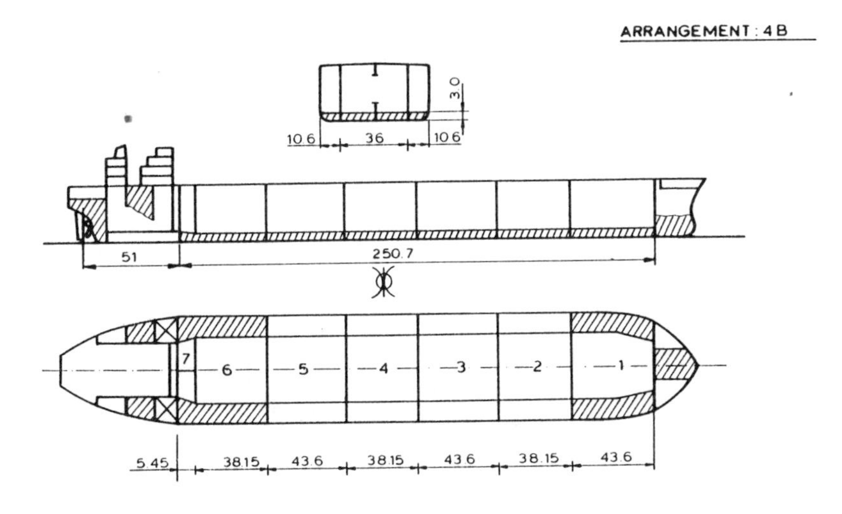

	IMO	PP1	PP2
REQUIRED	8875	13312	20708
ACTUAL	20920	21356	21356

Fig.3.8. Double bottom, single side (4B)

ARRANGEMENT: 5

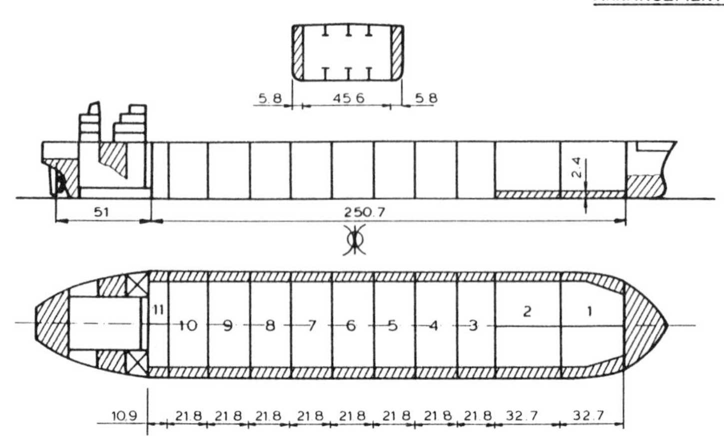

	IMO	PP1	PP2
REQUIRED	8875	13312	20708
ACTUAL	21093	22354	22354

Fig.3.9. Double side, partial double bottom (5)

ARRANGEMENT: 6

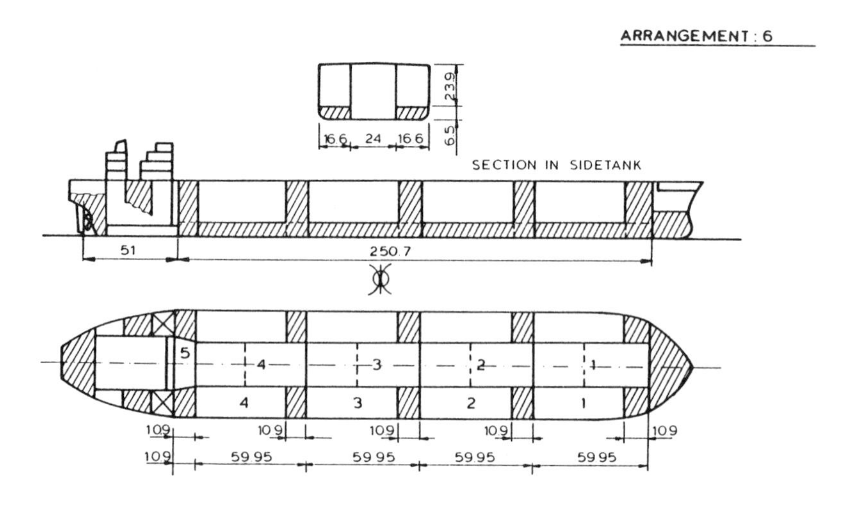

	IMO	PP1	PP2
REQUIRED	8875	13312	20708
ACTUAL	14186	14387	14387

Fig.3.10. Double bottom side tanks, single bottom centre tank (6)

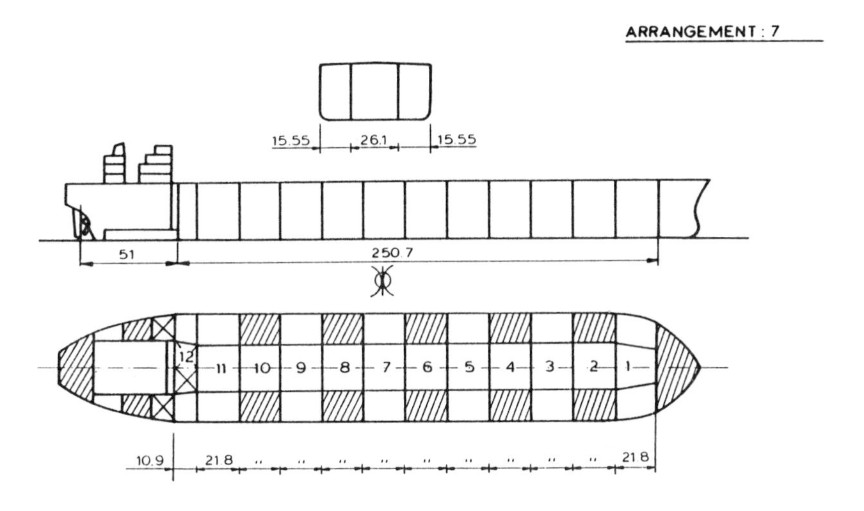

	IMO	PP1	PP2
REQUIRED	5917	11833	19229
ACTUAL	10017	10375	10375

Fig.3.11. Tank size half of MARPOL requirements (7)

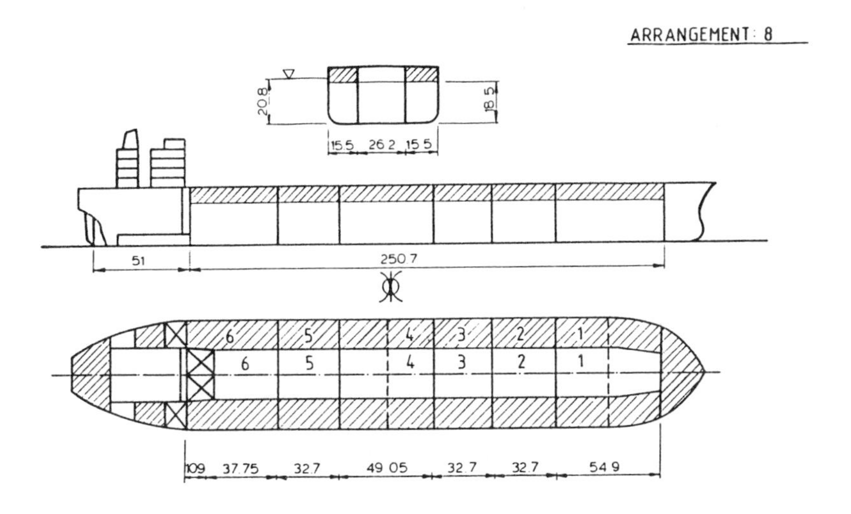

	IMO	PP1	PP2
REQUIRED	5917	11833	19229
ACTUAL	0	0	0

Fig.3.12. Intermediate Oil Tight Deck(8)

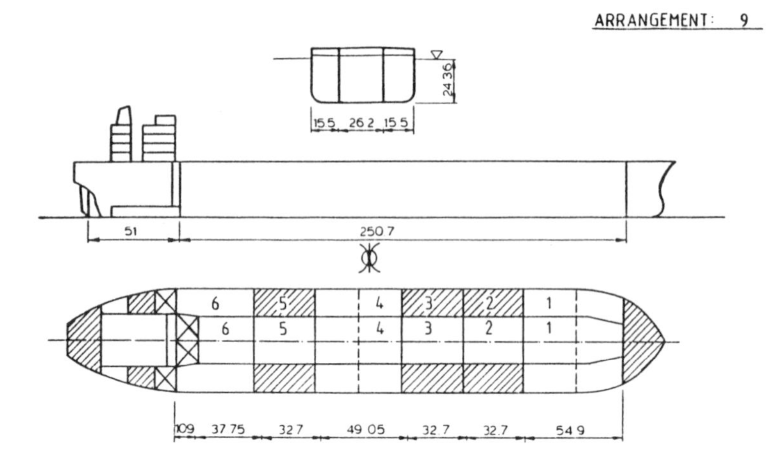

	IMO	PP1	PP2
REQUIRED	5917	11833	19229
ACTUAL	0	0	0

Fig.3.13. Hydrostatically Balanced Loading(9)

Intact stability, still water bending moments and shear forces have been calculated for designs 1A, 3 and 7. For 1A the damage stability calculations have been performed as well. The results are presented in Appendix 3.

An estimation of the steel weight has been carried out for all VLCC designs. The results are given in Appendix 3.

3.2 Ranking of VLCC Designs

Due to the simplifications and generalisations made above, and to shortcomings in the statistical information, the results from this study should be used for ranking the different VLCC designs only.

In the analysis the specific gravity for crude oil is assumed to be 0.9 t/m³. At this gravity the ship will float deeper than the design draught. The influence of various specific gravities on oil outflow has been examined.

3.2.1 Ranking in Collision

The results from the PROBAN analysis are shown in Fig.3.14-3.16. The cumulative probability on the ordinate in Fig.3.14-3.15

gives the probability for the escaping amount of oil to be equal or less the amount shown on the abscissa.

Double side as protective barrier against oil outflow is clearly demonstrated in Fig.3.14. In appr. 40% of all collisions resulting in rupture of the hull plating, no oil will escape at all for designs 3B and 5 as the inner side remains intact. Similarly, designs 1A and 1B will not leak oil in appr. 20% of all collisions resulting in rupture of hull plating.

Once the double side is ruptured, however, rather much oil is expected to escape in both 3B and 5 due to short tanks: the probability of penetrating several cargo tanks in collision is imminent.

The influence of cargo tank size on the potential oil outflow in collision for double side designs is also illustrated when comparing curves for 1A and 1B with each other: in 75% of all collisions design 1A with longer tanks is likely to leak significantly less oil than the same design 1B with short tanks.

As for single side VLCCs(Fig.3.15), 4B performs rather well and the oil outflow is quite limited for up to 60% of all collisions. For the rest of collisions the potential for oil outflow is quite high. This particular design has ballast side tanks immediately aft of the collision bulkhead which, considering the collision statistics used, explains the good performance in most collisions.

Best performance in all collisions is shown by design 7, with small cargo and ballast tanks. The ballast tanks provide a very effective protective barrier reducing the amount of oil likely to escape.

Considering the Intermediate Oil Tight Deck VLCC, this design performs similar to design 2 in about 70% of collisions, but clearly poorer for the rest of collisions. The reason for the difference is that at higher probability level the centre tanks are likely to be punctured, and as these contain only crude, then the potential oil outflow will increase much compared to design 2.

When loading the original VLCC SBT hydrostatically the potential oil outflow is reduced with an amount which corresponds to the difference in cargo quantity carried. The volume escaping remains the same as in the original VLCC SBT.

In order to rank all designs w.r.t. oil outflow in collision, the curves in Fig.3.14 and 3.15 have been integrated for accumulated probability of 0.0 to 0.9, and compared with the original VLCC (index=100). The reason for not integrating up to 1.0 is that the curves flatten out between 0.9 and 1.0, and the smoothening of the curves may introduce excessive errors influencing the index. In addition, the uncertainties associated with the different outflow model parameters will have a rather strong effect at this probability level reducing the validity of results.

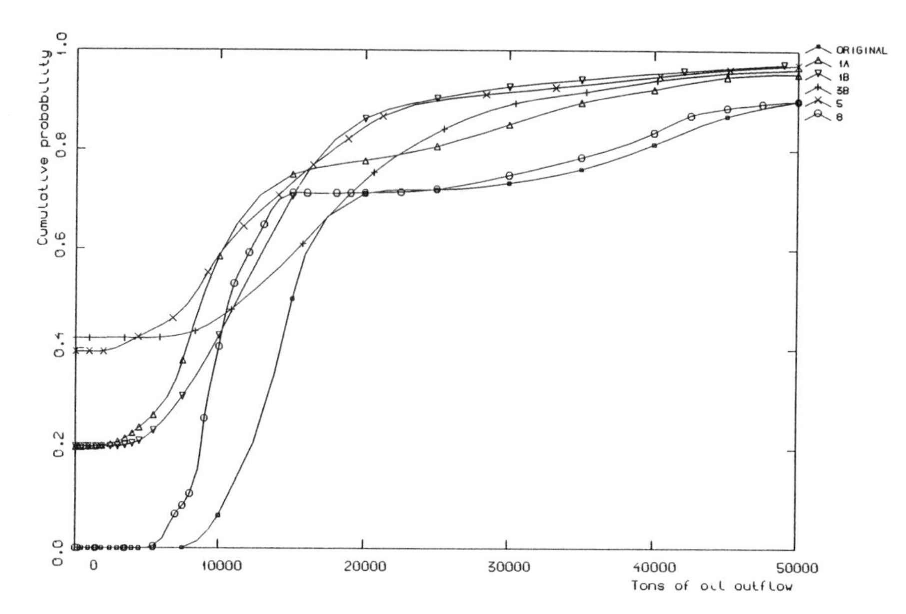

Fig.3.14. Cumulative Probability for Oil Outflow in Collision - Double Side Designs vs. Conventional VLCC

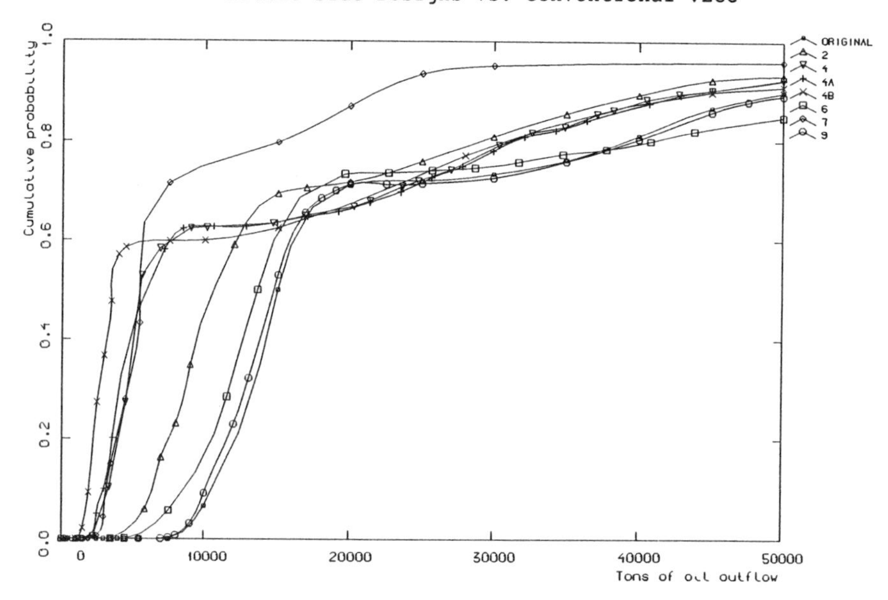

Fig.3.15. Cumulative Probability for Oil Outflow in Collision - Single Side Designs vs. Conventional VLCC

The resulting ranking is shown in Fig.3.16. It shows that designs 5(double side, partial double bottom) and 7(small tanks) leak about 25% less oil than 1A and 1B(double side, double bottom), and about only a third of the oil escaping from the conventional modern VLCC. Designs 6, 8 and 9 do not perform that well.

It is strongly stressed that the potential amount oil shown escaping in the results, is a weighed mean of the total amount oil contained in the cargo tanks damaged. In reality, the amounts are likely to be significantly lower as tanks might not wholly be ruptured from bottom to top due to actual bow forms, and as possible pollution preventive measures such as cargo transfer to other tanks may be initiated by the crew.

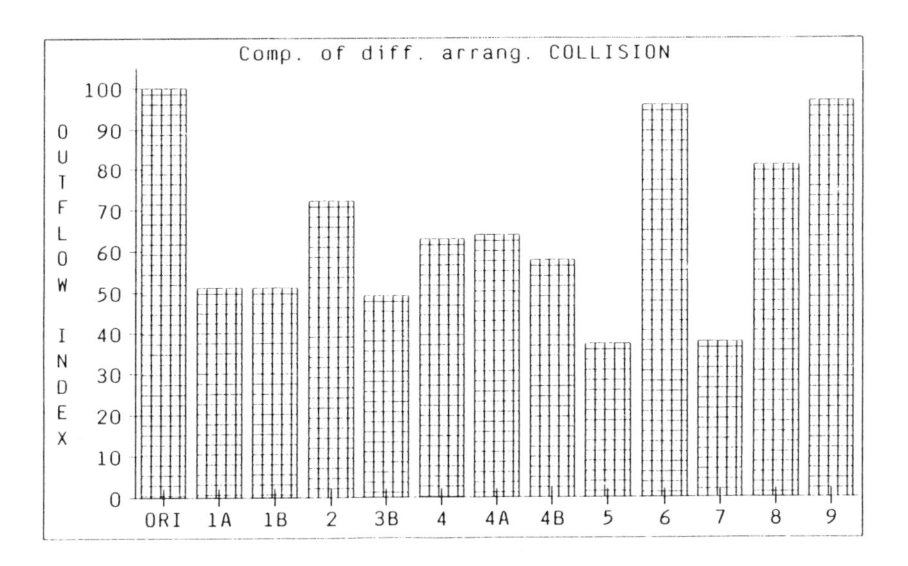

Fig.3.16. Ranking of VLCC Designs w.r.t. Oil Outflow in Collision

3.2 Ranking in Grounding

Extrapolating the calculated damage height mean for B/15(= 3.813) gives the damage height mean used in the PROBAN runs VLCCs as 2.308 with a standard deviation of 1.452.

As no supporting values on the damage height for grounded VLCCs have been available, the extrapolated damage height used in this analysis might be too high reducing the positive influence of double bottom height on the amount of oil escaping.

The results from the PROBAN analysis are shown in Fig.3.17-3.24. The cumulative probability on the ordinate in Fig.3.19-3.22 gives the probability for the escaping amount of oil in grounding to be equal or less than the amount shown on the abscissa at a grounding speed of 5 kn(Fig.3.19-3.20) and 10 kn(Fig.3.21-3.22).

Fig.3.19 and Fig.3.21 clearly show the merits of a double bottom w.r.t. reducing the amount of escaping oil. In some 85% of all groundings no oil is likely to escape from designs 1A, 1B, 4 and 4A having a double bottom of 3.9 m. For design 2, with a double bottom height of 6.6 m, this figure is not less than 99.8% !

The intermediate oil tight deck does not completely stop oil outflow in grounding; - groundings damaging the ship bilge and side will always result in oil outflow. A double bottom will protect against most of these damages.

The same conclusion applies to a hydrostatically loaded VLCC, due to possible bilge and side damage in grounding it will always leak some oil. Should design 3B be hydrostatically loaded, virtually no oil would leak in grounding as the probability for the bilge damage extending to the cargo tank is extremely small.

On the other hand, should the double bottom be penetrated, rather much oil will escape due to rather long damage lengths even at moderate grounding speeds. This is shown by the horisontal part of the oil outflow curves in 3.19 and 3.21.

A double bottom VLCC with hydrostatically filled cargo tanks will perform well in grounding. The double bottom will protect against bilge damage, and should the bottom be punctured, then the bottom will be filled with water. If the tanks are filled to a level corresponding to hydrostatic balance plus double bottom height, then some oil may escape in grounding due to mixing of water and oil.

In Fig. 3.17 - 3.18 the damage lengths have been calculated based on Eq. 3.3 and 3.4 for a grounding speed of 5 and 10 kn respectively. An increase in the grounding speed to 10 kn will result in the damage length reaching 100 m and more; - several tanks will be damaged in single bottom ships, or when the double bottom height is insufficient.

From the figure the influence of increasing the double bottom height on the damage length is evident: - as the height goes up, so goes the damage length i.e. the less influence will the inner bottom have on absorbing kinetic energy. For double bottom heights of 5 m and above, no kinetic energy is absorbed by the inner bottom given the damage heights used in this study.

This is illustrated in Fig.3.20 and Fig.3.22. All VLCCs without a double bottom seem to leak quite much oil should they run aground. Designs 3B and 5 perform both much worse than the conventional VLCC in grounding. This is due to wide and relatively short cargo tanks in 3B and 5; - once running aground much crude oil will escape as the ballast side tanks are quite narrow.

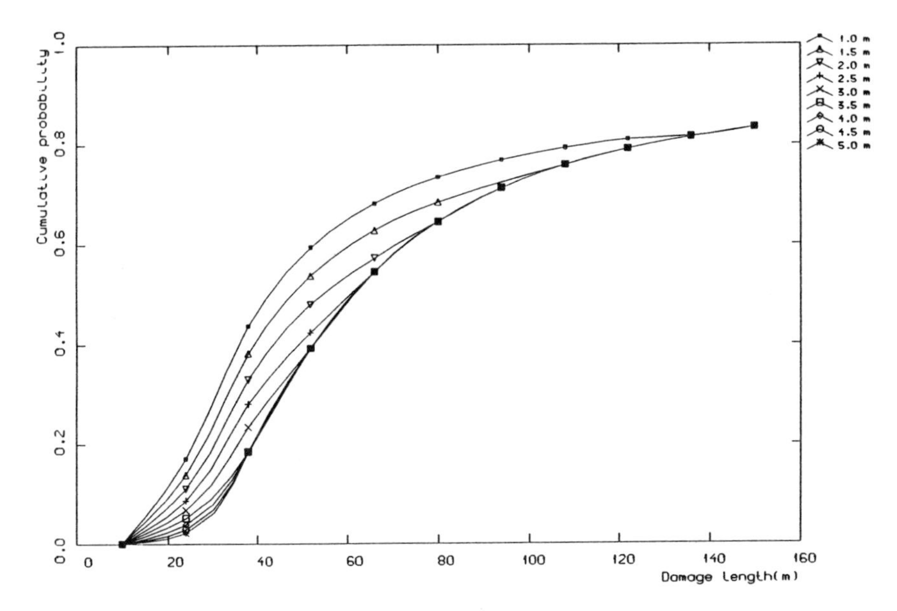

Fig.3.17. Damage Length vs. Double Bottom Height, 5 kn

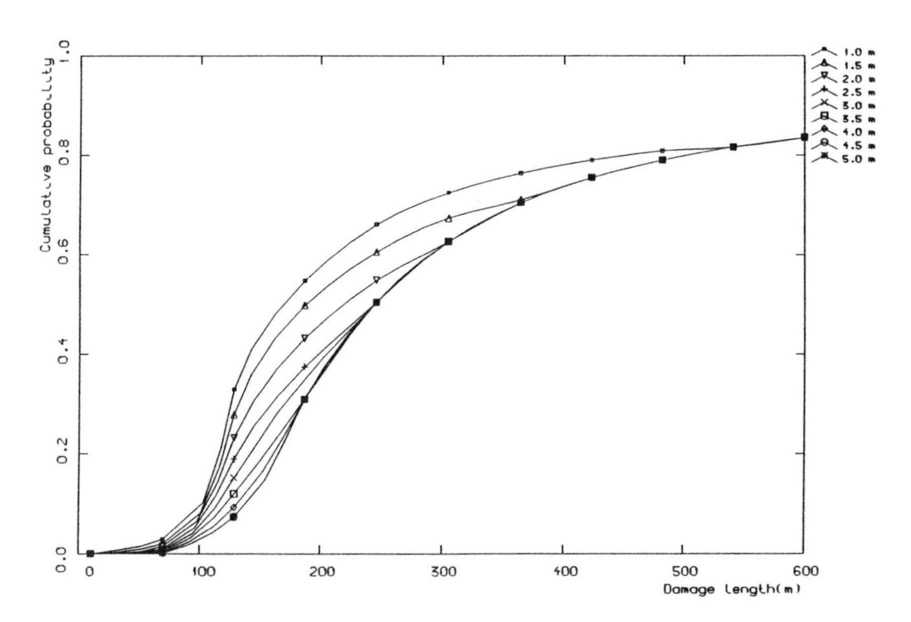

Fig.3.18. Damage Length vs. Double Bottom Height, 10 kn

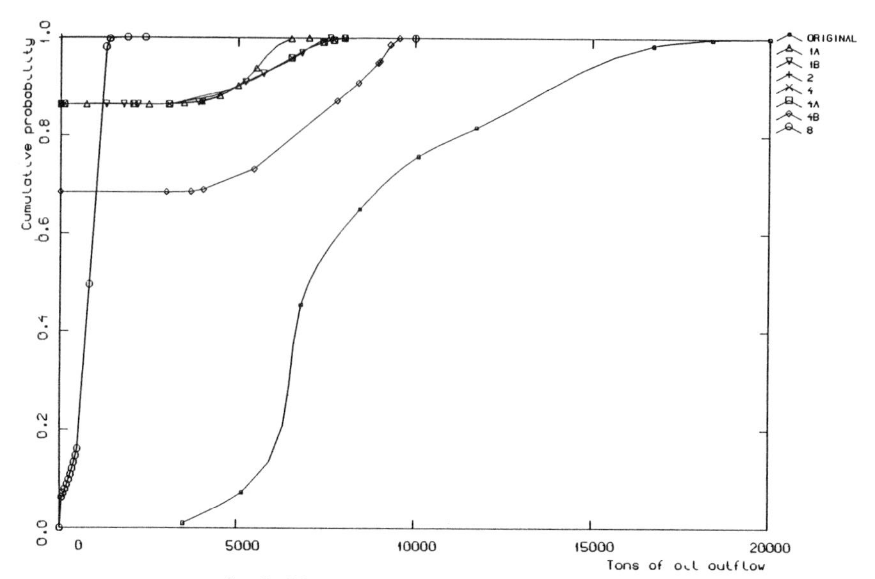

Fig.3.19. Double Bottom, 5 kn

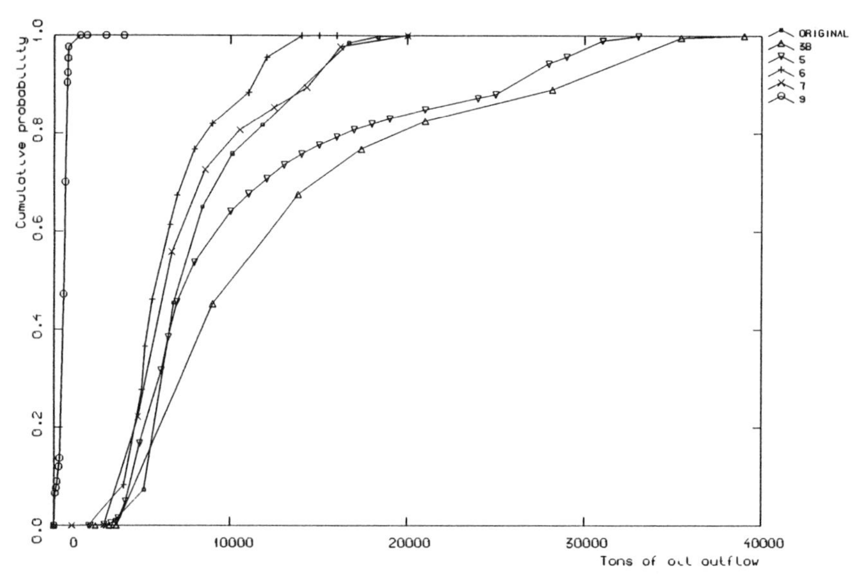

Fig.3.20. Single Bottom, 5 kn

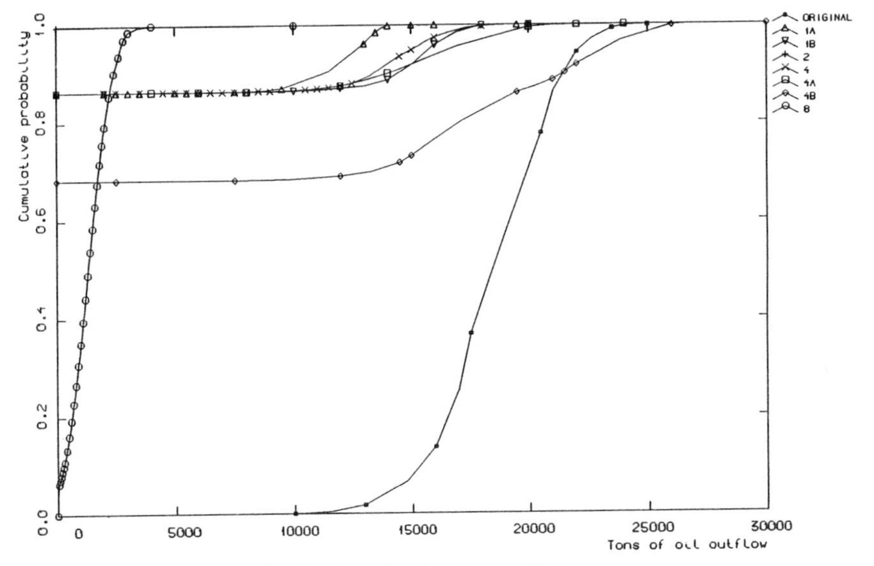

Fig.3.21. Double Bottom, 10 kn

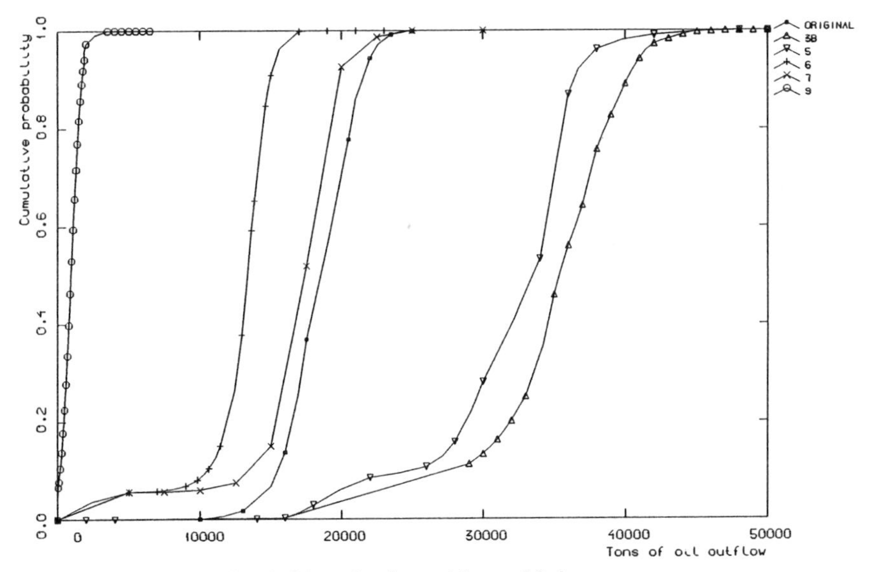

Fig.3.22. Single Bottom, 10 kn

The analysis shows that the partial double bottom in design 5 seems to be somewhat short and too low in order to be effective against oil outflow in grounding.

Considering the influence of speed at grounding on the oil outflow, the curves in Fig.3.19 and 3.21, and in Fig.3.20 and 3.22 respectively, indicate a doubling in the potential amount of oil escaping in grounding as the speed increases from 5 to 10 kn at the same probability level.

A sensitivity analysis has been performed in order to assess the influence of rock shape on damage extent showing that if the breadth of damage is halfed, the damage length at 5 kn will be approach the results obtained above for 10 kn.

In Fig.3.23 the average oil outflow index is given for groundings at 5 kn. The index is for each design is obtained by integrating the curves in Fig. 3.19-20 from 0.0 to 0.9, and comparing with the original VLCC.

Again, the double bottom designs 1A, 1B, 2, 4 and 4A show their good overall resistance against oil outflow in grounding. The intermediate oil tight deck design and the hydrostatically loaded tanker perform almost as well. The inferior designs 3B and 5 are likely to leak appr. 60% and 25% more oil respectively than the conventional VLCC in grounding.

Fig.3.24 gives the corresponding oil outflow index for 10 kn. As speed is increased, the cumulative probability for no leakage remains the same but because the damage length is increased so is the amount of oil likely to escape. One observes that the intermediate oil tight deck design as well as the hydrostatically balanced tanker now perform slightly better than designs 1A, 1B and 4, 4A. This is due to the more pronounced speed sensitivity of the latter designs(see Fig.3.21).

The influence of specific gravity on potential oil outflow has been studied for the original VLCC in grounding at 10 kn. When the gravity is 0.85 t/m^3, the potential oil outflow index is reduced by some 20%.

3.2.3 Combined Ranking - Collision and Grounding

An analysis of serious collision and grounding casualties carried out by DnVC for tankers above 20,000 tdw shows that the probability for collision casualties is appr. 40% and for groundings appr. 60% on a world wide basis.

Using these percentages, the average oil outflow for different designs in collision and grounding respectively have been combined in Fig.3.25 for 5 kn grounding speed, and in Fig.3.26 for 10 kn grounding speed. In the figures, the total index has been given as well as the parts from collision and grounding.

From Fig.3.25 it may be concluded that the double side/double bottom designs 1A and 1B provide the best overall protection

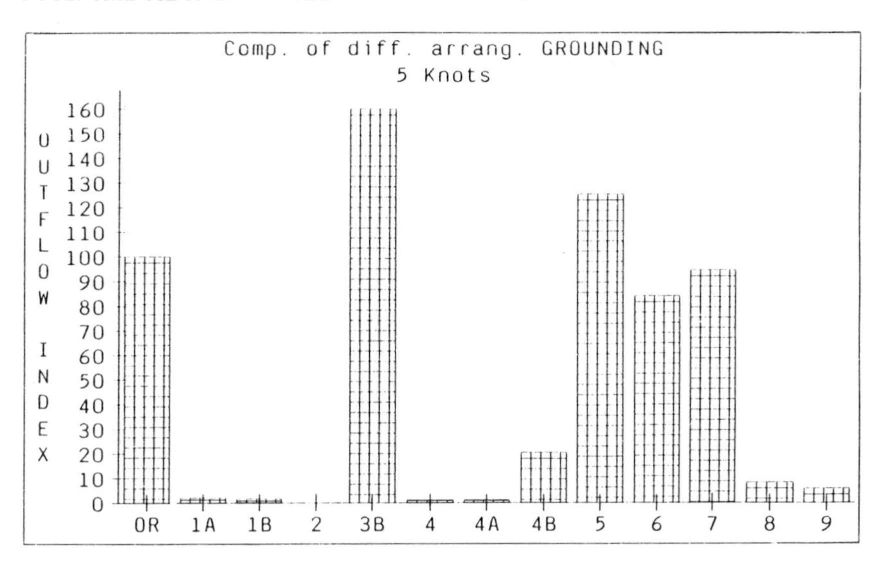

Fig.3.23. Ranking of Designs, Grounding 5 kn

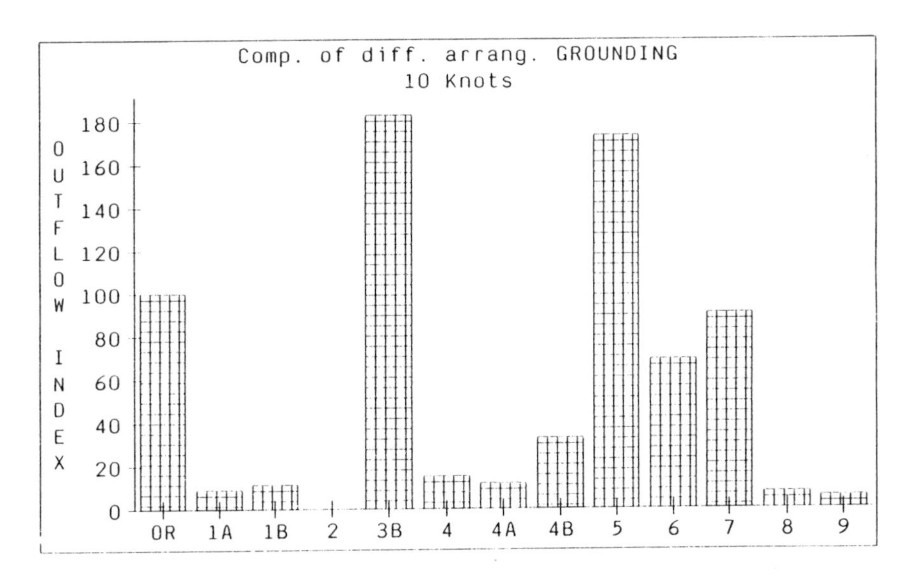

Fig.3.24. Ranking of Designs, Grounding 10 kn

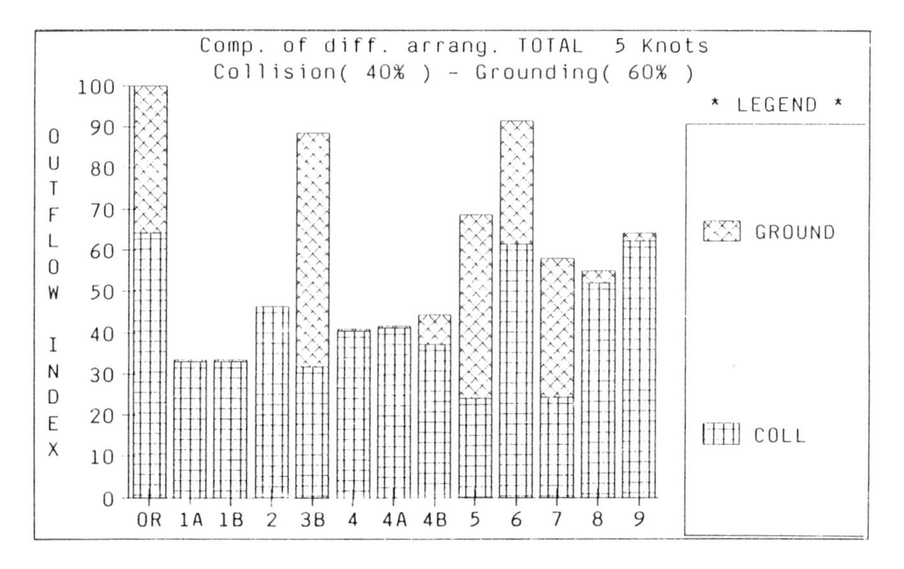

Fig.3.25. Combined Ranking, 5 kn

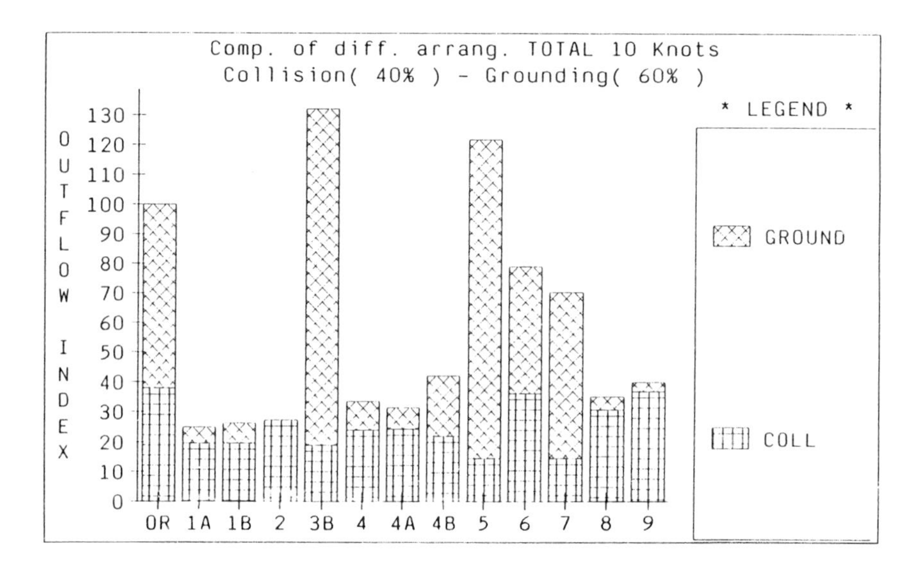

Fig.3.26. Combined Ranking, 10 kn

against oil outflow. Designs 4 and 4A are almost as good as designs 1A and 1B.

The wide side tank, single bottom design 3B performs worse than the conventional VLCC at high grounding speeds as does the partial double bottom design 5. At low grounding speeds, the performance of the partial double bottom design is much improved.

Design 7 with 50% MARPOL tank sizes is likely to leak some 35-40% less oil on the average in case of collision and grounding.

The combined index for the intermediate oil tight deck design suffers from bad collision performance. At 10 kn, this design leaks only some 35% oil compared to the original VLCC but 40% more than designs 1A, 1B and 2.

3.2.4 Impact of a Vacuum System on Oil Outflow in Grounding

The impact of a vacuum system[1)] on the oil outflow has been calculated for the conventional VLCC and the single bottom designs 3B and 7 respectively at a grounding speed of 5 kn.

It is assumed that the vacuum system is able to keep the vacuum in 40% of the total tank volume. Should more than 40% of the total tank volume be damaged, it is assumed that the first 40% of the damaged tanks keep the vacuum, and that the oil contained in subsequent tanks damaged will escape until hydrostatic balance is reached in the tanks.

The results are shown in Fig.3.27. For the conventional VLCC and design 7, the oil escaping is reduced to some 11%, for design 3B to about 29%.

The average oil outflows in collision and grounding with vacuum have been combined using above weights; - the results are shown in Fig.3.28. Comparing these values with corresponding values in Fig.3.25 shows that design 7 with vacuum is likely to perform as well as the double skin designs 1A and 1B, except that in 85% of the groundings no oil will escape from 1A and 1B at all whereas there always will be some oil leakage from a single bottom VLCC with the vacuum system fitted in the cargo tanks.

It is interesting to note that a conventional VLCC fitted with a vacuum system is likely to reduce its overall average oil outflow by half as compared with the basic design without the vacuum system. However, even with the vacuum system fitted, the VLCC is likely to leak about twice the amount of oil when compared with the double skin designs 1A and 1B.

---

1) A negative pressure - vacuum - is created in a damaged cargo tank provided the tank is quickly sealed off after the incident. The negative pressure may be established by closing the PV valves, and maintained by using an ejector which prevents pressure equalization and further outflow.

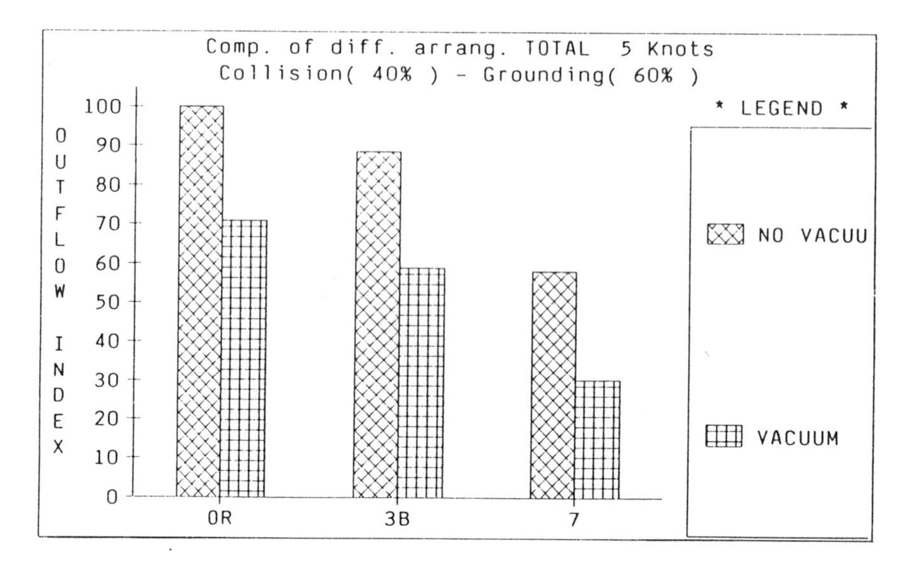

Fig.3.27. Vacuum vs. No Vacuum, Grounding 5 kn

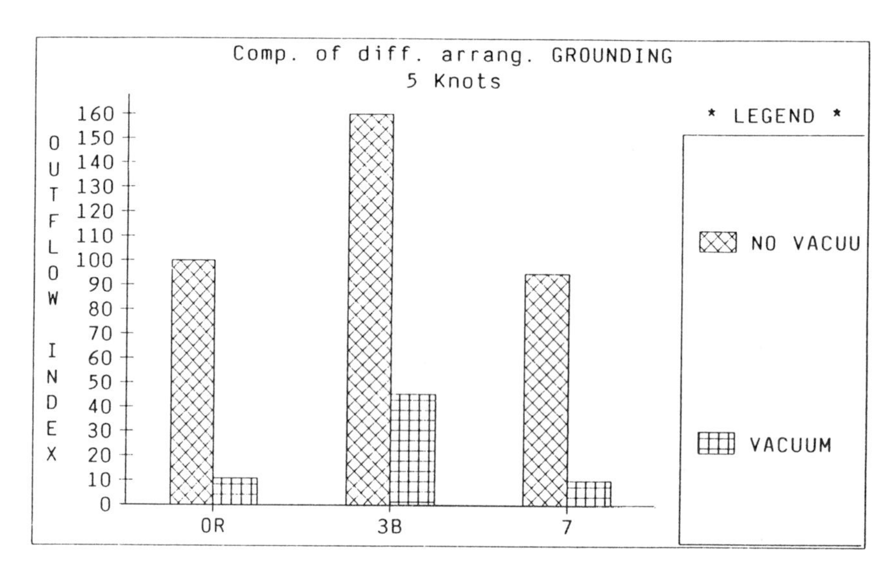

Fig.3.28. Vacuum vs. No Vacuum, Combined 5 kn

3.3 Conclusions and recommendations

The study shows that the double side/double bottom VLCC designs (1A,1B) have the smallest potential for oil outflow in collision and grounding with the given assumptions regarding damage location and extent respectively. Compared with a modern conventional VLCC the double side/double bottom VLCC is likely to leak on the average only about 33% of the corresponding amount of oil.

Comparing double side/double bottom designs, the design with two longitudinal bulkheads in addition to the double side(1A), is likely to leak less oil in about 75% of all collisions than the design with only one longitudinal bulkhead at the centreline(1B) although both designs have the same overall performance.

The amount of oil likely to escape from a VLCC with double bottom for ballast and with segregated ballast side tanks(4,4A and 4B), is about 40% of the amount oil a modern conventional VLCC may leak under the same conditions. The positive influence of locating the ballast tanks forward in the side tanks on potential oil outflow in collision is evident(4B vs. 4A).

As for single bottom designs with wide wing tanks for ballast (3B), their overall performance i.e. combined ranking, is quite poor due to large amount of oil likely to escape in grounding. In fact, they perform only slightly better than modern conventional VLCCs.

Reducing the tank size to half of MARPOL requirements(7) reduces the potential average oil outflow to about 60% as compared with the conventional VLCC.

The partial double bottom VLCC(5) analysed does not perform too well in grounding; - the basic reason is that the double bottom height is too low in order to be effective against oil outflow in grounding under the assumptions made in the study. Increasing the double bottom height, and reducing the side tank width, brings the results for design 5 closer towards those of 1A and 1B. Considering mud or sand bottoms, design 5 might be attractive because of reduced damage lengths.

It should also be stressed that in about 85% of all groundings the double bottom designs studied leak no oil at all, whereas always some oil will leak from single bottom designs irrespective of tank size and the ship having a vacuum tank system or not.

A vacuum system reduces significantly the amount of oil escaping in grounding for the single bottom designs analysed. However, the total amount of oil escaping from the modern conventional VLCC with a vacuum system is still about twice the amount escaping from the double side/double bottom design in case of collision and grounding.

Comparing VLCCs with double sides to VLCCs with single sides, the the double side provides an effective barrier against oil outflow in 20% of the all collisions for designs 1A and 1B, and in 42% of all collisions for 3B. The protection is particularly efficient for low energy collisions with limited damage penetration.

The influence of increasing the double bottom height as compared with increasing the double side width on oil outflow is shown in Table 3.3.1.

From the Table one may observe that increasing the double bottom height from 2.0 to 3.9m(B/15) reduces the probability for oil outflow in grounding from 58% to 14%, and increasing the width of double side from 2.0 to 3.0 m reduces the probability for oil outflow in collision from 88% to 80% only. Introducing the above zero outflow values for 2.0 m double bottom and double side into Fig. 3.14 and 3.24, and integrating assuming the curves follow those for design 1A and 1B, one obtains the outflow index for collision as ~59 and grounding as ~61. Combined index is ~60, or about 2.4 times the index for 1A.

Table 3.3.1. Probability for NO oil leakage as function of double bottom height and distance between double sides

G R O U N D I N G

DIST. BETWEEN INNER/OUTER BOTTOM	PROBAB. FOR NO LEAKAGE
2.0 M	41.7 %
2.4 M	52.6 %
3.0 M	69.0 %
3.9 M	86.0 %
6.6 M	99.8 %

C O L L I S I O N

DIST. BETWEEN INNER/OUTER SKIN	PROBAB. FOR NO LEAKAGE
2.0 M	12.1 %
3.0 M	20.4 %
5.8 M	39.4 %
6.3 M	42.0 %

CONCLUDING FROM THE ABOVE ANALYSIS ON VLCC DESIGNS :

* narrow and long cargo tanks, and increased double bottom height clearly reduces the oil outflow as does reduced tank volumes.

* ballast forward in side tanks will reduce the potential oil spill in collision, and in low speed groundings

* VLCCs should have a double bottom height approaching B/15 in order to be effective against pollution in grounding

* in addition, double sides having a width of at least B/20 should be considered.

* single bottom designs should preferrably be fitted with a vacuum system in order to reduce the amount of oil escaping in grounding.

* hydrostatic loading may be an alternative to vacuum systems, the lost cargo capacity is not prohibitive at A-freeboard. The increased draught may create problems in coastal waters and loading/discharging terminals.

* intermediate oil tight decks may be effective against oil pollution if adequate collision protection is provided.

* low L/D ratios($\approx$ 10) contribute to excessive pollution in grounding due to high freeboard. A higher L/D ratio($\approx$ 12) would significantly reduce the amount of oil escaping in grounding.

SEE COMMITTEE COMMENTS ON DNV CONCLUSIONS FOR VLCCS ON PAGE 300.

4. PROBABILISTIC RANKING OF 40,000 DWT TANKERS

4.1 The 40,000 dwt Designs Analysed

The 40,000 dwt designs chosen for the analysis are essentially similar to the VLCCs the objective being to detect trends in the results obtained which would be valid for other tanker sizes as well.

Of particular interest has been to investigate the influence of double side width on oil outflow as several 40,000 dwt product tankers have been built with narrow sides.

4.1.1 General Features

All designs analysed have following main particulars:

L_{pp} = 190.1 m
B = 27.4 m
D = 17.8 m
T = 12.9 m
C_B = 0.82

Following parameters have been kept constant:

* the length of the cargo area = 143.0 m
* tank lengths(except design 7) = 28.0 m
* ballast capacity outside the cargo area ≃ 3000 m^3
* slop tank volumes ≃ 800 m^3
* deadweight ≃ 46,500 tons.

The specific gravity for oil is taken as 0.9 t/m^3. For designs with less ballast in the cargo area than the original 40,000 dwt tanker, the increased cargo volume has not been utilized due to constant deadweight assumption. Resulting possible slack in oil tanks(if oil level in tank $< .98D$) has been assumed equally distributed on all crude oil tanks. At lower specific oil gravities the excess tank volume may be utilized without exceeding the design draught.

MARPOL/2/ requirements to tank lengths have been considered as have requirements to tank sizes.

4.1.2 Particulars of 40,000 dwt Tankers

The numbers below refer to the designs shown in Fig.4.1-4.8, and to graphs showing the potential oil outflow which has been calculated for different designs. A fold out page showing the designs has been enclosed at the end of the report.

The box in the lower right hand corner in Fig.4.1-4.8 shows IMO requirements, and DNVC PP1,PP2 and PP3 requirements. The actual values have been included.

1. Modern 40,000 tdw SBT

Cargo capacity for this design is ~ 52700 m^3. The ballast capacity in the cargo area is ~ 17,000 m^3.

The ballast/cargo ratio is 0.38 which is rather high for a single skin tanker.

This design barely meets the IMO requirements and is short of meeting the DNVC PP1,PP2 and PP3 requirements.

2. Double sides and single bottom

All ballast in the cargo area is carried in the double side. The width of the double side is 3.0 m(~B/9). Wide tanks.

The ballast capacity in the cargo area is ~15,100 m^3.

3. Double bottom and single sides

All ballast is carried in the double bottom which has a height of 3.9 m(~B/7). A centreline bulkhead has been fitted.

The ballast capacity in the cargo area is ~ 15,500 m^3.

4. Narrow double sides and double bottom

This design complies with the IMO Chemical Code Type 2/9/ ships w.r.t. the width of double side i.e 0.76 m. Double bottom height is 2.6 m. Centreline bulkhead is fitted.

The ballast capacity in the cargo area is ~ 13,600 m^3.

5. Double sides and double bottom

The double sides are 1.2 m apart. The height of the double bottom is 2.0 m. Centreline bulkhead.

The ballast capacity in the cargo area is ~13,600 m^3.

6. Wide double sides and double bottom

The double side width is 2.0 m. The height of the double bottom is 1.83 m(B/15). Centreline bulkhead.

The ballast capacity in the cargo area is ~16,300 m^3.

7. Wide double sides and double bottom, short tanks

This design features the same double side width and double bottom height as design 6. The difference is that design 7 has no centreline bulkhead and short tanks(20 m).

8. Intermediate Oil Tight Deck

An intermediate oil tight deck has been fitted 8.9 m above the baseline. The side tanks have the same width and length as the original 40,000 dwt tanker.

4.2 Probabilistic Ranking of 40,000 dwt Tankers

The same procedure for estimating oil outflow in collision and grounding has been adopted for 40,000 dwt tankers as for the VLCCs. The same statistics have been used but weighed for the smaller tanker i.e. the extent of damage depth, breadth and height have been related to the main particulars of the 40,000 dwt.

4.2.1 Ranking in Collision

The results from the PROBAN analysis are shown in Fig.4.9-4.11.

The very good performance of the original design is not surprising due to the location of the long and wide ballast tanks in the bow. These tanks provide an extremely efficient protection against collision.

The influence of double side width on oil outflow is evident from Fig.4.9 - in 42% of all collisions will no oil escape from the 40,000 dwt with 3 m wide double tanks carrying only ballast. Once the inner side is punctured rather much oil will escape however. This is the drawback of having no centreline bulkhead and long cargo tanks.

This is supported by studying the curves for oil outflow from designs 6 and 7 having the same double side width. Design 6 with long tanks and a centreline bulkhead performs much better after the inner side has been punctured; - at 0.8 probability level the short tank design is likely to leak ~66% more than the centreline bulkhead design.

In 9 % of all collisions resulting in structural damage, the narrow double side of 0.76 m will not leak any oil. The narrow double side provides protection against oil outflow especially at low velocity quay contacts. The impact of the centreline bulkhead on oil outflow is obvious when compared to the wide double side design above.

Comparing the original tanker with double side tankers one observes from Fig.4.9 that although the original tanker will always probably spill some oil in collision, the tanker will spill relatively little in more than 60% of all collisions.

The poor collision behaviour of the double bottom tanker is expected.

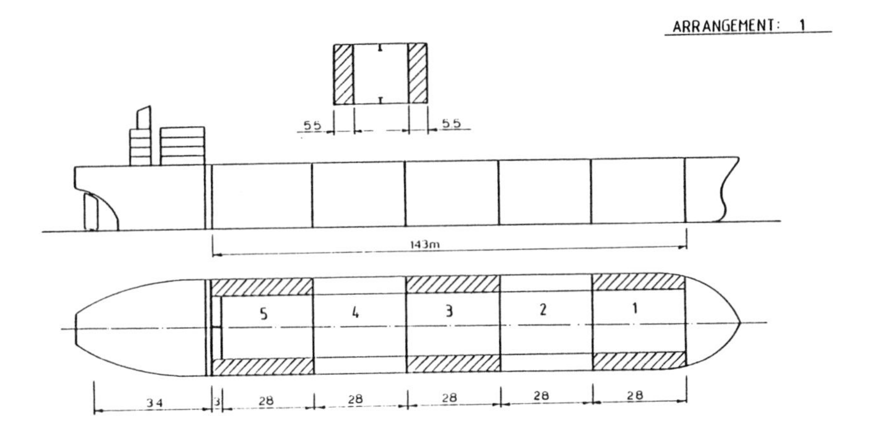

BALLAST 20500 TOTAL
BALLAST 17500 WITHIN CARGOAREA

	IMO	PP1	PP2	PP3
REQUIRED m^2	3906	4857	7110	8923
ACTUAL m^2	4054			

Fig.4.1 Modern 40,000 dwt SBT tanker(1)

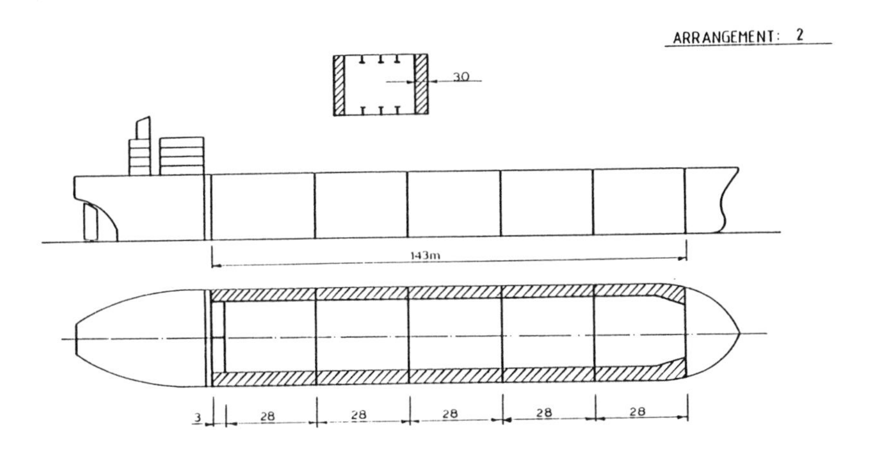

BALLAST 18500 TOTAL
BALLAST 15500 WITHIN CARGOAREA

	IMO	PP1	PP2	PP3
REQUIRED m^2	3906	4857	7110	
ACTUAL m^2		5949		

Fig.4.2 Double sides and single bottom(2)

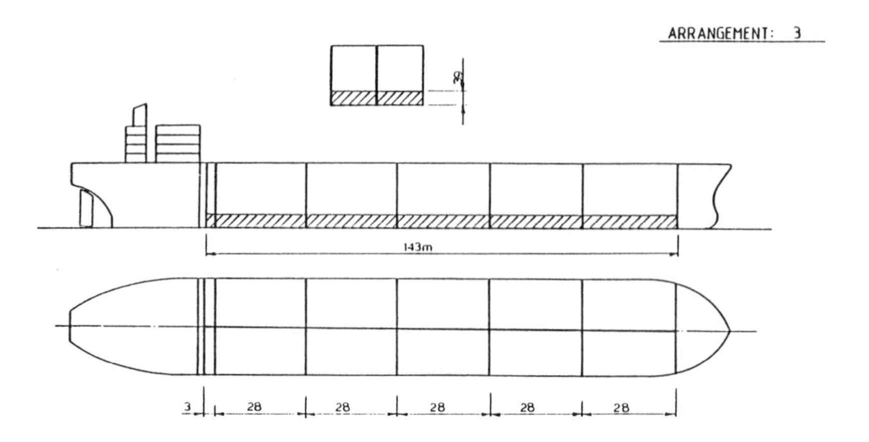

BALLAST 18500 TOTAL
BALLAST 15500 WITHIN CARGOAREA

	IMO	PP1	PP2	PP3
REQUIRED m^2	3906	4857	7110	
ACTUAL m^2	4780			

Fig.4.3 Double bottom and single side(3)

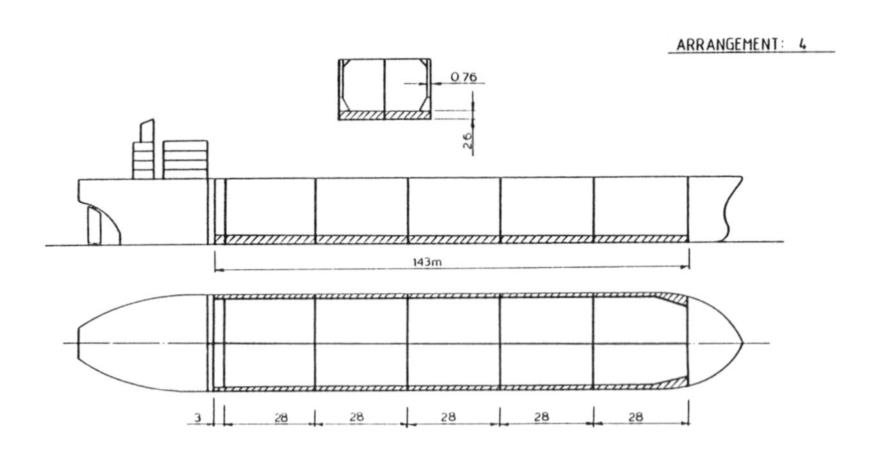

BALLAST 17000 TOTAL
BALLAST 14000 WITHIN CARGOAREA

	IMO	PP1	PP2	PP3
REQUIRED m^2	3906	4857	7110	8923
ACTUAL m^2	4666			

Fig.4.4 Narrow double sides and double bottom(4)

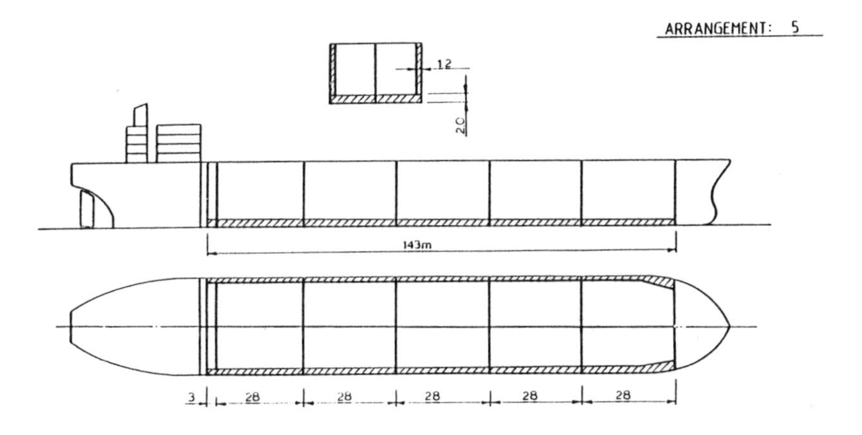

BALLAST 17000 TOTAL
BALLAST 14000 WITHIN CARGOAREA

	IMO	PP1	PP2	PP3
REQUIRED m^2	3906	4857	7110	8923
ACTUAL m^2	4494			

Fig.4.5 Double sides and double bottom(5)

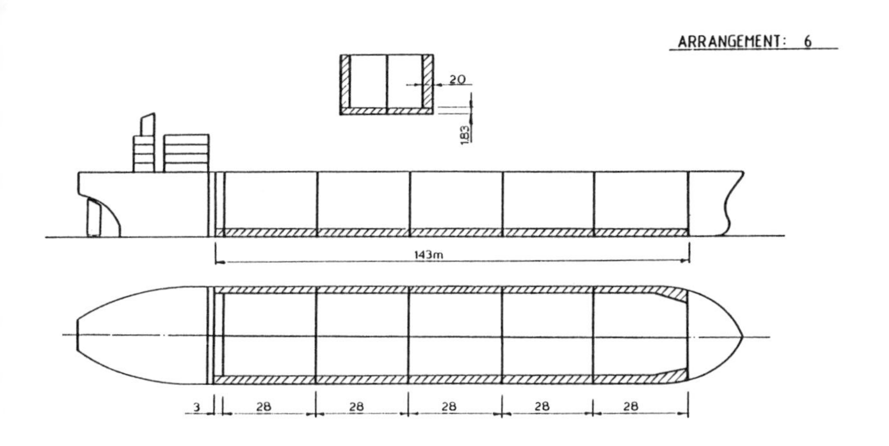

BALLAST 19700 TOTAL
BALLAST 16700 WITHIN CARGOAREA

	IMO	PP1	PP2	PP3
REQUIRED m^2	3906	4857	7110	8923
ACTUAL m^2				8923

Fig.4.6 Wide double sides and double bottom(6)

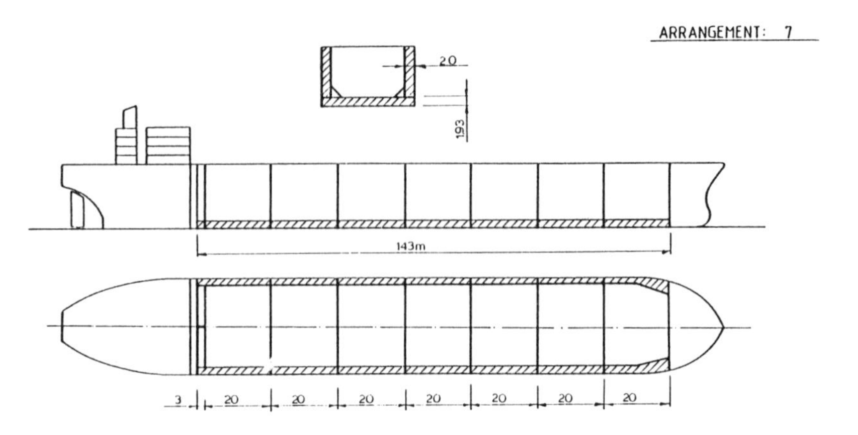

BALLAST 19700 TOTAL
BALLAST 16700 WITHIN CARGOAREA

	IMO	PP1	PP2	PP3
REQUIRED m^2	3906	4857	7110	8923
ACTUAL m^2				8923

Fig.4.7 Wide double sides and double bottom, short tanks(7)

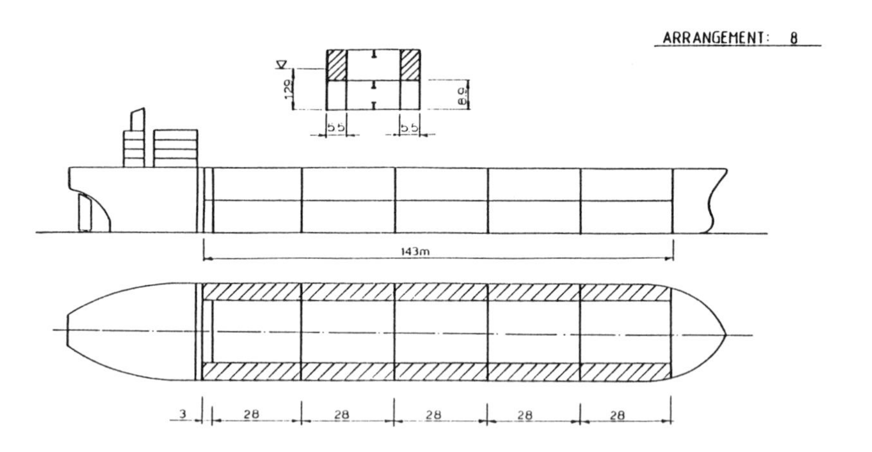

BALLAST 17500 TOTAL
BALLAST 14000 WITHIN CARGOAREA

	IMO	PP1	PP2	PP3
REQUIRED m^2	3906			
ACTUAL m^2	2545			

Fig.4.8 Intermediate Oil Tight Deck(8)

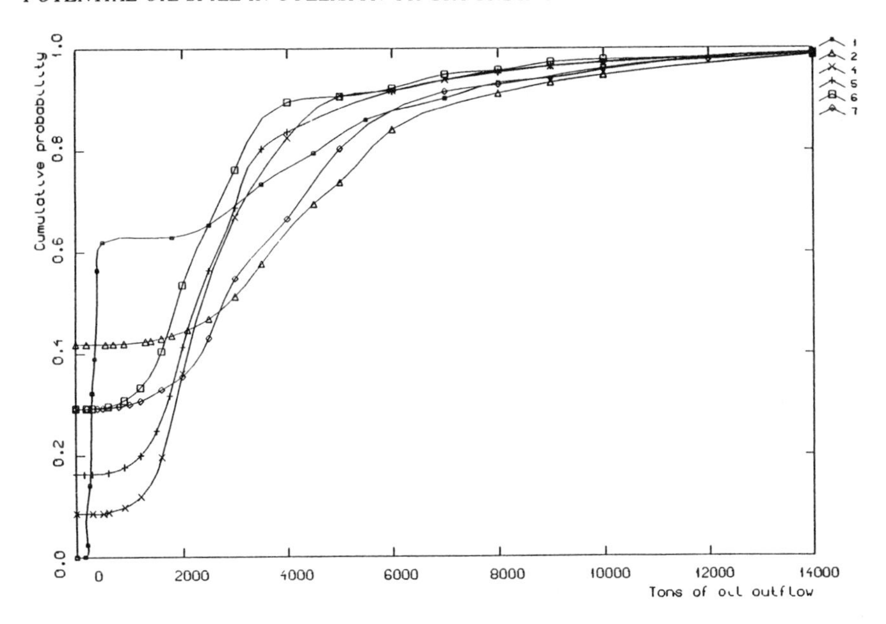

Fig.4.9. Cumulative Probability for Oil Outflow in Collision - Double Side 40,000 dwt Tankers

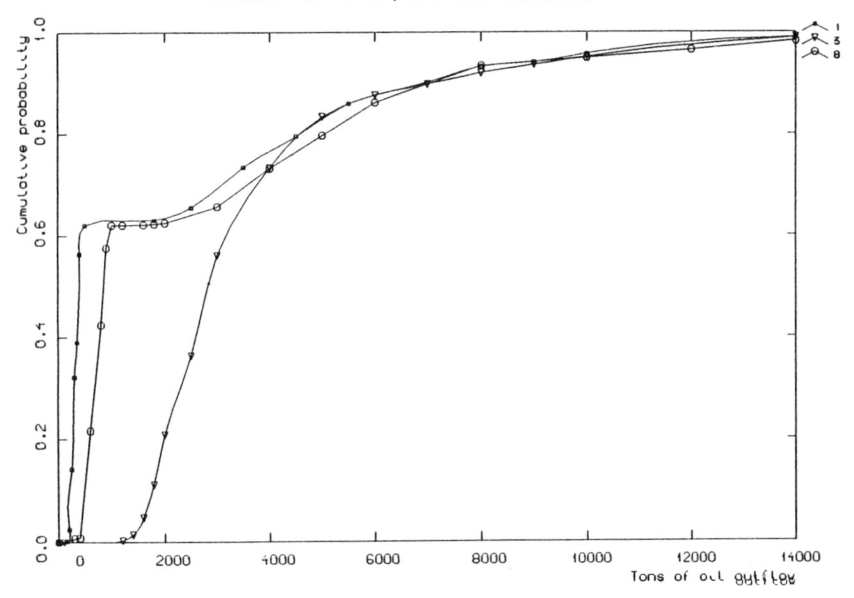

Fig.4.10. Cumulative Probability for Oil Outflow in Collision - Single Side 40,000 dwt Tankers

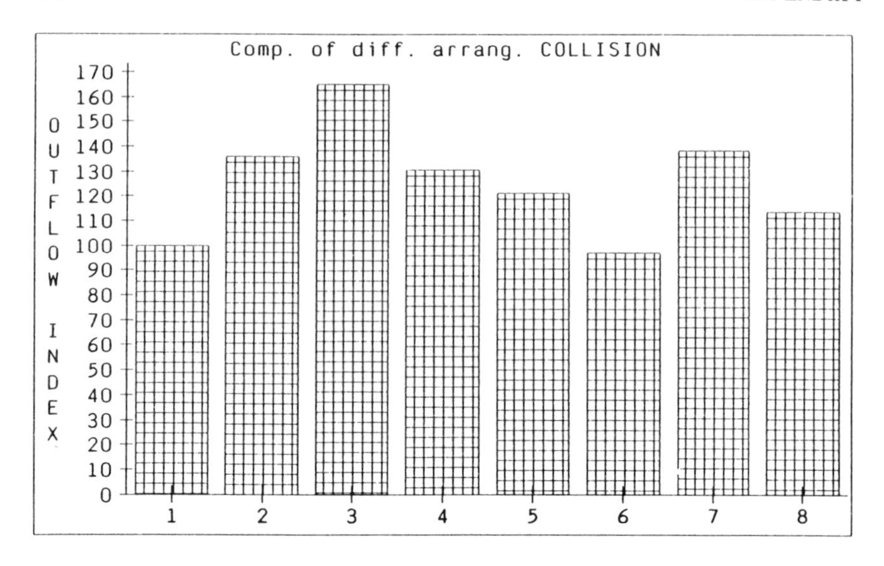

Fig.4.11. Ranking of 40,000 dwt Tankers in Collision

The 40,000 tdw tanker with an intermediate oil tight deck performs overall slightly worse than double side tanker with centreline bulkhead(design 6) and the original. Under the assumptions made regarding the vertical extent of collision damage, it will always leak some oil in collsion.

Integrating the outflow curves in Fig.4.9-4.10, and comparing with the original design shows(Fig.4.11) that the only design that can match the original design is design 6 - the wide double side and double bottom design with long tanks and centreline bulkhead(outflow index 98). In comparison, the wide double side design(7) with short tanks has an index of 140.

The narrow double side design(4) has an index of 130, this design performs better than design 7 which has about three times wider side tanks. Tank length seems obviously to have larger influence on oil outflow in collision than the width of the double side.

4.2.2 Ranking in Grounding

The results from the PROBAN analysis are shown in Fig.4.12-4.17.

The original 40,000 dwt tanker performs quite badly in grounding even at low speeds. This is because L/D is quite low, - in case of grounding(assuming hydrostatic equilibrium in a damaged tank) about 25% of the oil in the damaged tank will escape. The tanker with ballast in side tanks only, design 2, performs even worse.

Comparing the oil outflow curves for the original 40,000 dwt for 5 and 10 kn respectively, one observes that at 5 kn the curve is rather steep. This indicates that ballast in forward side tanks combined with long cargo tanks reduce the oil outflow at low speed grounding. At 10 kn, the oil outflow curve is rather horisontal at the beginning implying that there is a low probability for the damage to be restricted to the bow area only.

The performance of the original 40,000 dwt tanker and design 2 would improve much should they be fitted with a vacuum system, or be hydrostatically loaded. An alternative for design 2 may be to fit a double bottom in the cargo area only.

All double bottom designs perform quite well in grounding. Designs 3 and 4, with a double bottom of 3.9 and 2.6 m respectively, are likely to leak no/almost no oil at all in grounding. In more than 80% of all groundings, the B/15 double bottom of 1.83 m will provide protection against oil outflow(designs 6 and 7). Double bottoms restrict the potential oil outflow in case of bilge and side damage due to grounding.

Due to lack of time, the influence of double bottom height and ship speed on damage length in grounding has not been established for 40,000 dwt tankers.

As expected the intermediate oil tight deck tanker performs quite well in grounding.

Integrating the curves in Fig.4.12-4.13 gives the average oil outflow index in grounding at 5 kn. All designs, except design 2, have an index below 10(Fig.4.14).

The effect of increasing grounding speed on potential oil outflow is similar for 40,000 dwt tankers as for VLCCs. The results are shown in Fig.4.15-4.17. At 10 kn, all designs except design 2, have an index below 10.

Carrying oil with a lower specific gravity would result in the ship draught either being reduced or remain about the same when utilising the excess tank capacity. Assuming a specific gravity of 0.7 t/m^3, the new draught for the original 40,000 dwt would be in the order of 10.5 m. This would cause the potential oil outflow to be reduced by some 35% in grounding. For the other designs volume spilled may increase when filling up tanks with light products; - the difference in potential oil outflow has not been calculated.

The hydrostatically loaded 40,000 dwt tanker would, due to same damage statistics as for the VLCCs, leak some oil in grounding because the bilge and side may be damaged in grounding.

In case of a tanker with double bottom being hydrostatically loaded, the hydrostatic balance level may be increased with an amount equal to the double bottom height. Should the double bottom be damaged, then water and oil will fill the double bottom. Some oil outflow may occur to oil mixing with water in the damaged double bottom.

Should the tanker be part loaded then the oil outflow in grounding would be reduced significantly.

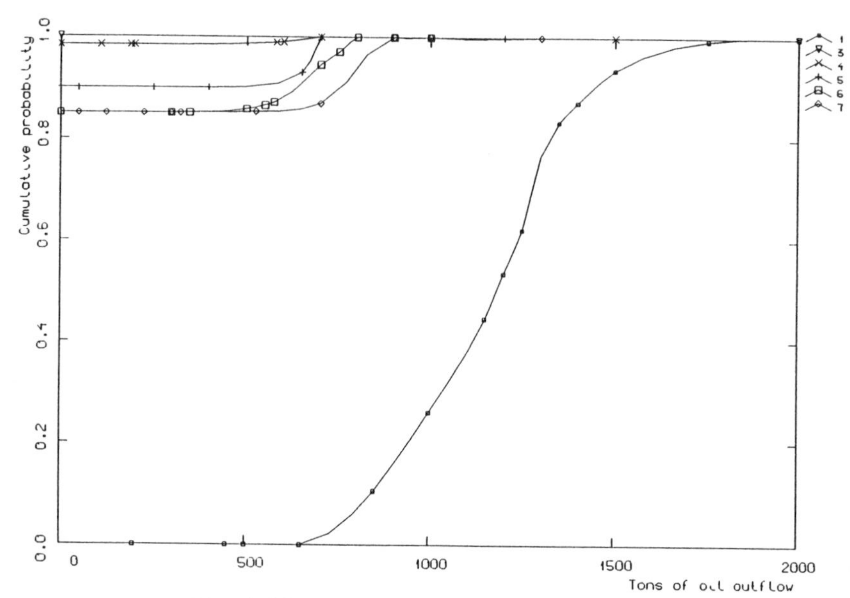

Fig.4.12. Double Bottom, 5 kn

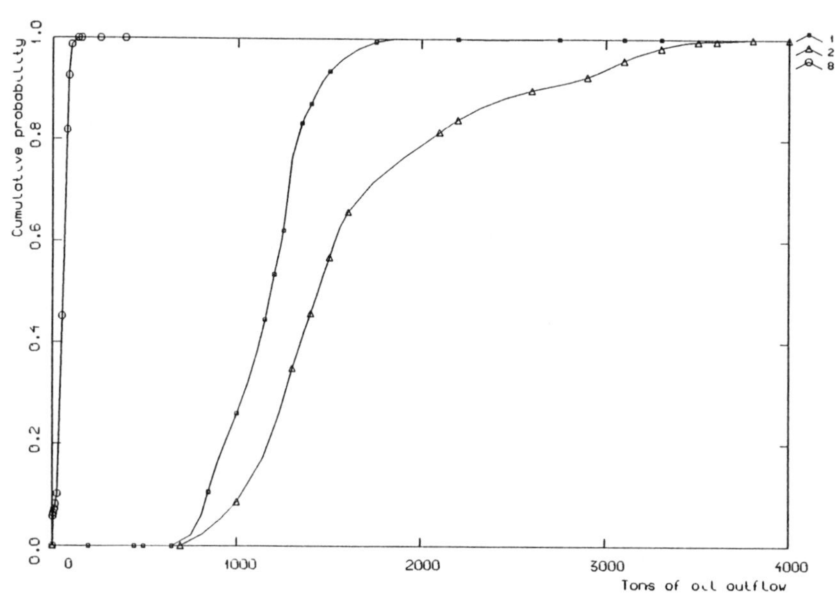

Fig.4.13. Single Bottom, 5 kn

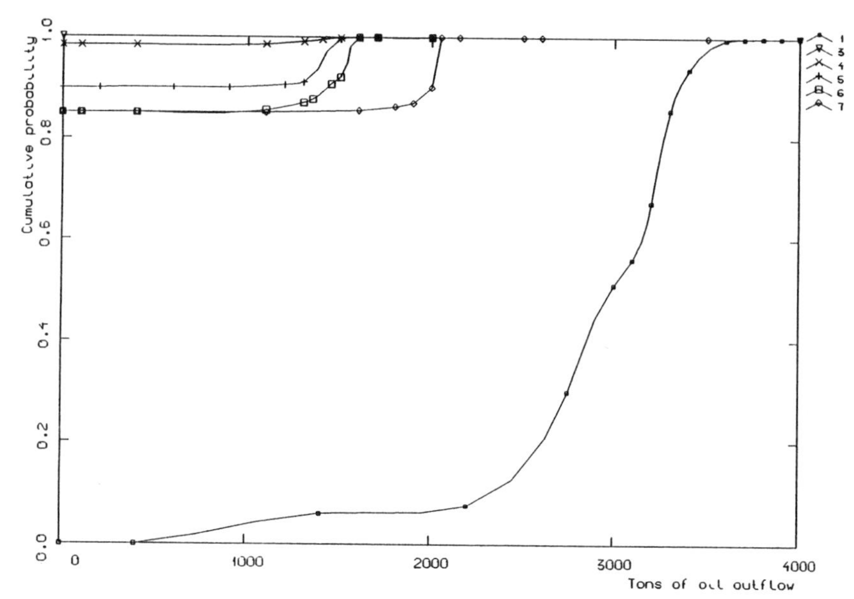

Fig.4.14. Double Bottom, 10 kn

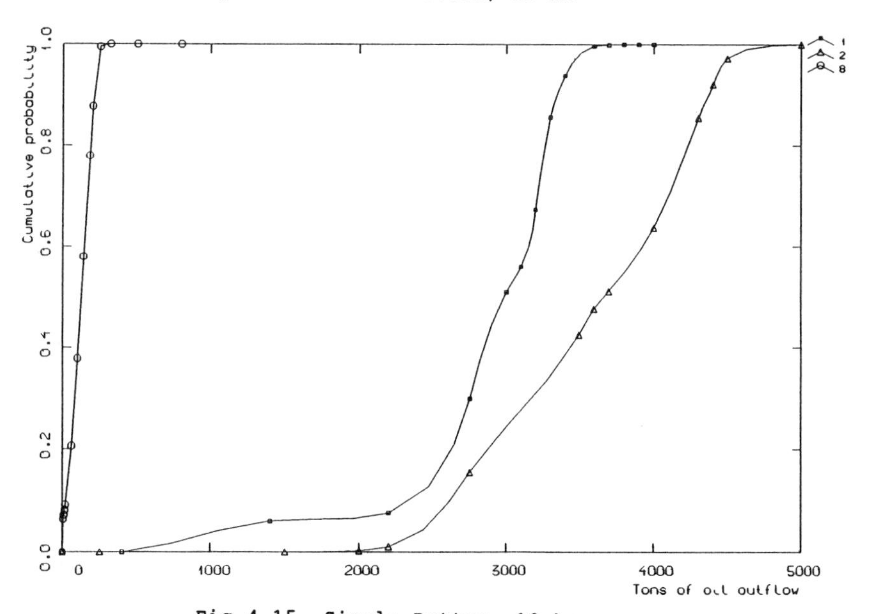

Fig.4.15. Single Bottom, 10 kn

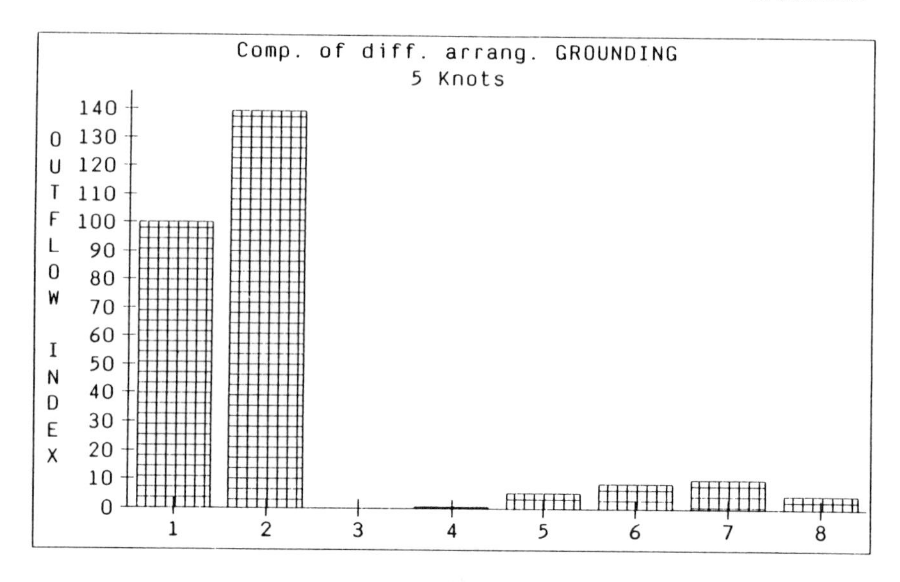

Fig.4.16. Ranking of Designs, 5 kn

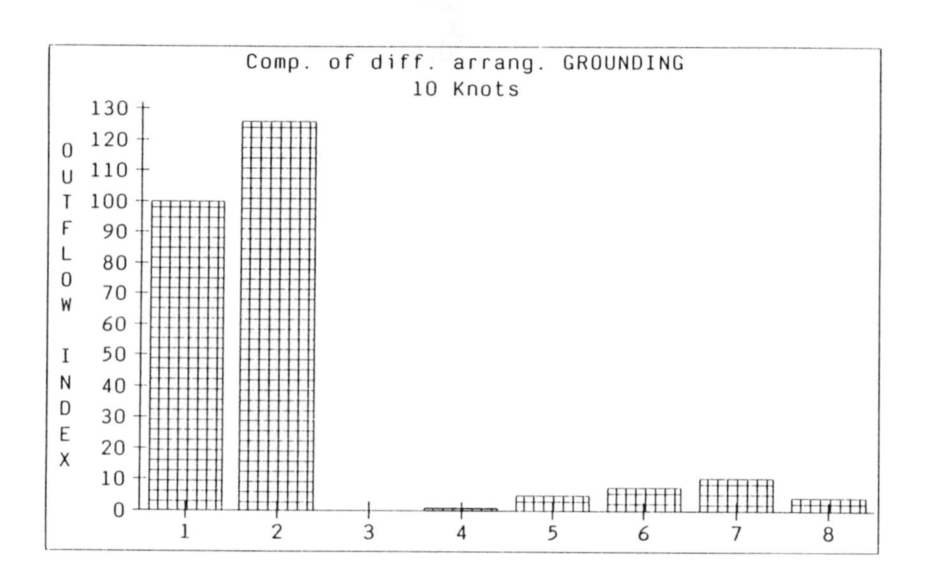

Fig.4.17. Ranking of Desins, 10 kn

4.2.3 Combined Ranking

Combining the above indices for collision, and grounding at 5 and 10 kn respectively, using a 40-60 collision/grounding weighing gives the combined ranking shown in Fig.4.18 for 5 kn grounding speed and in Fig.4.19 for 10 kn grounding speed.

At 5 kn, design 6 with wide double side and low double bottom obtains an index of ~57 followed by design 8 having an index of ~65. Only design 2, the one without double bottom, has an index above the original 40,000 dwt tanker.

The ranking does not change for 10 kn; - design 6 is still best with an index of ~38 followed by designs 8, 4 and 5.

4.3 Conclusions and Recommendations

The above analysis for 40,000 dwt tankers shows the importance of double bottom on the potential oil outflow. In Table 4.3.1, the effect of increasing the double bottom height and width of double side on the probability for leaking no oil is given respectively.

The Table shows that the probability for oil leaking is reduced rapidly as double bottom height is increased.

For collisions, increasing the width of the double side has not the same dramatic effect on the probability for no oil outflow.

Table 4.3.1. Probability for no oil outflow in collision and grounding

G R O U N D I N G

DIST. BETWEEN INNER/OUTER BOTTOM	PROBAB. FOR NO LEAKAGE
1.83 M	85.1 %
2.0 M	90.0 %
2.6 M	98.4 %
3.9 M	99.99%

C O L L I S I O N

DIST. BETWEEN INNER/OUTER SKIN	PROBAB. FOR NO LEAKAGE
0.76 M	8.6 %
1.2 M	16.3 %
2.0 M	29.2 %
3.0 M	41.9 %

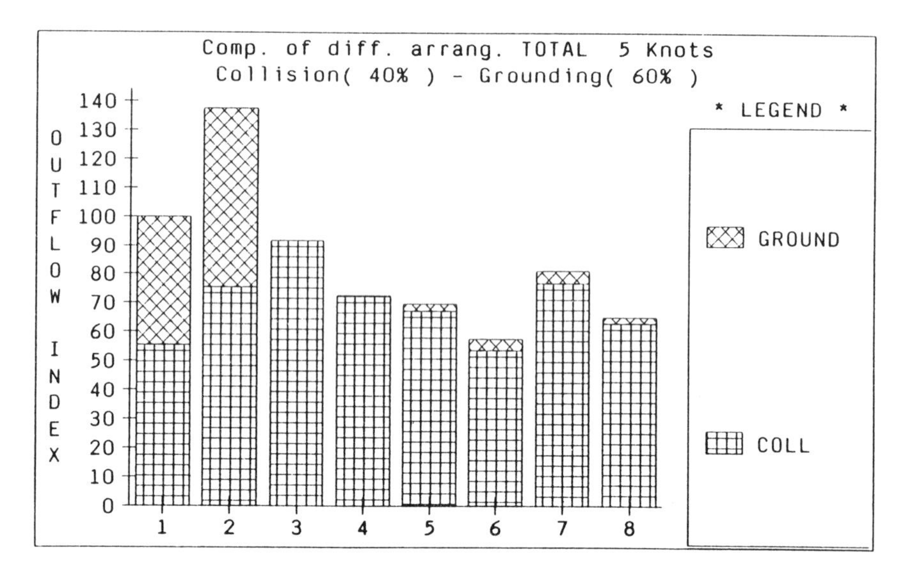

Fig.4.18. Combined Ranking, 5 kn

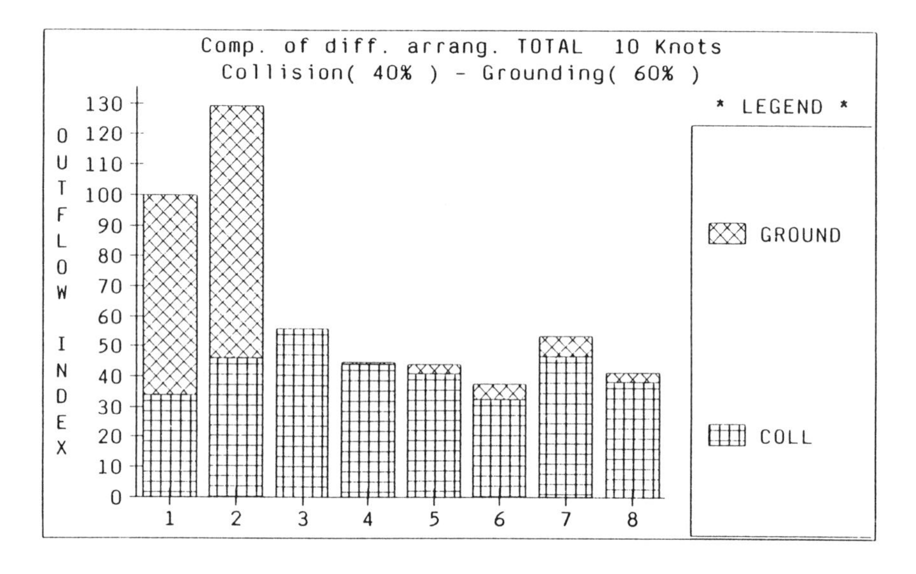

Fig.4.19. Combined Ranking, 10 kn

CONCLUDING FROM THE ABOVE ANALYSIS ON 40,000 DWT TANKERS:

* narrow and long side tanks reduce the potential oil outflow in collision and grounding as does reduced tank volumes.

* introducing a centreline bulkhead reduces the potential oil outflow in collisions when compared with short wide cargo tank designs.

* for ships < 50,000 dwt, the width of double side could be related to B/25 with a minimum of 760 mm according to the Chemical Code Type 2 ships in order not to reduce the cargo capacity(ref. limited influence of increased width on oil outflow), and allowing for centreline bulkheads.

* double bottom provides an effective barrier against oil pollution for the 40,000 dwt tankers analysed. The height of double bottom should not be less than B/15.

* double bottoms reduce oil outflow in groundings causing bilge and side damage

* single bottom designs should be fitted with a vacuum system, alternatively loaded only to a hydrostatic balance level to reduce oil outflow in grounding.

* the intermediate oil tight deck design performs quite well but will always leak some oil in collision and grounding

SEE COMMITTEE COMMENTS ON DNV CONCLUSIONS FOR 40,000 DWT TANKERS ON PAGE 301.

5. ESTIMATED OIL OUTFLOW FROM A 80,000 DWT TANKER

Based upon above studies for VLCCs and 40,000 dwt tankers, the potential oil outflow from a 80,000 dwt is discussed in this Chapter. No supporting calculations have been carried out with PROBAN.

5.1 Comparison of Performance for VLCCs vs. 40,000 dwt Tankers

Original VLCC and 40,000 dwt tanker designs - single skin:

The smaller tanker with large ballast side tanks forward performs relatively much better than the VLCC design which is penalised for not having ballast in the same side tanks.

The high ballast/cargo volume ratio for the 40,000 dwt necessarily improves the performance of the original design in relation to other 40,000 dwt designs.

Comparing the potential oil spill volume to total cargo volume ratio, the difference is only ~ 1.3% in ratios for the VLCC and the 40,000 dwt tanker.

In grounding, the overall performance is rather poor for both the VLCC and the 40,000 dwt tanker. Again ballast in forward side tanks reduces the oil spill, but due to long damage lengths at 10 kn, the effect is not very pronounced.

Ballast in forward side tanks reduce also the potential for oil spill in bilge and side damages caused by grounding; - at low grounding speeds the reduction is considerable.

The volume of oil spill in grounding in relation to total cargo volume is virtually the same for the VLCC and the 40,000 dwt tanker at 5 and 10 kn respectively. At 10 kn, the total amount oil both ships leak is 2.5-3 times more than at 5 kn.

A vacuum system will reduce the oil outflow in grounding by ~90% at 5 kn provided a vacuum can be maintained in 40% of the cargo tanks. At 10 kn the relative reduction will be smaller because damage lengths increase and more tanks are damaged. Increasing the vacuum capacity will reduce the oil outflow correspondingly.

Hydrostatically balanced loading of cargo tanks will reduce oil outflow in grounding to less than 10% of the original amount.

For a 80,000 dwt SBT tanker with conventional tank arrangement, much the same behaviour is expected.

Double side and double bottom designs:

Oil outflow from VLCCs with double sides and double bottom is radically lower compared with the VLCC SBT - the overall oil outflow is ~33% of the outflow from the VLCC SBT at 10 kn. For the 40,000 dwt tanker the corresponding percentage is ~37 at 10 kn.

The same reduction in overall oil outflow may be expected for a 80,000 dwt tanker provided the ship meets with the B/20 double side and B/15 double bottom recommendations.

Double sides and single bottom designs:

In the VLCC and the 40,000 dwt tanker with double sides and single bottom, the double side width is appr. B/9. The probability for no oil outflow in collision is for both ships ~42.0%.

Both ships perform rather poorly in grounding. The VLCC is likely to leak 80% more than the VLCC SBT at 10 kn. The 40,000 dwt tanker 25 % more than the original 40,000 dwt tanker.

A vacuum system reduces the potential oil spill in grounding at 5 kn for the VLCC by ~70 %. Increasing the speed results in relatively more oil spill due to increased damage lengths.

Hydrostatically balanced loading of cargo tanks would virtually eliminate all oil outflow in grounding as the probability for a bilge damage to extend into the cargo tanks is very low(assuming double side width B/9 as above).

For the 80,000 dwt tanker, the oil spill in grounding may be 35-40% in comparison. This value depends on the L/D ratio for the design.

Single side and double bottom designs:

In collision, the 40,000 dwt tanker with ballast in double bottom only, leak considerably more than the original design. The increase in total oil outflow is some 65 % in spite of the ship having a centreline bulkhead and long tanks.

In case of the VLCC, the single side/double bottom design is likely to leak ~30% less than the original VLCC. As stated above, the reason for poor behaviour of the original VLCC is due to no ballast in forward side tanks.

In grounding neither the VLCC(B/dbh = 8.6) nor the 40,000 dwt tanker(B/dbh=7.0) is expected to leak any oil.

For a similar 80,000 dwt tanker no difference in performance is expected.

Intermediate Oil Tight Deck Designs:

In collision, the performance of the intermediate oil tight deck design is a little better for the VLCC and a little inferior for the 40,000 dwt tanker.

In grounding, the performance is good for both sizes.

For the 80,000 dwt, the same behaviour is expected.

5.2 Conclusions and Recommendations:

Following general conclusions may be drawn regarding oil outflow from a 80,000 dwt tanker:

* long and narrow tanks are likely to leak less than short and wide tanks. A centreline girder should be fitted to reduce oil outflow in collisions

* a double bottom of B/15 should be fitted, an alternative may be a vacuum system with high capacity or hydrostatic loading of cargo tanks

* a double side should be considered to restricted pollution in low energy collisions; preferred width B/20

* high L/D reduces oil outflow in grounding

SEE COMMITTEE COMMENTS ON DNV CONCLUSIONS FOR THE 80,000 DWT TANKER ON PAGE 302.

6. SUMMARY OF RESULTS

The results should strictly be used only for ranking the VLCC and 40,000 dwt tanker designs as the statistics for damage extent used contain all types and sizes of ships. In addition, some simplifications have been made in the study influencing structural response in collision and grounding.

The adopted procedure for assessing the potential oil outflow in collision and grounding seems promising, however. Provided more detailed statistics is applied for tanker damages, the results would improve.

Summing up conclusions in Chapters 3-5, following measures should be considered for reducing the probability for oil outflow in collision and grounding:

* a low L/D ratio will increase oil outflow in grounding
* wide tanks are likely to spill more oil than long narrow tanks
* location of ballast forward in side tanks provides good collision protection
* double side protects against oil outflow in low energy collision
* width of double side should tentatively be B/25 for ships < 50,000 dwt, and B/20 for ships above, in order to be effective in ship/ship collision
* double bottom reduces oil outflow in grounding
* double bottom protects against bilge and side damage in grounding
* the double bottom should tentatively be B/15
* vacuum systems reduce oil outflow in grounding
* hydrostatic loading of cargo tanks reduces oil outflow in collision and grounding
* intermediate oil tight deck reduces oil outflow in grounding

REFERENCES

/1/ PROBAN Theory Manual, Veritas Research 1986

/2/ IMO International Conference on Marine Pollution 1973

/3/ IMO Regulations on Subdivision and Stability of Passenger Ships, Resolution A.265(VIII)

/4/ Valsgård, S. - Jørgensen, L.: Evaluation of Ship/Ship Collision Damage Using a Simplified Nonlinear Finite Element Procedure, Int. Symp. on Practical Design in Shipbuilding PRADS'83, Tokyo/Seoul, 1983

/5/ Køhler, P.E. - Jørgensen, L.: Ship Ice Impact Analysis, 4 th Int. Symp. on Offshore Mechanics and Arctic Engineering OMAE, Dallas, 1985

/6/ Vaughan, H.: Bending and Tearing of Plate with Application to Ship Bottom Damage, The Naval Architect, May 1978

/7/ Wierzbicki, T. et.al.: Damage Estimates in High Energy Grounding of Ships, MIT, June 27, 1990

/8/ Køhler, P.E.et.al.: Potential Oil Spill from Tankers in Case of Collision and/or Grounding - A Comparative Study of Different VLCC Designs, DNVC Report 90-0074, May 1990

/9/ IMO International Code for the Construction and Equipment of Ships Carrying Dangerous Chemicals in Bulk, London 1986

INTERPRETATIONS AND COMMENTS OF THE COMMITTEE ON TANK VESSEL DESIGN CONCERNING SECTIONS OF THE DET NORSKE VERITAS REPORT

Following are the committee's interpretations of the most significant assumptions in the collision analysis. Statements drawn from the DnV report are in italics (some excerpts have been edited for brevity and style), and committee comments are in standard type.

DNV Collision Assumptions (Section 2.1.1)

Several simplifying assumptions were made with respect to the collision assessment. To avoid overly optimistic results, the assumptions were conservative. Therefore, the committee feels that outflow may be overstated on a consistent basis.

The study assumes that oil begins to escape from a cargo tank when the bow of the striking ship touches a side or corner of the tank. No large deformations resulting in yielding of the tank sides were required. Thus, a tank is considered penetrated at contact.

The striking bow is assumed to be wedge-shaped, and it remains in shape after penetrating hull plating. As the vertical bow penetrates a tank, the full height of the tank is ruptured, from bottom to top. This assumption was made to conform with the MARPOL assumption that all oil will escape from a damaged tank over time.

One consequence of these assumptions is that the results apply strictly to the final condition — that is, after all oil has escaped from a holed tank. To obtain more accurate estimates on oil outflow during the collision, it would be necessary to conduct detailed modeling of bow crushing behavior and ship side structural response during the collision process, taking into account the changes in ship speeds and headings, and added mass effects. This is beyond the scope of the present study. Furthermore, the committee feels that additional research would be required even to develop the capability to obtain these estimates.

In the present study, only damage in way of the collision contact point is considered. Damage elsewhere, such as possible tearing in weld seams due to excessive tension (potentially resulting in oil leakage), have not been considered.

As for possible explosions due to friction heat and sparks generated in a collision, the study assumes that the inert gas system is effective and no fires ensue.

Grounding Assumptions (Section 2.2.1)

Following are the committee's interpretations of the most significant assumptions in the DnV grounding analysis. Statements drawn from the DnV report are in italics (some excerpts have been edited for brevity and style), and committee comments are in standard type.

The simplifying assumptions relate to extent of damage and oil outflow. Again, the assumptions are conservative, so outflow may be overstated consistently.

The analysis assumes that the grounding ship has forward speed, so that the damage is initiated at the bow and develops towards the stern. Damage caused by grounding while the ship is adrift, turning, or going astern, has not been considered.

The bottom surface is defined as a solid rock, which does not crush during the grounding process. The rock is assumed to be wedge-shaped (triangular in the transverse plane), with a constant breadth. The Vaughan method outlined in Chapter 3 was used in this analysis.

For purposes of this study, the hull plating is considered penetrated when the rock comes into contact with a tank side, bottom, or corner. Oil escapes until hydrostatic equilibrium is achieved in the damaged tank. Should the ship side be damaged, then all oil below the intact side plating will escape, to be replaced by water. Due to the reduced hydrostatic pressure, the oil level will drop proportionally to the specific gravities of water and oil, and the height of side damage.

Statistics on maximum vertical extent of grounding damage were applied to the full length of damage. This conservative assumption was made due to lack of detailed information on vertical damage extent, or penetration, as a function of damage length. Consequently, the damage lengths calculated may be too short, due to overstatement of the energy absorbed vertically.

Possible tearing of welds well away from the direct grounding damage have not been considered. Nor was the influence of ship motions during the grounding process considered in detail. During a grounding, a ship may develop a forward trim due to either downward suction of the bow, or sudden loss of buoyancy as the forward bottom plating is peeled off. Or, the ship may run up on the rock, lifting the bow. In this analysis, a variation in ship draft (+ 0.5 m) was included, partly to account for these factors.

Tidal effects on oil outflow were not considered. At low tide, more oil may escape from a ship sitting on a rock, due to reduced draft. Tides also may result in excessive hogging moments, which might result the hull breaking apart, with uncontrolled pollution. This matter requires a separate study, using available computer programs for ultimate strength analysis of the hull girder.

DNV Conclusions for VLCCs (Section 3.3)

Following are the committee's comments on each of DnV's conclusions for the VLCC designs. DnV conclusions are in italics, and the committee's comments are in standard type.

Narrow and long cargo tanks and increased double bottom height clearly reduces the oil outflow as does reduced tank volumes. The committee agrees.

Ballast forward in side tanks will reduce the potential oil spill in collision, and in low speed groundings. The committee agrees.

VLCCs should have a double bottom height approaching B/15 in order to be effective against pollution in grounding. The committee agrees, as one approach.

In addition, double sides having a width of at least B/20 should be considered. The committee prefers wider double sides.

Single bottom designs should preferably be fitted with a vacuum system in order to reduce the amount of oil escaping in grounding. The committee feels that the vacuum system needs further evaluation.

Hydrostatic loading may be an alternative to vacuum systems, the lost cargo is not prohibitive at A-freeboard. The increased draft may create problems in coastal waters and loading/discharge terminals. While the committee sees some benefit to hydrostatic loading, as noted in this statement, it perceives additional problems and feels hydrostatic loading should be considered for existing but not new vessels (except in conjunction with structural designs employing double hulls or double sides).

Intermediate oil-tight decks may be effective against oil pollution if adequate collision protection is provided. The committee feels that the intermediate oil-tight deck, when combined with double sides, shows promise and deserves further consideration.

Low L/D ratios (=10) contribute to excessive pollution in grounding due to high freeboard. A higher L/D ratio (=12) would significantly reduce the amount of oil escaping in grounding. The committee feels that high freeboard, not L/D ratio, is the relevant issue.

DNV Conclusions for 40,000 DWT Tankers (Section 4.3)

Following are the committee's comments on each of the DnV conclusions for the 40,000 DWT tankers. DnV conclusions are in italics, and committee comments are in standard type.

Narrow and long side tanks reduce the potential oil outflow in collision and grounding as does reduced tank volume. The committee agrees.

Introducing a centreline bulkhead reduces the potential oil outflow in collisions when compared with short wide cargo tank designs. The committee agrees but points out that centerline bulkheads may introduce possible asymmetric damage scenarios.

For ships over 50,000 DWT, the width of double side could be related to B/25 with a minimum of 760 mm according to the Chemical Code Type 2 ships in order not to reduce the cargo capacity (ref. limited influence of increased width on oil outflow), and allowing for centreline bulkheads. The committee prefers wider double sides.

Double bottom provides an effective barrier against oil pollution for the 40,000 DWT tankers analysed. The height of double bottom should not be less than B/15. The committee agrees that double bottoms are beneficial in terms of protection from groundings.

Double bottoms reduce oil outflow in groundings causing bilge and side damage. The committee agrees that double bottoms, as well as double sides, would reduce outflow.

Single bottom designs should be fitted with a vacuum system, alternatively loaded only to a hydrostatic balance level to reduce oil outflow in grounding. The committee feels that the vacuum system needs further evaluation and that hydrostatic loading should be considered for existing but not new vessels except in conjunction with structural designs employing double hulls or double sides. It does concur with the DnV conclusion that some oil leakage will occur if the hull is penetrated. The intermediate oil-tight deck with double sides will provide a reduction of outflow in high energy groundings.

The intermediate oil-tight deck design performs quite well but will always leak some oil in collision and grounding. The committee feels that the intermediate oil-tight deck, when combined with double sides, shows promise and deserves further consideration.

DNV Conclusions for the 80,000 DWT Tanker (Section 5.2)

Long and narrow tanks are likely to leak less than short and wide tanks. A centreline girder should be fitted to reduce oil outflow in collisions. The committee agrees with the benefits of long and narrow tanks and centerline bulkheads, but points out that centerline bulkheads may introduce possible asymmetric damage scenarios.

A double bottom of B/15 should be fitted, an alternative may be a vacuum system with high capacity or hydrostatic loading of cargo tanks. The committee agrees with the benefits of double bottoms in terms of grounding protection, but feels that the vacuum system needs further evaluation, and recommends that hydrostatic loading be considered for existing but not new tank vessels except in conjunction with structural designs employing double hulls or double sides.

A double side should be considered to restrict pollution in low-energy collisions; preferred width B/20. The committee prefers wider double sides.

High L/D reduces oil outflow in groundings. The committee feels that freeboard, not L/D ratio, is the relevant issue.

APPENDIX

G

Transport Cost Model Calculations

This appendix presents the calculations performed using the transport cost spreadsheet model described in Chapter 6.

Annual cost differentials for the alternative tank vessel designs are presented below (Table G-1). These costs are based on:

- Oil transport cost (dollars per ton) for each vessel size and design alternative, taken from Tables G-2 to G-10.
- Average transport costs for each alternative, weighting the three representative sizes and route distances, as noted in Chapter 6.
- Difference between the weighted average transport cost for each alternative and the cost of the base case, MARPOL single-skin ship.
- Annualization of these transport cost differentials, using a representative annual volume of 600 million tons of oil moved by tank vessels in U.S. waters.

Calculations of representative transport costs for each vessel design are presented in the following order:

Table	Vessel Design
G-2	MARPOL (Single Hull)
G-3	MARPOL + Hydrostatic Control
G-4	Double Hull
G-5	Double Hull + Hydrostatic Control
G-6	Double Side
G-7	Double Side + Hydrostatic Control
G-8	Double Side + Intermediate Oil Tight Deck
G-9	Double Bottom
G-10	Small Tanks

TABLE G-1 Transport Cost Model

Ship Size, kDWT	40	80	240		Wtd	Annual
Voy Dist RT, nm	2000	8000	4000	Wtd	Avg	Cost
Weighting, %	25	58	17	Avg	Diff	@ 600M T/yr
Design Alternative	---- Transport Cost, dollars per ton ----					$ Mill.
Marpol	4.33	10.30	3.65	7.68	0.00	0
Marpol + HB	5.34	12.71	4.51	9.48	1.80	1080
DH	5.01	11.87	4.26	8.86	1.19	712
DH + HB	5.76	15.19	4.93	11.09	3.41	2047
DS	4.69	11.02	3.97	8.24	0.56	339
DS +HB	5.34	12.76	4.56	9.51	1.84	1102
DS + MHD	5.19	12.23	4.35	9.13	1.45	872
DB	4.77	11.31	4.06	8.45	0.77	462
ST	4.74	11.25	4.01	8.39	0.72	430

TABLE G-2 MARPOL (Single Hull)

Vessel DWT		40000	80000	240000
Capital Cost				
Cost, $M		34.0	49.7	89.6
CRF @ 12% 20 yrs	0.13388			
Annualized Cost		4.55	6.65	12.00
Non-Voyage Optg Costs				
Manning		0.81	0.89	1.15
Stores & Lubes		0.25	0.27	0.32
M & R		0.29	0.33	0.47
Insurance				
H & M		0.34	0.50	0.90
P & I		0.08	0.10	0.23
Admin & Other		0.21	0.23	0.29
Voyage Costs				
Fuel				
Fuel cons, T/d		25	45	75
Fuel cost, $/T	150			
Fuel cost		1.02	2.15	2.92
Port costs		0.49	0.21	1.14
Total Costs, $M		8.04	11.34	19.40
Cargo Del'd, MT/yr		1.86	1.10	5.31
Trans Cost, $/T		4.33	10.30	3.65
cents/gal		1.47	3.50	1.24
Memo Items				
Voyage R/T, n mi		2000	8000	4000
Speed avg, kts		15.0	15.0	14.6
Port Time				
Load, days		0.75	1	1.5
Disch, days		0.75	1	1.5
Delays, days		0.1	0.2	1
Voyage days		7.2	24.4	15.4
Voys/yr		48.9	14.3	22.7
Cargo DWT/Total DWT		0.95	0.96	0.975

TABLE G-3 MARPOL and Hydrostatic Control

Vessel DWT		40000	80000	240000
Capital Cost				
Cost, $M		34.0	49.7	89.6
CRF @ 12% 20 yrs	0.13388			
Annualized Cost		4.55	6.65	12.00
Non-Voyage Optg Costs				
Manning		0.81	0.89	1.15
Stores & Lubes		0.25	0.27	0.32
M & R		0.29	0.33	0.47
Insurance				
H & M		0.34	0.50	0.90
P & I		0.08	0.10	0.23
Admin & Other		0.21	0.23	0.29
Voyage Costs				
Fuel				
Fuel cons, T/d		25	45	75
Fuel cost, $/T	150			
Fuel cost		1.02	2.15	2.92
Port costs		0.49	0.21	1.14
Total Costs, $M		8.04	11.34	19.40
Cargo Del'd, MT/yr		1.51	0.89	4.30
Trans Cost, $/T		5.34	12.71	4.51
cents/gal		1.82	4.32	1.53
Memo Items				
Voyage R/T, n mi		2000	8000	4000
Speed avg, kts		15.0	15.0	14.6
Port Time				
Load, days		0.75	1	1.5
Disch, days		0.75	1	1.5
Delays, days		0.1	0.2	1
Voyage days		7.2	24.4	15.4
Voys/yr		48.9	14.3	22.7
Cargo DWT/Total DWT		0.95	0.96	0.975

TABLE G-4 Double Hull

Vessel DWT		40000	80000	240000
Capital Cost				
Cost, $M *		39.4	58.2	105.7
CRF @ 12% 20 yrs	0.13388			
Annualized Cost		5.27	7.79	14.15
Non-Voyage Optg Costs				
Manning		0.81	0.89	1.15
Stores & Lubes		0.25	0.27	0.32
M & R *		0.50	0.58	0.83
Insurance *				
H & M		0.39	0.58	1.06
P & I		0.07	0.09	0.20
Admin & Other		0.21	0.23	0.29
Voyage Costs				
Fuel				
Fuel cons, T/d		25	45	75
Fuel cost, $/T	150			
Fuel cost		1.02	2.15	2.92
Port costs		0.49	0.21	1.14
Total Costs, $M		9.02	12.80	22.05
Cargo Del'd, MT/yr		1.80	1.08	5.18
Trans Cost, $/T		5.01	11.87	4.26
cents/gal		1.70	4.04	1.45
Memo Items				
Voyage R/T, n mi		2000	8000	4000
Speed avg, kts		15.0	15.0	14.6
Port Time				
Load, days		0.75	1	1.5
Disch, days		0.75	1	1.5
Delays, days		0.1	0.2	1
Voyage days		7.2	24.4	15.4
Voys/yr		48.9	14.3	22.7
Cargo DWT/Total DWT *		0.92	0.94	0.95

TABLE G-5 Double Hull and Hydrostatic Control

Vessel DWT		40000	80000	240000
Capital Cost				
Cost, $M *		47.3	80	126.7
CRF @ 12% 20 yrs	0.13388			
Annualized Cost		6.33	10.71	16.96
Non-Voyage Optg Costs				
Manning		0.81	0.89	1.15
Stores & Lubes		0.25	0.27	0.32
M & R *		0.61	0.80	1.00
Insurance *				
H & M		0.47	0.80	1.27
P & I		0.07	0.09	0.21
Admin & Other		0.21	0.23	0.29
Voyage Costs				
Fuel				
Fuel cons, T/d *		27.5	49.5	82.5
Fuel cost, $/T	150			
Fuel cost		1.12	2.36	3.21
Port costs		0.49	0.21	1.14
Total Costs, $M		10.36	16.37	25.54
Cargo Del'd, MT/yr		1.80	1.08	5.18
Trans Cost, $/T		5.76	15.19	4.93
cents/gal		1.96	5.17	1.68
Memo Items				
Voyage R/T, n mi		2000	8000	4000
Speed avg, kts		15.0	15.0	14.6
Port Time				
Load, days		0.75	1	1.5
Disch, days		0.75	1	1.5
Delays, days		0.1	0.2	1
Voyage days		7.2	24.4	15.4
Voys/yr		48.9	14.3	22.7
Cargo DWT/Total DWT *		0.92	0.94	0.95

TABLE G-6 Double Side

Vessel DWT		40000	80000	240000
Capital Cost				
Cost, $M *		35.7	52.2	95.9
CRF @ 12% 20 yrs	0.13388			
Annualized Cost		4.78	6.99	12.84
Non-Voyage Optg Costs				
Manning		0.81	0.89	1.15
Stores & Lubes		0.25	0.27	0.32
M & R *		0.46	0.52	0.75
Insurance *				
H & M		0.36	0.52	0.96
P & I		0.07	0.09	0.21
Admin & Other		0.21	0.23	0.29
Voyage Costs				
Fuel				
Fuel cons, T/d		25	45	75
Fuel cost, $/T	150			
Fuel cost		1.02	2.15	2.92
Port costs		0.49	0.21	1.14
Total Costs, $M		8.45	11.88	20.58
Cargo Del'd, MT/yr		1.80	1.08	5.18
Trans Cost, $/T		4.69	11.02	3.97
cents/gal		1.60	3.75	1.35
Memo Items				
Voyage R/T, n mi		2000	8000	4000
Speed avg, kts		15.0	15.0	14.6
Port Time				
Load, days		0.75	1	1.5
Disch, days		0.75	1	1.5
Delays, days		0.1	0.2	1
Voyage days		7.2	24.4	15.4
Voys/yr		48.9	14.3	22.7
Cargo DWT/Total DWT *		0.92	0.94	0.95

TABLE G-7 Double Side and Hydrostatic Control

Vessel DWT		40000	80000	240000
Capital Cost				
Cost, $M *		42.8	63.7	115
CRF @ 12% 20 yrs	0.13388			
Annualized Cost		5.73	8.53	15.40
Non-Voyage Optg Costs				
Manning		0.81	0.89	1.15
Stores & Lubes		0.25	0.27	0.32
M & R *		0.55	0.63	0.90
Insurance *				
H & M		0.43	0.64	1.15
P & I		0.07	0.09	0.21
Admin & Other		0.21	0.23	0.29
Voyage Costs				
Fuel				
Fuel cons, T/d *		26.25	47.25	78.75
Fuel cost, $/T	150			
Fuel cost		1.07	2.26	3.06
Port costs		0.49	0.21	1.14
Total Costs, $M		9.61	13.75	23.62
Cargo Del'd, MT/yr		1.80	1.08	5.18
Trans Cost, $/T		5.34	12.76	4.56
cents/gal		1.82	4.34	1.55
Memo Items				
Voyage R/T, n mi		2000	8000	4000
Speed avg, kts		15.0	15.0	14.6
Port Time				
Load, days		0.75	1	1.5
Disch, days		0.75	1	1.5
Delays, days		0.1	0.2	1
Voyage days		7.2	24.4	15.4
Voys/yr		48.9	14.3	22.7
Cargo DWT/Total DWT *		0.92	0.94	0.95

TABLE G-8 Double Side and Intermediate Oil Tight Deck

Vessel DWT		40000	80000	240000
Capital Cost				
Cost, $M *		41.4	60.7	108.8
CRF @ 12% 20 yrs	0.13388			
Annualized Cost		5.54	8.13	14.57
Non-Voyage Optg Costs				
Manning		0.81	0.89	1.15
Stores & Lubes		0.25	0.27	0.32
M & R *		0.53	0.60	0.86
Insurance *				
H & M		0.41	0.61	1.09
P & I		0.07	0.09	0.20
Admin & Other		0.21	0.23	0.29
Voyage Costs				
Fuel				
Fuel cons, T/d		25	45	75
Fuel cost, $/T	150			
Fuel cost		1.02	2.15	2.92
Port costs		0.49	0.21	1.14
Total Costs, $M		9.33	13.18	22.52
Cargo Del'd, MT/yr		1.80	1.08	5.18
Trans Cost, $/T		5.19	12.23	4.35
cents/gal		1.76	4.16	1.48
Memo Items				
Voyage R/T, n mi		2000	8000	4000
Speed avg, kts		15.0	15.0	14.6
Port Time				
Load, days		0.75	1	1.5
Disch, days		0.75	1	1.5
Delays, days		0.1	0.2	1
Voyage days		7.2	24.4	15.4
Voys/yr		48.9	14.3	22.7
Cargo DWT/Total DWT *		0.92	0.94	0.95

TABLE G-9 Double Bottom

Vessel DWT		40000	80000	240000
Capital Cost				
Cost, $M *		37.7	55.7	101.3
CRF @ 12% 20 yrs	0.13388			
Annualized Cost		5.05	7.46	13.56
Non-Voyage Optg Costs				
Manning		0.81	0.89	1.15
Stores & Lubes		0.25	0.27	0.32
M & R *		0.40	0.46	0.66
Insurance *				
H & M		0.377	0.557	1.013
P & I		0.07	0.09	0.21
Admin & Other		0.21	0.23	0.29
Voyage Costs				
Fuel				
Fuel cons, T/d		25	45	75
Fuel cost, $/T	150			
Fuel cost		1.02	2.15	2.92
Port costs		0.49	0.21	1.14
Total Costs, $M		8.68	12.32	21.26
Cargo Del'd, MT/yr		1.82	1.09	5.23
Trans Cost, $/T		4.77	11.31	4.06
cents/gal		1.62	3.85	1.38
Memo Items				
Voyage R/T, n mi		2000	8000	4000
Speed avg, kts		15.0	15.0	14.6
Port Time				
Load, days		0.75	1	1.5
Disch, days		0.75	1	1.5
Delay, days		0.1	0.2	1
Voyage days		7.2	24.4	15.4
Voys/yr		48.9	14.3	22.7
Cargo DWT/Total DWT *		0.93	0.95	0.96

TABLE G-10 Small Tanks

Vessel DWT		40000	80000	240000
Capital Cost				
Cost, $M *		36.2	53.8	97.0
CRF @ 12% 20 yrs	0.13388			
Annualized Cost		4.85	7.20	12.99
Non-Voyage Optg Costs				
Manning		0.81	0.89	1.15
Stores & Lubes		0.25	0.27	0.32
M & R *		0.46	0.54	0.76
Insurance *				
H & M		0.36	0.54	0.97
P & I		0.08	0.10	0.22
Admin & Other		0.21	0.23	0.29
Voyage Costs				
Fuel				
Fuel cons, T/d		25	45	75
Fuel cost, $/T	150			
Fuel cost		1.02	2.15	2.92
Port costs		0.49	0.21	1.14
Total Costs, $M		8.53	12.13	20.75
Cargo Del'd, MT/yr		1.80	1.08	5.18
Trans Cost, $/T		4.74	11.25	4.01
cents/gal		1.61	3.83	1.36
Memo Items				
Voyage R/T, n mi		2000	8000	4000
Speed avg, kts		15.0	15.0	14.6
Port Time				
Load, days		0.75	1	1.5
Disch, days		0.75	1	1.5
Delays, days		0.1	0.2	1
Voyage days		7.2	24.4	15.4
Voys/yr		48.9	14.3	22.7
Cargo DWT/Total DWT *		0.92	0.94	0.95

APPENDIX

H

Expert Judgment Technique

The method chosen by the committee to corroborate its findings is a rating technique based on expert judgments. This process functioned as an aid in finalizing the committee's conclusions, encouraging committee members to synthesize disparate information and make difficult choices. The committee already had reached general agreement on the preferred designs and the justifications, but the rating process provided a means of clarifying these ideas and offered a degree of quantitative support for the overall consensus.

The rating, it must be emphasized, represents a compilation of the committee members' individual conclusions and a check on the consensus. It was not a decision-making tool, but rather illustrative of the process. Fortunately, the ratings results generally reflected the same general conclusions indicated by the pollution-effectiveness study (detailed generally in Chapter 5): No one design is clearly superior, and none is a panacea for the oil spill problem.

METHODOLOGY OVERVIEW

The rating method made use of a matrix, which relates the alternative designs and design combinations (plus the reference or base case design) to important performance attributes. Ten attributes were used, including performance in four major accident scenarios, susceptibility and containment of fires/explosions, structural failure, salvageability, personnel safety, and cost. A typical expert-judgment rating matrix is shown in Figure H-1.

Each committee member ("expert") contributed a set of ratings, assigning a value for each attribute of each design. Each committee member also

Design Option (X_i) \ Attribute (Y_j)	Y_1	Y_2 ----	Y_j ---	Y_m
X_1	R_{11}	R_{12}	R_{1j}	R_{1m}
X_2	R_{21}	R_{22}	R_{2j}	R_{2m}
⋮				
X_i	R_{i1}	R_{i2}	R_{ij}	R_{im}
⋮				
X_n	R_{n1}	R_{n2}	R_{nj}	R_{nm}
Weighting Factors	W_1	W_2 --------	W_j ----	W_m

FIGURE H-1 Typical rating matrix.

assigned a weighting factor to each attribute, according to its judged importance; the same set of weighting factors was used for all design options. For each expert, attribute ratings then were combined, to estimate an overall rating for each design option. Finally, all of the expert judgments were aggregated, by arithmetic averaging. The computing process was facilitated by use of spreadsheet software.

To provide additional perspective, a sensitivity analysis was conducted, using ratings values and methodology that were identical in every way except that all of the attributes were assigned equal weighting factors (importance).

RULES FOR THE EXPERT JUDGMENTS

The first step was to design the rating matrix, which includes the important performance attributes discussed by the committee. The matrix, which was designed during a committee meeting, is shown in Figure H-2. The nine design options (X1, X2, . . . X9) and 10 attributes (Y1, Y2, . . . Y10) are described in Table H-1. (All of the options and attributes have been discussed previously in this report.) Each expert thus provided a total of 90 ratings, or best estimates.

The rules for the rating system were as follows:

<---------- POLLUTION EFFECTIVENESS ------------>

Attribute / Design Options	Y1	Y2	Y3	Y4	Y5	Y6	Y7	Y8	Y9	Y10
	LEG	HEG	LEC	HEC	F&E S	F&E C	SF	SA	PS	C
X1 = MARPOL										
X2 = MARPOL+HC										
X3 = DH										
X4 = DH+HC										
X5 = DS										
X6 = DS+HC										
X7 = DS+IOTD										
X8 = DB										
X9 = ST										
WEIGHTING FACTORS $0<W<1$	W1	W2	W3	W4	W5	W6	W7	W8	W9	W10

Note: The reference design option (i.e., X1=MARPOL) represents the design that the committee members are most familiar with. However, it does not necessarily represent an average design with a Rating of 5 with respect to all attributes. Whether it is an average design or not, it has to be judged by the experts.

FIGURE H-2 Rating matrix used in the decision analysis.

• Ratings could be any whole number from 1 to 9, where 1 was the *least* desirable (lowest rated) design option with respect to the attribute in question, and 9 was the *most* desirable (most highly rated). Thus, for the cost attribute, 1 represents the most costly design option, and 9 the least expensive. For personnel safety, 1 represents the least safe option, and 9 the safest.

• For each attribute, at least one option had to be assigned a 1, and at least one option had to be assigned a 9. This ensured a broad range of ratings, such that any differences among the design alternatives would stand out.

TABLE H-1 Descriptions of Design Options and Attributes

Design Option Descriptions:
X_1 = MARPOL Ship
X_2 = MARPOL + HC = MARPOL ship & hydrostatic control
X_3 = DH = double hull
X_4 = DH + HC = double hull & hydrostatic control
X_5 = DS = double sides
X_6 = DS + HC = double sides & hydrostatic control
X_7 = DS + IOTD = double sides & intermediate oil-tight deck
X_8 = DB = double bottom
X_9 = ST = small tanks

Attributes Descriptions:
Y_1 = LEG = low-energy groundings
Y_2 = HEG = high-energy groundings
Y_3 = LEC = low-energy collisions
Y_4 = HEC = high-energy collisions
Y_5 = F&E S. = fires and explosions (susceptibility)
Y_6 = F&E C. = fires and explosions (containment)
Y_7 = SF = structural failures
Y_8 = SA = salvageability
Y_9 = PS = personnel safety
Y_{10} = C = cost

• If attributes were judged to be of *unequal* importance, then an estimated weighting factor between 0 and 1 (a fraction) was assigned to each attribute. The higher the weighting factor, the more important the attribute, relative to the others listed. If attributes were judged equal in importance, then no weighting factor was assigned. The conditions are:

$$0 < W_j < 1 \text{ and } \sum_{j=1}^{10} w_j = 1$$

In summary, for the design options, 1 represents the worst rating, 2-4 poor ratings, 5 an average rating, 6-8 good ratings, and 9 the best rating. For each attribute, a weighting factor at or near zero indicates low relative importance, while a higher weighting factor indicates greater importance.

Two important assumptions were made regarding the double-bottom, double-side, and double-hull designs: 1) mandatory minimum outside hull thickness, and 2) mandatory minimum/maximum spacing between hulls. The significance of these factors was discussed in Chapter 4.

DATA ANALYSIS METHOD

The expert judgment estimates for each rating, R_{ijk} (for *i*th design option, *j*th attribute, and *k*th expert), and each weighting factor, W_{jk} (for *j*th attribute given by *k*th expert), were entered into a PC-based spreadsheet. The ratings for various attributes were combined, to estimate each expert's overall rating for each design option. Mathematically, this process is expressed as:

$$R_{iK} = \sum_{j=1}^{m} R_{ijk} \cdot W_{jk} \qquad m = \text{number of attributes}$$

Finally, the expert judgment estimates were aggregated, to find order in the series of collective ratings for the various designs. The arithmetic averaging method was used. Mathematically, this can be expressed as:

$$R_i = (1 / K)\sum_{k=1}^{K} R_{ik} \qquad K = \text{number of experts}$$

This process produced the combined final results. For sensitivity analysis, all of the calculations were repeated with equal weighting factors of 0.1.

The *range* of the 12 expert inputs on ratings and weighting factors, for each attribute for all of the design options, is shown in Table H-2.

As part of the learning process for committee members unfamiliar with the expert judgment rating method, and to achieve some consistency in methodology, the entire rating process (inputs and calculations) was conducted three times. The third round of inputs, the results of which are included here, was provided independently by each committee member.

RESULTS AND CONCLUSIONS

The individual average ratings given by all 12 experts were averaged to produce aggregate ratings for the various design options. The first analysis included the weighting factors assigned by each expert, and the sensitivity analysis was conducted with equal weighting factors of 0.1. Based on the aggregate ratings, the design options were ranked with respect to desirability. The results are shown in Table H-3 (with weighting factors) and Table H-4 (with equal weighting factors). The rank order changed only slightly in the sensitivity analysis. The principal conclusions are as follows:

- The design options tended to cluster into three groupings. The most desirable overall (from Table H-3) were the double hull with hydrostatic control, the double hull, double sides with intermediate oil-tight deck and double sides with hydrostatic control. The other four options were deemed somewhat less desirable.
- The differences within each grouping should be viewed as marginal. However, with the help of the weighting factors, the experts expressed a slight preference for the two double-hull options. Of these, the double-hull with hydrostatic control was viewed as the most desirable.

TABLE H-2 Range of Inputs to Rating Matrix by Experts

<----------- POLLUTION EFFECTIVENESS ------------>

Attribute / Design Options	Y1 LEG	Y2 HEG	Y3 LEC	Y4 HEC	Y5 F&E S	Y6 F&E C	Y7 SF	Y8 SA	Y9 PS	Y10 C
X1 = MARPOL	1-2	1-2	1-5	1-5	1-9	1-7	1-9	1-7	1-9	8-9
X2 = MARPOL+HC	2-6	2-8	1-5	1-8	1-9	1-9	1-9	1-9	1-9	2-9
X3 = DH	7-9	5-9	8-9	2-8	1-9	1-9	1-9	1-9	1-9	1-7
X4 = DH+HC	9	7-9	8-9	2-9	1-9	1-9	1-9	2-9	1-9	1-5
X5 = DS	1-4	1-3	7-9	6-9	2-6	2-7	3-7	1-7	3-7	3-7
X6 = DS+HC	2-8	2-8	8-9	7-9	1-8	1-8	3-8	3-8	2-8	2-7
X7 = DS+IOTD	3-9	3-9	7-9	6-9	1-6	1-7	1-9	1-8	1-8	1-6
X8 = DB	3-9	4-9	1-5	1-6	1-9	1-9	1-9	1-9	2-7	3-7
X9 = ST	2-7	2-6	2-8	2-7	3-9	3-9	2-9	3-9	3-9	3-9
WEIGHTING FACTORS 0<W<1	0.04 - 0.35 W1	0.05 - 0.3 W2	0.04 - 0.2 W3	0.05 - 0.2 W4	0.01 - 0.12 W5	0.01 - 0.08 W6	0.01 - 0.15 W7	0.03 - 0.2 W8	0.01 - 0.26 W9	0.01 - 0.3 W10

Note: The reference design option (i.e., X1=MARPOL) represents the design that the committee members are most familiar with. However, it does not necessarily represent an average design with a Rating of 5 with respect to all attributes. Whether it is an average design or not, it has to be judged by the experts.

• None of the designs was judged to be clearly superior. This seems to corroborate the results of the pollution-effectiveness study detailed in Chapter 5.

• The lowest rated design was the reference design, the existing MARPOL vessel. This appears to confirm that existing design standards should be upgraded.

TABLE H-3 Ranked Design Options (with weighting factors)

Rank #	Aggregate Rating	Design Option
1*	6.4	X4 = DH+HC
2	6.0	X3 = DH
3	5.8	X7 = DS+IOTD
4	5.7	X6 = DS+HC
5	5.0	X8 = DB
5	5.0	X9 = ST
6	4.8	X2 = MARPOL+HC
6	4.8	X5 = DS
7**	3.8	X1 = MARPOL

TABLE H-4 Ranked Design Options (with equal weighting factors)

Rank #	Aggregate Rating	Design Option
1*	5.8	X4 = DH+HC
2	5.7	X3 = DH
2	5.7	X6 = DS+HC
2	5.7	X9 = ST
3	5.4	X7 = DS+IOTD
4	5.1	X2 = MARPOL+HC
4	5.1	X5 = DS
5	4.7	X8 = DB
6**	4.4	X1 = MARPOL

NOTES ON THE RATING METHOD

The results of the matrix rating method confirmed what committee members already sensed, apparently fulfilling the stated purpose of clarifying and corroborating the conclusions. However, several limitations of the process, which of course reflect judgment rather than fact, should be noted.

First, as stated earlier, the ratings represent expert estimates, rather than precise actual values. The credibility of the ratings rests on the expert qualifications of committee members and on the information provided and available to the committee.

Second, committee members may have arrived at their ratings by different routes. Some may have assigned values in a more systematic way, and the degree to which their minds were made up may have varied. As the same directions were given to each committee member, this is a reflection of personal style. The differing degrees of expertise in some areas was surely a factor, and differences in interpretation likely contributed as well. At the same time, the diversity of the committee may have been a positive influence on the overall outcome; the committee was structured deliberately to provide a balance of viewpoints, with the aim of arriving at a collective opinion that was as close to objective as possible.

BACKGROUND MATERIAL

Electric Power Research Institute. 1986. Seismic Hazard Methodology for the Central and Eastern United States, Vol. 1: Methodology. Palo Alto, Calif.: EPRI. NP-4726.

Keeney, R.L., and D. Von Winderfeldt. 1988. Probabilities are Useful to Quantify Expert Judgments. Reliability Engineering and System Safety 23(4):293-298.

National Research Council. 1981. Reducing Tankbarge Pollution. Washington, D.C.: National Academy Press.

Ortiz, N. R., T. A. Wheeler, R. L. Keeney, and M. A. Meyer. 1989. Use of Expert Judgment in NUREG-1150. Paper presented at American Nuclear Society/European Nuclear Society International Topical Meeting on Probability, Reliability, and Safety Assessment, Pittsburgh, Pennsylvania, April 1989.

APPENDIX

I

Biographies of Committee Members

HENRY S. MARCUS, Chairman, holds the positions of associate professor of marine systems in the Ocean Engineering Department and chairman, Oceans Systems Management Program at the Massachusetts Institute of Technology. He has also served as a transportation consultant to maritime industries and government. Dr. Marcus holds a B.S. degree in naval architecture from Webb Institute, two M.S. degrees from M.I.T. (one in naval architecture and the other in shipping and shipbuilding management), and a D.B.A. degree from Harvard University. Dr. Marcus' research interests include ocean system logistics and marine environmental protection. Dr. Marcus was a member of the NRC's Maritime Transportation Research Board during the late 1970s; more recently he has served as a member of the Marine Board's Committee on Productivity of Marine Terminals, and of the Committee on Control and Recovery of Hydrocarbon Vapors from Ships and Barges.

WILLIAM O. GRAY, Vice Chairman, is President, Skaarup Oil Corporation. He received his B.E. (Mechanical Engineering) from Yale University and B.S.E. (Naval Architecture and Marine Engineering), with honors, from the University of Michigan. He served on the NAS Committee on Maritime Safety and frequently served on committees and panels with industry, national and international organizations, and professional associations. In 1985, he was appointed as Chairman of the International Chamber of Shipping's Tanker Safety Sub-Committee, representing tanker owners from 32 nations before the International Maritime Organization (IMO). Mr. Gray had an extended career with Exxon Corporation, where he served progressively as marine designer, planning section head, Manager of Tanker Research and the Arctic Tanker Project, Senior Advisor for Industry and Government, and Operations Coordinator for Marine and Marine Terminals. Mr. Gray is a member of the Society of Naval Architects and Marine Engineers and has represented International Chamber of Shipping and other professional trade associations.

DAVID M. BOVET is a Vice President at Temple, Barker and Sloane, Inc. He received his B.S. (Naval Architecture and Marine Engineering) from the Webb Institute of Naval Architecture. He was awarded his M.B.A., with highest honors, and M.A. (Economics) from Stanford University. He earned his M.S. (Ocean Engineering) from George Washington University. Mr. Bovet previously served in marine safety positions in the Coast Guard, and held professional positions with Booz-Allen Applied Research, Standard Oil of California, and Litton Systems. He was a staff economist with the World Bank, where he directed macroeconomic and project-level studies. Mr. Bovet is a member of the Society of Naval Architects and Marine Engineers, the Transportation Research Forum, and the Maritime Law Association.

J. HUNTLY BOYD, JR. is a senior associate at Booz-Allen & Hamilton, Inc. He received his B.S. from the Naval Academy. He earned his M.S. (Naval Architecture and Marine Engineering) and professional degree of Naval Engineer from the Massachusetts Institute of Technology. Mr. Boyd, a member of the Society of Naval Architects and Marine Engineers, chairs their Panel on Salvage and Rescue Systems, and serves as a committee chairman for the Marine Technology Society. He previously served in the Navy as Director of Ocean Engineering and Supervisor of Salvage in the Naval Sea Systems Command, and commanded a naval shipyard. Prior to joining Booz-Allen, he was a private consultant conducting marine engineering studies and casualty analyses. Mr. Boyd completed 28 years of Navy service as Captain and is an honorary life member of the American Society of Naval Engineers.

JOHN W. BOYLSTON is President of Argent Marine Operations, Inc. He received his B.S. (Marine Transportation) from the U.S. Merchant Marine Academy and his B.S. (Naval Architecture and Marine Engineering) from the University of Michigan. He holds a Coast Guard license as Third Officer, Unlimited Tonnage. He worked with the NRC Panel on Response to Casualties Involving Ship-Borne Hazardous Cargoes in their analyses of operational simulations conducted under the auspices of the Marine Board. He worked on three other NAS panels including Human Error in Navigation and Research Needs to Prevent Collisions, Rammings and Groundings. Previously he was Vice President of Seaworthy Systems, Inc., naval architect consultants. Earlier, as Vice President, Engineering, Giannotti & Associates, Inc., he managed the T-5 tanker design project for the Military Sealift Command. He directed marine operations for El Paso Marine Company. As Chief Naval Architect for Sea-Land Service, Inc., he was responsible for the design and construction of 13 container ships and conversion of 35 others. Mr. Boylston is a life member of the Society of Naval Architects and Marine Engineers where he serves on various panels and committees, and is a member of the Board of the American Bureau of Shipping.

JOHN M. BURKE is Vice President and Manager, Marine Technical Division, Mobil Shipping and Transportation Company where he is responsible for worldwide marine transportation objectives related to technical and new building strategies. He received his B.S. (Naval Architecture and Marine Engineering) from the University of Michigan. His previous positions include Chartered Fleet Coordinator for Exxon International Company, Marine Engineering Manager for Exxon Company USA, and naval architect for ESSO International Company in London. Earlier, he was employed by Newport News Shipbuilding in design and construction programs for naval vessels. Mr. Burke, a Chartered Engineer, holds membership in the Society of Naval Architects and Marine Engineers, the American Society of Mechanical Engineers, and the Royal Institute of Naval Architects.

THOMAS D. HOPKINS is Gosnell Professor of Economics, Rochester Institute of Technology. He received his B.A. (Economics), with Highest Honors, from Oberlin College. He earned his M.A., M.Phil., and Ph.D. (Economics) from Yale University. He held positions with the American University, the University of Maryland, and Bowdoin College. Dr. Hopkins held the position of Deputy Administrator for Regulatory and Statistical Analysis, Office of Management and Budget and was Acting Director, Council on Wage and Price Stability, conducting and reviewing regulatory analyses. He has served as a consultant to the Office of Technology Assessment on benefit charges for financing infrastructure and has studied benefits valuation in hazard regulation for the Administrative Conference of the United States.

JAMES HORNSBY completed service as Director General, Ship Safety, Canadian Coast Guard, and as Chairman of the Board of Steamship Inspection, retiring in 1988. He graduated from the Royal Technical College, Glasgow, Scotland. Mr. Hornsby headed the Canadian delegation from 1981-88 at all Marine Safety Committee sessions as well as several other IMO sessions and was a member of the Technical Committee, American Bureau of Shipping. He had extensive experience with the Canadian Coast Guard as ship surveyor and marine inspector, directing and administering all ship inspections in the Maritime Provinces. He was Canada's first On-scene Commander for a major oil tanker stranding, subsequently serving as senior technical advisor for salvage and pollution response for various other major marine casualties. As Director General, he held prime responsibility for all ship safety legislation, administrative and functional direction of field activities.

VOJIN JOKSIMOVICH is President and Chief Executive Officer of Accident Prevention Group, Inc. He has directed studies of risk management, assessment and human reliability for the Electric Power Research Institute, U.S. utilities, the Nuclear Regulatory Commission, as well as for European and Japanese utilities. Prior to founding Accident Prevention Group, Inc., he directed reliability/risk assessment programs for NUS Corporation, and was manager, safety and reliability of General Atomic Company. He has authored more than 100 papers technical papers and periodicals presented at various international conferences. Dr. Joksimovich has an engineering degree from the University of Belgrade, Yugoslavia, and a Ph.D. (nuclear engineering) from Imperial College, London University. Dr. Joksimovich is a member of the American Nuclear Society and the Society for Risk Analysis.

SALLY ANN LENTZ is a staff attorney with The Oceanic Society and Friends of the Earth. She received her B.A. (Sociology and Anthropology), with highest honors, from Oberlin College, and her J.D., Order of the Coif., from the University of Maryland. She also holds a Diploma in European Integration awarded by the University of Amsterdam, and is a member of the District of Columbia and Maryland Bars. Ms. Lentz served extensively as private sector advisor to U.S. delegations to IMO's Legal and Marine Environment Protection Committees, representing Friends of the Earth (FOE) and other private interest groups at national and international meetings on ocean dumping and marine environmental protection issues. She headed the FOE delegation to the joint meeting of the London Dumping Convention and Oslo Commission on Ocean Incineration in 1987. She is also a member of the National Committee for the Prevention of Marine Pollution. Prior to her present position, Ms. Lentz was a research attorney with a private law office.

ROBERT G. LOEWY is Institute Professor, School of Engineering, Rensselaer Polytechnic Institute. He received his B.A.E. (Aeronautical Engineering) from Rensselaer and his M.S. (Aeronautical Engineering) from the Massachusetts Institute of Technology. He earned his Ph.D. (Mechanical Engineering) from the University of Pennsylvania. Dr. Loewy is a member of the National Academy of Engineering. He is a member of

NRC's Naval Studies Board, a Director of Mohasco Industries, and General Partner, Advanced Technology Ventures. Dr. Loewy previously was Dean of the College of Engineering and Applied Science, University of Rochester, and held positions as Chief Scientist, U.S. Air Force, and Chief Technical Engineer, Vertol Division, Boeing Company.

TOMASZ WIERZBICKI is Professor of Applied Mathematics, Department of Ocean Engineering, Massachusetts Institute of Technology. He received his M.S. from Warsaw Technical University. He earned his Ph.D. and D.Sc. from the Institute of Fundamental Technological Research, Warsaw. He was Co-Chairman of an international symposium on structural crashworthiness. Dr. Wierzbicki has taught on Advanced Analysis and Design of Ocean Engineering Structures, Structural Mechanics, Crushing Mechanics of Thin-Walled Structures, and Plastic Structural Dynamics. He held appointments with national and international academic institutions. He is a consultant to Det Norske Veritas (ship classification society), Oslo, Norway, and is involved in analysis of collision protection of offshore platforms. He has also been a consultant to automobile manufacturers and is consulting for VOLVO of Sweden on the development of crash energy management systems.

APPENDIX

J

Presentations to the Committee

American Bureau of Shipping. Donald Liu. Tank vessel design process and key elements. March 26, 1990.

American Petroleum Institute. E.J. Roland. Existing tank ships are adequate. November 6, 1989.

American Waterways Operators. Thomas Allegretti. Safety record spills and causes. June 6, 1990.

A.P. Moller Co. Bent E. Hanson. European operator's response to special hull requirements on behalf of International Chamber of Shipping, International Association of Independent Tanker Owners, and oil companies International Marine Forum. March 26, 1990.

Bethlehem Steel Corp. Frank Slyker. An Approach to Tanker Retrofit. March 27, 1990.

CHEVRON OREGON tour. San Pedro, California. January 20, 1990.

Cleary, W.A. Jr., consultant. Tank Vessel Design Alternatives. November 6, 1989.

E. I. Du Pont de Nemours and Company Protection Systems. Keither R. Watson. Flexible Composite Liner System for Preventing Oil Spills. July 17, 1990.

Exxon Shipping Co. Frank Iarossi. Views regarding tanker safety and improved safety. January 18, 1990.

EXXON VALDEZ tour. National Steel and Shipbuilding Company, San Diego, California. January 18, 1990.

Foss Maritime Co. Steven Scalzo. Alternatives for hull design to reduce pollution risk. June 6, 1990.

Glosten Associates. Duane H. Liable. Influences on construction, barges vs. tankers. June 6, 1990.

Hughes, Owen, Professor. Aerospace and Ocean Engineering Department, Virginia Polytechnic Institute and State University. Discussion of capability of detailed analytical techniques for determining hull deformation from grounding. June 6, 1990.

International Association of Independent Tanker Owners. Philip Embiricos. Double hulls and double bottoms—maintenance and inspection concerns. June 6, 1990.

International Chamber of Shipping. J. M. Joyce. European operator's response to special hull requirements. March 26, 1990.

Jones, Norman. University of Liverpool. Estimating Hull Penetration from Collisions. June 6, 1990.

Lloyd's Register of Shipping. J. M. Ferguson. Crude Oil Carriers in the 1990s. June 6, 1990.

Majestic Shipping Services Corp. Jack Devanney. Mechanisms governing oil spill from tank vessels. January 19, 1990.

Marinex International Inc. Robert D. Goldbach. Double hulls and double bottoms—maintenance and inspection concerns. June 6, 1990.

Maritime Overseas Corp. George Blake. Double hulls and double bottoms—maintenance and inspection concerns. June 6, 1990.

Mitsubishi Heavy Industries. N. Aikawa. Double Sided Hull with Mid-Height Deck (Alternative Idea to Double Hulls). July 17, 1990.

National Oceanic and Atmospheric Administration. Charles N. Ehler. Estimating the cost of spills—the Exxon Valdez experience. June 6, 1990.

National Steel and Shipbuilding Co. Al Lutter. Repair Process for the EXXON VALDEZ. January 18, 1990.

National Wildlife Federation. Eric Olson. The Prince William Sound precedent for future costs. June 6, 1990.

Platzer Shipyard. Neal Platzer. Influences on construction, barges vs. tankers. June 6, 1990.

Shipbuilders Council of America. Richard Thorpe Jr. U.S. Shipyard response and capability. March 27, 1990.

Stolt-Nielson. Stefan Nystrom. Double hulls and double bottoms—maintenance and inspection concerns. June 6, 1990.

Tanker Advisory Center Inc. Arthur McKenzie. Higher standards are needed. November 6, 1989.

U.S. Coast Guard. Captain James M. MacDonald. Charge to Committee. November 6, 1989.

U.S. Coast Guard. Capt. James M. MacDonald and Stephen Shapiro. Updates on IMO Marine Environment Protection Committee 30 and Oil Pollution Act of 1990. December 13, 1990.

APPENDIX

K

Committee Estimates of Oil Outflow Relative to MARPOL Standard (100%)

Outflow estimates were provided by Det norske Veritas (DnV) for several design alternatives; the committee made its own estimates for other alternatives using the DnV individual estimates of grounding and/or collision performance. This performance rating is stated as an outflow index relative to the conventional MARPOL tanker (designated by DnV as "original" in their report) whenever possible.

The cases for which the committee made its own estimates are marked with an "x" in Table K-1. The figures and calculations leading to the committee estimates are on the following pages.

TABLE K-1 Ratings Derived by Committee

Design Alternative	Tanker Size	
	VLCC (Table 5-5)	40,000 DWT (Table 5-6)
Intermediate oil-tight deck with double sides	x	x
Double sides with hydrostatic control	x	x
Double hull with hydrostatic control	x	x
Hydrostatic control	Calculated by DnV	x
Small tanks	Calculated by DnV	x

VLCC

		Reference: DnV Outflow Figures from Chapter 5:	
		Figure 5-13 Speed: 5 Knots	Figure 5-14 Speed: 10 Knots
Intermediate Oil Tight Deck with Double Sides	**DnV Arrangement Number**		
Double sides with hydrostatic control collision	(3 B)	30	19
Intermediate oil tight deck grounding	(8)	2	4
Total		32	23
Double Sides with Hydrostatic Control	**DnV Arrangement Number**		
Double sides collision	(3 B)	32 x .9[1] = 29	19 x .9 = 17
Hydrostatic control grounding	(9)	3	4
Total		32	21

Double Hull with Hydrostatic Control	DnV Arrangement Number	Reference: DnV Outflow Figures from Chapter 5: Figure 5-13 Speed: 5 Knots	Figure 5-14 Speed: 10 Knots
Double hull collision	(1A or 1B)	33 x .9 = 30	20 x .9 = 18
Double hull grounding	(1A or 1B)[2]	$\frac{0 + 1}{2} = 0$	$\frac{6 + 3}{2} = 4$
Total		30	22

40,000 DWT TANKER

Hydrostatic Control + Base	DnV Arrangement Number	Reference: DnV Outflow Figures from Chapter 5: Figure 5-15 Speed: 5 Knots	Figure 5-16 Speed: 10 Knots
Base collision	(1)	55 x .9 = 49	34 x .9 = 31
Intermediate oil-tight deck grounding	(8)	3	3
Total		52	34

Reference: DnV Outflow Figures from Chapter 5:

Smaller Tanks

Assume base ship (1) with twice as many transverse bulkheads. This would make collision index about 0.5 and grounding index about 0.9 compared to MARPOL base.

	Figure 5-15 Speed: 5 Knots	Figure 5-16 Speed: 10 Knots
Base grounding x 0.5	55 x .5 = 27.5	34 x .5 = 17
Base collision x 0.9	45 x .9 = 40.5	66 x .9 = 59
Total	68	76

Intermediate Oil-tight Deck with Double Sides	**DnV Arrangement Number**		
Double sides collision	(6)[3]	55	34
Intermediate oil-tight deck grounding	(8)	2	2
Total		57	36

Double Sides with Hydrostatic Control			
Double sides collision	(2)	75 x .9 = 67	44 x .9 = 40
Hydrostatic control grounding	(Use VLCC Values)	3	4
Total		70	44

Reference: DnV Outflow Figures from Chapter 5:

Double Hull with Hydrostatic Control	**DnV Arrangement Number**	**Figure 5-15 Speed: 5 Knots**	**Figure 5-16 Speed: 10 Knots**
Double hull collision	[Avg. of (5+6+7)] x .9	= 60	= 37
Hydrostatic control	Average hydrostatic control + "Perfect" - or - $\frac{3+0}{2}$	= 1.5	= 1.5
Total		61	39

[1] 0.9 assumes all cargo tanks are 90% filled for the hydrostatic control design.

[2] Grounding averages 1A + 1B (double hull) and 9 (hydrostatic control).

[3] Uses DnV arrangement number 6 (double hull with centerline bulkhead) as being the more representative of double side that would be DnV arrangement number 2 (with no centerline bulkhead).

Glossary

These definitions, provided for readers not familiar with nautical terms, are those used for purposes of the present study. Some definitions may not be applicable in other contexts.

ballast: non-cargo load (generally sea water) used to make a vessel without cargo heavier and more stable.
barge: a tank vessel lacking onboard means of propulsion.
bending moment: the summation of forces (ship and cargo weight, buoyancy, and dynamic) acting on a vessel's hull that tend to bend the hull.
bulkhead: longitudinal or transverse structure dividing cargo tanks.

class: classification society.
combination carrier: a vessel designed and built to carry dry, bulk, or liquid cargo.
crude oil washing (COW): a method of washing or rinsing out cargo tanks, using high-pressure jets of crude oil as the washing medium.

deadweight: a measure of the carrying capacity of a vessel (the weight of cargo, fuel, fresh water, and stores).
draft: the depth of water a vessel draws, especially when loaded.
DWT: deadweight tons.

Exclusive Economic Zone: area generally considered to extend 200 nautical miles from shore.

flag of convenience: flag state selected (by non-resident ship owners) on the basis of favorable conditions for non-residents in terms of commercial flexibility and tax treatment.

flag state: a nation where ships can be registered.

freeboard: the distance from the waterline to a vessel's deck.

free surface effect: cargo movement resulting from light loading of tanks that tends to reduce vessel stability.

girders: large structural support framing in tanker hull. Girders can be transverse or longitudinal, and vertical or horizontal.

green water: heavy seas washed on deck.

gross registered tonnage (GRT): a volumetric measure of both earning spaces (for cargo) and non-earning spaces, such as the engine room, bridge, and accommodations.

heel: the extent of a vessel's incline or tilt to one side.

hydrostatic balance: the level of oil in a cargo tank such that the oil pressure is equal to the exterior sea water pressure.

IGS: inert gas system (a safety feature that prevents combustion by using non-flammable, or "inert", gas to exclude oxygen from tanks).

IMO: International Maritime Organization, the United Nations agency responsible for maritime safety and environmental protection of the seas.

lightening: same as lightering.

lightering: the process of transferring cargo at sea, from one vessel to another.

lightweight: weight of a ship without cargo, crew, fuel, or stores (same as lightship weight).

list: a ship's tilt to one side in a state of equilibrium (as from unbalanced loading).

load on top (LOT): a cargo loading system that minimizes operational pollution: when oil and water are left standing, the heavier water sinks and can be drawn off and returned to the sea, and cargo "loaded on top" of remaining oil/water residues.

LOOP: Louisiana OffShore Oil Port (18 miles off the coast of Louisiana in the U.S. Gulf of Mexico).

MARPOL: The International Convention for the Prevention of Pollution from Ships, adopted in 1973 and amended in 1978. It constitutes the basic international law for limiting all ship-source pollution, including structural and operational provisions for tank vessel pollution control; the term is used in this study to describe the current standard for vessel design.

net registered tonnage: a measure of the earning capacity of a vessel, based on cubic capacity of revenue-earning spaces.

ocean-going: a vessel designed and certificated (authorized by the U.S. Coast Guard) for operation on ocean routes but not on inland (except for port entrances) or Great Lakes waters. In the present study, the term refers to vessels over 10,000 DWT.

peak tanks: tanks at the bow and stern, often used for ballast.

PL: a MARPOL provision for protectively located ballast tanks (refers to strategic placement of SBT, to afford some protection in a grounding or collision).

retrofit: major structural alteration of an existing vessel.

SBT: segregated ballast tanks (tanks dedicated to carriage of seawater ballast, never cargo). For purposes of this study, refers to MARPOL requirements for SBT.

scantlings: the dimensions of the structural members of a vessel.

SOLAS: the international convention for the Safety of Life at Sea. It constitutes the basic international law governing ship safety features for ships and their crew.

tanker: a tank vessel with onboard means of propulsion.

territorial waters: each flag state may establish its territorial waters up to 12 nautical miles from baselines (determined in accordance with the 1982 Law of the Sea Convention).

transshipment: re-shipment of crude oil or petroleum products.

trim: the position of a vessel with reference to the horizontal, or the difference in draft forward and aft.

ullage: the space in a cargo tank above the cargo.

VLCC: Very Large Crude Carrier (can refer to vessels over 160,000 DWT or 200,000 DWT).

webs: structural support framing in tanker hull (see girders).

Index

T